Cities and Natural Process

This book is a discussion of a fundamental conflict in the perception of nature in the city, an expression of the essential need for a view that is grounded in natural processes that will inform the theory and practice of urban design at the scale of the city and its urban region. It is, therefore, a search for a sustainable urban future. With public concern for energy conservation, the protection of natural systems and environmental awareness growing, the desire to nurture a rewarding environment where nature and urbanism co-exist is the challenge for the twenty-first century.

Revisiting many of the cases originally studied in *City Form and Natural Process* and the first edition of *Cities and Natural Process*, this new edition has been extensively revised to take account of recent theoretical and practical developments. The author discusses the processes of nature at work and how they are altered in urban environments, and suggests a design framework for achieving an alternative, environmentally informed, view of cities. By examining natural and human processes as they operate in balance with nature, the book reveals how alternative values can tip the balance in favour of a constructive relationship between the two.

Cities and Natural Process is an essential addition to the study of urban design and environmental planning and of the potential of cities to be environmentally, economically and socially sustainable.

Michael Hough is a landscape architect and principal and founder of his firm ENVISION the hough group, and Adjunct Professor in the Faculty of Environmental Studies at York University, Ontario, Canada. His previous books include *City Form and Natural Process* (1984), *Cities and Natural Process* (1995) and *Out of Place* (1990).

Cities and Natural Process

A basis for sustainability

Second edition

Michael Hough

LONDON AND NEW YORK

First published 1995
by Routledge
2 Park Square, Milton Park, Abingdon, Oxon, OX14 4RN

Simultaneously published in the USA and Canada
by Routledge
270 Madison Ave, New York, NY 10016

Second edition published 2004

Reprinted 2006

Routledge is an imprint of the Taylor & Francis Group, an informa business

Typeset in Times New Roman by
Florence Production Ltd, Stoodleigh, Devon
Printed and bound in Great Britain by
TJ International, Padstow, Cornwall

British Library Cataloguing in Publication Data
A catalogue record for this book is available from the British Library

Library of Congress Cataloging in Publication Data
Hough, Michael.
Cities and natural process: a basis for sustainability/Michael Hough.
– 2nd ed.
p. cm.
Includes bibliographical references and index.
1. City planning – Environmental aspects. 2. Urban ecology. I. Title.
HT166.H664 2004
307.1′216–dc22 2003015520

ISBN10: 0–415–29854–7 (hbk)
ISBN10: 0–415–29855–5 (pbk)

ISBN13: 978–0–415–29854–4 (hbk)
ISBN13: 978–0–415–29855–1 (pbk)

Contents

Preface

This edition has been written with the knowledge of the continuing change that has occurred since *Cities and Natural Process* was first published in 1995. At a grass-roots level there is an increasingly sophisticated understanding of environmental and social issues, the power of citizen action and involvement that has been gathering momentum. Dedicated groups in Europe and North America are finding new ways of effecting change by initiating projects, whether these are constructing wetlands, restoring wild-flower meadows or bringing rivers and watersheds back to health, many in collaboration with various government agencies and municipal departments. Greening cities and sustainability have become major environmental and social movements that have grown to include the participation of international organizations. There is also an increasing understanding of some larger concerns of urban growth, the impacts of sprawl on air, land and water, and the need for approaches that address regional issues. Thus, while the fundamental message of this edition remains pertinent, I have incorporated much new material. A new chapter has been added that discusses different ways of understanding the landscape in its regional context and how this addresses the control of urban growth and the maintenance of environmental health. I have also included some discussion on the economic value of the landscape as this relates to parks and open spaces, and the significance of tree canopies in relation to storm drainage and air quality. A number of new case studies have been added and I have continued the practice, initiated in the first edition, of revisiting projects I discussed in the mid-1980s and reporting on the changes that have since occurred.

Embracing an ecological view of cities is an ongoing and evolving process. My intent, therefore, is to continue the broadening and enrichment of the book's scope to bring it into line with contemporary values and the issues that need to be addressed in the twenty-first century.

Acknowledgements

Many friends and colleagues have helped and advised me in the writing of this new edition and I am much indebted to them all. A few, however, must be acknowledged with particular gratitude. Suzanne Barrett for her willingness to read and comment in considerable detail on initial drafts; Arifa Hai who spent many months researching new areas of investigation both in Toronto and Dhaka, her home city; Sybrand Tjallingii for taking me to many Dutch cities and significant projects; Simon Miles for spending much time in discussion of urban issues and reviewing drafts; Jae Cheon for his very considerable help with graphics and photographs; Alex Murray for sharing his research on ecological footprints; and my colleagues at ENVISION the hough group, with whom I have continued to collaborate for many years in exploring and realizing creative ideas.

Also for specifically granting permission to incorporate material into this book. Peter Milroy (UBC Press), Photographers Steve Frost, Jaekyung Kim, Dr S. M. A. Rashid, Andrew Beddingfield, Ute Schmidt, Karen Yukich and Gera Dillon, Arifa Hai, Gord Macpherson (Toronto and Region Conservation Authority), Bruce Ferguson, Willem van Leeuwen, Herbert Dreiseitl, Richard Scott, Jae Cheon, National Capital Commission (Ottawa), Creekside Education Trust, David Satterthwaite (Environment and Urbanism), Mick Magennis (Kentish Town City Farm), Dr Jurgen Baumuller (Office of Environmental Protection: Urban Climate and Planning, Stuttgart), Michael Schwarz-Rodrian, Professor Alex Murray and Eric Krause, Marlaine Koehler (Waterfront Regeneration Trust), Justin Houk (Metropolitan Portland Planning Department), Pat Guy (Thompson Publishing Services), Rob Buffler (Y2Y Initiative), Rick Boychuk (Canadian Geographic), Robert Paehlke, Nicole Hudon (Environment Canada), (Ottawa).

Many others also contributed generously with information, maps and photographs and include Sue Cornwell, Chris Baines, Jeremy Iles, Irene Lucas, Mike Houck, Ethan Seltzer, Peter Koddermann, Bill Sleeth, Beth McEwen, Jamie Laut, Jim Barlow. I should also like to thank Routledge and specifically Andrew Mould and Melanie Attridge for their considerable help in the book's publication.

Credits

The illustrations for this edition were drawn by Ian Dance, Tim Hough and Jay Cheon. Photographs for the new edition were taken by Werner Hannappel, Peter Leidtke, Arifa Hai, Richard Scott, Steve Frost, Andrew Beddington and others and have been credited to them. Except where noted the remainder were taken by the Author.

Introduction

This book is about natural processes, cities and design. Its overall thesis is that the traditional design values which have shaped the physical landscape of our cities have contributed little to their environmental health, or to their success as civilizing, enriching places to live in. My purpose is to find alternative approaches to thinking about the physical environment of cities in ways that promote sustainability. This book has been written with two purposes in mind. The first is to offer a conceptual, philosophical base on which urban design can rest, one that has received, until recently, only scant attention in the literature. The second is to illustrate with examples drawn from real life how the practical application of theory is both relevant and useful to the urban designer. The book deals with five general areas of concern.

First, there is the alienation of urban society from environmental values that embrace both the cities and their larger regions. The technologies that sustain the modern city have touched every corner of human life, every landscape and wilderness, no matter how remote, and reinforce this isolation. This fact was forcefully demonstrated to me on a journey I made to the Hudson Bay lowlands many years ago. As a newcomer to Canada, I searched for an image of the great unspoiled Canadian wilderness, free from the sights, sounds and pressures of the urbanized south. Armed with hip waders and binoculars, I spent many days tramping through a landscape of water, muskeg and granite boulders, the dome of the sky creating a feeling of extraordinary wildness and beauty. Forget the dense clouds of hungry mosquitoes and blackfly (known as the scourge of the north), the discomforts of permanently wet boots; here was wild nature. Yet one day, a pink object, lying in the tangle of sedges at the edge of a pond, caught my eye. It was the rubber nipple from a baby's bottle, abandoned there by a passing group of Inuit. The rude shock of this relic of human society, so alien to the environment around me, brought me abruptly back to reality. The incident was a powerful reminder that the pervasive influences of the city are everywhere, even in the remoteness of the Hudson Bay lowlands. The nipple, my hip waders and binoculars, and the fact that I was there, transported by plane hundreds of kilometres from Toronto, verified that urbanism is a universal fact of life. This is so, not only for the white Canadian, but for the native individual today who uses the white Canadian's canned foods and machinery, hunts with a rifle, travels by skidoo and lives in permanent northern communities. And as I learned much later, evidence of human activities are evident in the presence of DDT that has been found in polar bears, the effects of greenhouse gases on polar ice, evidence of climate change and consequent implications to the northern landscape and to life systems. The perceptual distinction between cities and the larger landscapes beyond them has been a root cause of many social and environmental conflicts and the lack of attention to the environment of the cities where most problems begin.

Second, little attention has been paid to understanding the natural processes that have shaped the city's physical form and which in turn have been altered by it. In the presence of plentiful energy, the urban environment has been largely influenced by imperatives that are economic rather than environmental or social. Yet, it has become more and more apparent that the achievement of sustainable goals depends on the interdependence of environmental, social and economic determinants. The explosive growth of urban areas since the Second World War has brought about fundamental changes, not only to the physical landscape but also to people's perceptions of land and environment. An affluent and mobile urban society takes refuge in the countryside in search of fresh air and natural surroundings that are denied at home. Unsustainable pressures are placed on natural and cultural landscapes, degrading their ecological integrity and their inherent distinctiveness – the things that make one place different from another. At the same time, efforts to reclaim derelict land replace naturally regenerated sites for new, placeless environments.

Third, there are issues that concern urban processes and how we think about them. They include vast areas of inefficiently used land from urban sprawl, enormous water, energy and nutrient resources that are the by-products of urban drainage, sewage disposal and other functions of city processes. Having no perceived value, these contribute instead to the pollution loads of an overstressed environment.

Fourth, there are questions of aesthetic values from which the city's formal landscape has evolved. These values have little connection with the dynamics of natural process and lead to misplaced priorities. Horticultural science, not ecological processes, determines the development, form and management of the city's open spaces. At the same time, another landscape, the fortuitous product of natural and cultural forces, flourishes without care and attention. These two landscapes symbolize a fundamental conflict of values in the perceptions of nature: the desire to nurture the one and suppress the other in a perpetual and costly struggle to maintain order and control. At a time when natural processes are being recognized as central to landscape planning and design, they have until recently been virtually ignored in the urban context. It is clear that the conventional framework for the design of cities must be re-examined. In-built assumptions about traditional priorities and standards must give way to more unconventional approaches. How can a recognition of the essential relationships between natural processes, people and economy provide clues to the shaping of cities, to their sustainability? How has the familiar concept of the urban park changed, and what is its role and function today in creating healthy and dynamic places? What are the implications of evolving citizen initiatives on urban form and action?

Fifth, there are questions of environmental values and perceptions, and of how we respond to the environments around us. If it can be shown that there are less costly and more socially valuable ways of shaping urban landscapes than has traditionally been the norm, then we have a realistic and practical basis for action. The biologist and city planner Patrick Geddes once remarked that 'civics as an art has to do not with imagining an impossible no-place where all is well, but making the most and the best of each and every place, especially in the city in which we live'.[1] So utopian ideals of the perfect city set in bucolic landscapes that were once the fashion in planning and architectural philosophy are not relevant to our concerns. This book addresses urban design issues that focus on existing cities, since it is here that the opportunities lie and where the effort must be made.

The forthcoming chapters explore these and other issues in further detail. Chapter 1 suggests the design framework within which urban natural processes should be examined. It describes the general character and evolution of the urban landscape in both the pre-industrial and modern city, and examines the constraints of energy, environment and social necessity that have helped shape their form, character and use. The attitudes and values that pervade urban life are reviewed in terms of the environmental problems they have generated and the opportunities that exist for creating a rational basis for design. Some basic principles that derive from the application of ecology to the design process are suggested. These become the frame of reference for subsequent discussions of the city's physical and social environment.

The chapters on water, plants, wildlife, city farming and climate examine the various components of the natural and human environment in several ways: first, as they operate as natural systems, or in balance with nature; second, how they are affected or changed by urban processes, and the attitudes and cultural values that these changes have engendered. Some alternative values based on ecological insights are suggested that would tip the balance in favour of a constructive relationship to the urban environment. Such a change also reveals opportunities instead of problems and substitutes economy for high cost. Practical examples of opportunities that are often unrecognized, but occur everywhere in the city, serve to illustrate the potential that exists for beneficial change. The implications of an ecological view on urban form are then examined through various case studies as the foundation for a coherent philosophy of design. I have revisited many of them several times since the publication of *City Form and Natural Process* in 1984, for several reasons. First, much can be learned from revisiting places at different points in time. It is important to determine the level of success or failure of different projects in relation to the premises on which they were designed and implemented and the reasons why some failed, why others succeeded, and how others evolved and changed in unexpected ways. Second, the purpose of this approach is to reinforce the notion that we are, fundamentally, dealing with the principle of process – the evolution of natural processes, places and events over time. This will be explored in Chapter 1. Chapter 7 is concerned with cities in their regional landscape contexts. It discusses various ways in which urban growth can be understood and managed when the larger landscape is conceived of as an organizing framework giving shape to urban form, the implications of ecological footprints in relation to sprawl and containment. It also brings together the primary themes of the book and develops an integrated view that includes both an ecological and behavioural framework for urban design at local, regional scales, and in its larger world context.

1 Urban ecology: a basis for shaping cities

As we enter the twenty-first century, the environmental concerns and values that began to take root in the 1960s have brought into sharp focus an acute awareness of the earth's fragility as a natural system. We have begun to understand human beings as biological creatures immersed in vital ecological relationships within the biosphere; with the need to live within its limits, sharing the planet with non-human life. These perceptions are leading us to a view that there must be a transition from a society preoccupied with consumerism and exploitation, to one that gives priority to a more sustainable future. The Bruntland Commission interprets sustainability as 'meeting the needs of the present without compromising the ability of future generations to meet their own'. The World Business Council on Sustainable Development defines it as (involving) 'the simultaneous pursuit of economic prosperity, environmental quality and social equity.'[1] Such essentially homocentric interpretations, however, indicate the ultimate need for an ethic that recognizes the interdependence of all life forms and the maintenance of biological diversity. In this sense, therefore, sustainability becomes everyone's concern. When we consider a world population projected to be a possible 10 billion by the middle of this century; the relentless migration of people in developing countries to urban areas; and the massive impacts of human activities on world terrestrial and marine ecosystems, then it is clear that there are inseparable links between nature, cities and sustainability. The Ecological Footprint[2] of every facet of human activity has profound implications for survival.

Patrick Geddes, Ian McHarg, Philip Lewis and other eloquent voices concerned with bringing together nature and human habitat have shown that the processes which shape the land, and the limitless complexity of life forms that have been created over evolutionary time, provide the indispensable basis for shaping human settlements. The dependence of one life process on another, the interconnected development of living and physical processes of earth, climate, water, plants and animals, the continuous transformation and recycling of living and non-living materials, these are the elements of the self-perpetuating biosphere that sustain life on earth and which give rise to the physical landscape. They are the central determinants that must shape all human activities on the land.

If we consider the urban landscape in this context, some recognition of the links between natural processes and cities have begun to emerge. In a world concerned increasingly with the problems of a deteriorating environment, there are signs of changing values. We are beginning to understand that cheap energy, air and water pollution, vanishing plants, animals, natural or productive landscapes are issues intimately linked with the cities. At the same time, if urban design can be described as that art and science dedicated to enhancing the quality of the physical environment in cities, to providing civilizing and enriching places for the people who live in them, then much remains to be done.

Thus the basic premise on which this book rests is twofold. First, that an environmental view is an essential component of the economic, political, planning and design processes that shape cities. The often unrecognized natural processes occurring within them provide us with an alternative basis for their evolution and form. Second, that the problems facing urban and rural regions have their roots in the inner cities, and solutions must also be sought there. Thus the task is one of linking urbanism with nature at both local and regional scales. My purpose in this chapter is to identify current problems and examine the structure and principles for design that are central to this point of view.

The contradiction of values

The average urban dweller going about the business of daily living will experience the city through its patterns of streets, residential areas and pedestrian ways, office towers, shopping plazas and parking lots, civic squares, monuments, parks and gardens. However, there is another generally ignored landscape lying beneath the surface of the city's public places and thoroughfares. It is the landscape of railways, public utilities, vacant lands, urban expressway interchanges, abandoned industrial lands and derelict waterfronts. Thus two landscapes have long existed side by side in cities. The first is the nurtured 'pedigreed'[3] landscape of mown turf, flowerbeds, trees, fountains and planned places everywhere that have traditionally been the focus of civic design. Its basis for form rests in the formal design doctrine and aesthetic priorities of established convention. Its survival is dependent on high energy inputs and horticultural technology. Its image is that of the design solution independent of place: it can be found everywhere from Washington DC to Jakarta, Indonesia; from the city centre to the outlying suburbs. The second is the fortuitous landscape of naturalized urban plants and flooded areas left after rain that may be found in the forgotten places of the city. Urban 'weeds' emerge through cracks and gratings in the pavement, on rooftops, walls, poorly drained industrial sites or wherever a foothold can be gained. They provide shade and flowering groundcover and wildlife habitat at no cost or care and against all the odds of gasoline fumes, sterile or contaminated soils, trampling and maintenance men. There is also a humanized landscape hidden away in back alleys, rooftops and backyards of many an ethnic neighbourhood that may be described as the product of spontaneous cultural forces. It is here that one may find a rich variety of flourishing gardens and brightly painted houses. The turfed front yard of the well-to-do neighbourhood gives way to sunflowers, daisies, vegetable gardens, intricate fences, ornaments and religious icons of every conceivable variety, expressing rich cultural traditions, and the imperatives of necessity. The forces that shape the built vernacular, in fact, have remarkable parallels to the fortuitous landscape. Both have evolved in response to minimum interference from authority.

These two contrasting landscapes, the formal and the natural, the pedigreed and the vernacular, symbolize an inherent conflict of environmental values. The first has little connection with the dynamics of natural process; yet it has traditionally been held in high public value as an expression of care, aesthetic value and civic spirit. The second represents the vitality of altered but none the less functioning natural and social processes at work in the city; yet it is regarded as a derelict wasteland in need of urban renewal, the disorderly shambles of the poorer parts of town. If we make the not unreasonable assumption that diversity is ecologically and socially necessary to the health and quality of urban life, then we must question the values that have determined the image of nature in cities. A comparison between the plants and animals present in a regenerating vacant

(a)

(b)

Plate 1.1 Two urban landscapes. A formally landscaped boulevard, and an abandoned waterfront site. Which is the derelict site? (a) Has four or five species of plants and supports no wildlife. (b) Supports over 400 species of plants and is visited by 290 species of birds.

lot, and those present in a landscaped residential front yard, or city park, reveals that the vacant lot generally has far greater floral and faunal diversity than the lawn or city park. Yet all efforts are directed towards nurturing the latter and suppressing the former. The reclamation of 'derelict' areas, or the creation of new development at the city's edge where the native landscape is replaced and reshaped by a cultivated one, involves reducing diversity and sense of place, rather than enhancing them. The question that arises, therefore, is this: Which are the derelict sites in the city requiring rehabilitation? Those fortuitous and often ecologically diverse landscapes representing urban natural forces at work, or the formalized landscape created by design?

Such observations may be seen as a metaphor for larger environmental issues, but it is my contention that the formal city landscape imposed over an original natural diversity is the one in need of rehabilitation. While it will be obvious that such a landscape has a time-honoured place in the city, its *universal* application in the making of urban places is the most persuasive argument for considering it as a derelict landscape. Other paradoxes become apparent when we apply ecological insights to our observation of the city environment.

- Urban design has, with some exceptions, been more concerned with long held aesthetic conventions than with biophysical process as determinants of urban form. There are countless examples of cities and institutions whose design and form are wholly inappropriate to their regional climates.
- Traditional storm drainage systems, the conventional method of solving the problem of keeping the city's paved surfaces free of water, have until recently been unquestioned. As the established vocabulary of engineering, water drains to the catchbasin. Yet the benefit of well-drained streets and civic spaces is paid for by the environmental costs of eroded streams, flooding and impairment of water quality in downstream watercourses, a condition that is akin to environmental degradation by design.
- Waste disposal systems are seen as engineering rather than biologically sustainable solutions to the ultimate larger problems of the eutrophication of water bodies. Yet cities produce vast quantities of nutrient energy that recycling programmes are only beginning to reuse for productive purposes.

Humanity and nature have long been understood to be separate matters. Such a dichotomy has had profound influences on the way people have thought about themselves: the cities where people live and the non-urban regions beyond where nature lives. In the unique cultures from which the disciplines of intervention spring – civil engineering, building, planning and design – this perceived separation has also profoundly influenced the desire to control, not only nature, but also human behaviour. Thus the nature of pedigreed design has had little time or understanding for the natural and cultural processes that shape human environments, or the particular needs of multicultural communities that are the norm in most cities today.

The resolution of these contradictions must be found in an ecological view that encompasses the total urban landscape and the people who live there. This includes the unstructured spatial and social environments that are not currently seen to contribute to the city's civic image as well as those that do. To explore this point of view further we must examine the legacy that pre-industrial as well as modern cities have left.

Vernacular landscapes and the investment in nature

The vernacular has traditionally been described as urban and landscape forms that grow out of the practical needs of the inhabitants of a place and the constraints of site and climate.[4] The vernacular form of early city building was a response to the practical problems of shelter and security. Decisions to lay out a pattern of streets and squares, locate settlement on the tops of hills or at the bend of a river, were functional and pragmatic, rather than the result of preconceived doctrine. The tendency to romanticize the past embodied in winding streets, piazzas suddenly entered and ancient churches is typical of the average tourist who can wander through these places separated from the day-to-day struggle to survive that these places historically represented. So for the medieval builders, building towns on 'picturesque' hilltop locations was not a challenge to subdue nature, or to enjoy a view of the countryside, but to pose a challenge to their enemies and preserve precious sources of water and valley land for farming.[5] Yet at heart their shape, visual relationship to the countryside and harmonious siting are the historical product of economic and social forces and the physical constraints of the land. Early settlement, as Mumford has pointed out, could not grow beyond the limits of its water supply and food sources until better transportation and a more sophisticated administration could evolve.[6] The early association with food production maintained a connection between the country and the city in some form until the Industrial Revolution. For instance, fruit and vegetables consumed in New York and Paris came from nearby market gardens whose soil had been enriched with night refuse. A large part of the population had private gardens and practised rural occupations within the city, as American towns did up until 1890. Mumford also notes that the amount of usable open space within medieval cities throughout their existence was greater per head of population than any later form of city.[7]

From a design viewpoint, the most significant impression one receives of the pre-industrial town is that it made the most of what it had within the means and technology available at any point in time. In the absence of the technological means to ignore climate, built form has traditionally responded by developing direct and sometimes ingenious ways of moderating its effects. Since pre-industrial settlements were built and operated on solar power, they were limited by what stored energy was available from organic materials, running water and sunlight. Variations of climate, topography, agricultural soils and water supply shaped their form. Open spaces were functional, producing a variety of fruit and vegetables; the common and churchyard provided grass and were kept trim by livestock – a practice still kept up by many towns and cities in Europe. Groupings of houses around greens and courtyards were arranged on the basis of functional necessity to conserve heat, minimize winds and to provide sunlight and space.

Books on architecture have marvelled at the subtle sequence of spaces, the proportions of squares, formal avenues and powerful architectural statement of important buildings and other aesthetic urban qualities. Architectural history has been preoccupied with styles of building based on formal rules of design that commemorate power and wealth.[8] It excludes 'vernacular' building which responds to the environment, to social, economic and functional necessity. We can draw relevant parallels in the landscape. The literature on historical landscapes deals almost exclusively with the development of the artistic philosophies that produced the great parks and gardens of the times, from which much of our urban park tradition can be traced. It ignores the working vernacular landscape of town and country, created out of necessity and poverty that symbolized the investment in nature and the cultural landscape. But it is these landscapes that hold crucial lessons for us today in our search for a relevant basis for urban form in the twenty-first century.

(a)

(b)

Plate 1.2 Expressions of the vernacular landscape. (a) Visual connections to the countryside. (b) Places for walking, commerce and the local farmers' market.

The landscapes of the contemporary city

The patterns of space in the modern city are the product of market forces, transportation systems and design ideologies that are radically different from older building patterns. In the inner city, tower blocks rise from open plazas which are often windswept and shadowed in cold climates, or sun-baked in hot ones. Economic forces have created a landscape of uncontained plazas, parking lots, vehicular thoroughfares, highway interchanges and vacant sites. Aesthetic conventions have created a development landscape of parks, playgrounds and recreation spaces, whose character rests on a universal application of cultivated turf, asphalt and chain-link fences occasionally punctuated by an ornamental tree or exotic shrub. At the edges of the city, the expressway and arterial road determine suburban location and growth. The 'wet noodle' residential street system determines suburban form. The isolation of the component parts from each other – low-density residential housing, regional shopping centres, industrial estates, places of work, corner gas stations, hamburger franchises – determine a car-dependent living environment. What is missing is a semblance of order, integration, mix of uses, places that belong to the pedestrian – the kind of urbanism that has traditionally evolved in downtown areas. The native plant community disappears and is replaced by non-native, horticultural species, signifying a loss of bio-regional connections and the sense of being grounded in the native landscape. Today, in North America, more people live in the suburbs than in the inner city, depending on the car to shop, work and visit friends at enormous social cost in air quality, greenhouse gases, ecological degradation, loss of agricultural soils, and restrictive zoning that inhibits the natural growth and development of neighbourhoods over time. As a professional colleague of mine once observed, the suburb is a place where you need a litre of gasoline to buy a litre of milk.[9]

Of all the varied impressions that come to mind as one looks at the modern city, there are perhaps four that reveal the most about the subject at hand: a lack of visual connections to the countryside; the use of urban parks solely for leisure; the mutually exclusive nature of the relationship between town and countryside; and the abundant use of energy.

Visual connections to the countryside

The view to the countryside from the town that symbolized the pre-industrial dependence on the land has, with the exception of some rural towns and villages, all but disappeared. The traditional relationship between city and farmlands has been replaced by an industrialized agriculture that has no direct connections with the local place, but rather represents the dictates of international trade. Land that once produced crops and livestock is now producing real estate. What was once productive countryside immediately surrounding the city is now, in much of North America, the object of land speculation and sporadic development, defying planning solutions and perpetuating an unproductive landscape and the loss of ecologically diverse habitats.

Parks are for recreation

Since its inception, recreation has been the exclusive land use for the city's public open spaces. The migration of people from the countryside to the cities that began during the Industrial Revolution did more than create poverty and slums. The skills and knowledge of traditional patterns of rural life were replaced by the living and working patterns of the city. The psychological and physical separation between urban and rural environments widened as cities grew larger, more industrialized and more remote from the rural areas with which originally they had been connected. The urban park had an entirely different purpose from the countryside it replaced. The crops, orchards and livestock that had been a function of many open areas in the pre-industrial settlements were now replaced by those that catered exclusively to amenity and recreation.

Parks originated in the late eighteenth century in Britain as private residential squares at a time when some cities were becoming attractive places to live in for the upper classes. Among them were the famous Bloomsbury garden squares of London (1775–1850), and the crescents of Bath, developed by the Brothers Wood (1730–67).[10] The development of the public parks in the expanding cities of Europe and the US in the nineteenth century evolved out of the Romantic movement. They were created in the conviction that nature should be brought to the city to improve the health of the people, by providing space for exercise and relaxation. It was felt that the opportunity to contemplate nature would improve moral standards. A new preoccupation with the aesthetics of natural landscape led to the notion that parks would improve the appearance of cities.[11] The introduction of the Royal Parks in London, Olmsted's Central Park in New York, Boston Common and Mount Royal in Montreal, are testament to a period of extraordinary social convictions and purpose.

However, the continuing expansion of the city since the nineteenth century created new conditions. There was greater mobility, wealth and leisure among more people, and a desire to escape the city and renew contact with rural settings that the urban park was unable to satisfy. Work and play came to be perceived as separate and distinct activities. The perception that the countryside exists solely as an urban playground is borne out every weekend when its lakes, forests and farmlands are invaded by people who have little or no direct contact with the landscape as a place of work, or commitment to it as a natural environment. Thus recreation in the countryside contributes little to the land on which it occurs. Its social effects frequently create conflict between those who earn their living by the land and those who use it for leisure. Its environmental effects are often destructive of lakes and streams, soils and vegetation. At the same time the recreational needs of urban people are changing. The urban poor and culturally diverse groups without access to the countryside, a preoccupation with physical fitness and diversified recreational interests are changing conventional views of how the city's open spaces are used. Recreation, once confined to parks, now involves the entire city (see Chapter 3).

The city and countryside are mutually exclusive places

'[T]oday it is nature beleaguered in the country, too scarce in the city which has become precious.'[12] It is not hard to understand how this mental dissociation takes place. Perceptually we may miss the obvious evidence of natural surroundings, the woods, streams, marshes and fields. We fail, however, to see nature as an integrated connecting system that operates in one way or another regardless of locality, whether this be the rural or natural region, or within the city itself.

Water supply and disposal systems leave no indication that the water supplied to the kitchen tap had its origins in the forests and landscapes of upper watersheds, or that rain falling on rooftops and paved surfaces and disappearing without trace into catchbasins and underground sewers is part of a continuous hydrological cycle. The grass and specimen trees of the urban parks and gardens, their plants brought from Korea and the Himalayas, their turf maintained like a billiard table, are difficult to associate with the diverse community of plants that convert sunlight into energy, store carbon and produce the food and materials necessary for survival. The frozen and hermetically sealed plastic package one finds in the meat section of the local supermarket bears not the slightest resemblance to the animal from which it came (see Chapter 5). Park maintenance usurps natural succession. The regulated air-conditioned climate and tropical plantings of the interior shopping mall have replaced the cycle of the seasons. Sanitary sewers and the rubbish truck break the life cycle of nutrients and materials of natural systems. The occasional visit to the National Park or Protected Natural Area tells us little about its purpose, its natural systems, or its relation to the cities where most people live. Exacerbating the situation, the environment of the city, its mown lawns, golf-courses, horticultural displays and deteriorating infrastructure, is transported to the rural areas and protected spaces. The urban environment serves to isolate us from an awareness of the natural and human processes that support life.

Yet the essential creativity of nature, the processes that continue modified and often degraded, continue to function. Rich and diverse natural habitats, remnants from a pre-urban era, occur on quasi-public or private land where public access is restricted and where the gang mower has not penetrated. New communities of plant species, often alien to the region, establish themselves, flourishing in the warmer climate that northern cities afford. These plants, like the mosses, common dandelion, plantain, tree of heaven, buddleia, are, in some parts of the world, what we regard as weeds. Weeds, botanically speaking, are plants that colonize disturbed land. From a cultural viewpoint, they are plants growing where they are not wanted, but they none the less represent the fortuitous communities of the urban environment. Hydrological systems are in evidence in the rainfall impounded on the poorly drained parking lot and playing field, where the processes of evaporation and groundwater recharge continue the cycle, by accident rather than by intent. The sewage lagoon perpetuates the process of decay of organic material and release of nutrients and provides new urban marshland for large populations of shore birds. Flat-topped roofs provide nesting places for night hawks, and rubbish dumps and waste places attract small rodents that in turn attract the hawks and owls which feed on them. It is these natural systems operating within the city that are the ecological framework for urban design.

An abundance of energy

The availability of cheap fossil fuel has been an overriding determinant of urban form. The energy flow through a city, with its factories, automobiles, heating and cooling systems and high power consumption, is about a hundred times greater than the energy flow through a natural ecosystem.[13] Cities place enormous stresses on natural systems, depending on them for resource inputs and for the disposal of unwanted products. There is input of food, produced from oil-based synthetic fertilizers from agricultural regions, and output into the environment of heat and concentrated nutrient energy from sewage treatment plants. Industrial processes draw water from lakes and rivers for cooling and return waste heat energy. One hundred per cent of the water transported through the pipes to the filtration plants is treated to drinking-water standards, but only about 1 per cent is used for drinking. Solid waste and organic refuse are disposed of in landfill sites using up and often contaminating land and water, and generating large quantities of methane in the process of decomposition that contributes to greenhouse gases.

The problem of establishing natural elements in the hostile urban environment of the city centre has produced a landscape whose creation, maintenance and survival depends not on natural determinants, but on technology and high energy inputs. The international style of landscape design that one finds in the urbanizing city edge has little to do with the inherent characteristics of the place that was once there. It is established and maintained in isolation – predetermined design imposed on its site.

Perhaps the most striking aspect of the city is the amount of wasted energy and effort that is expended to create and maintain such an unrewarding environment. The wealth of opportunity that exists to create a better one is, in the second millennium, only beginning to be explored. It is therefore encouraging that architecture in Britain, through the Royal Institute of British Architects, has formally recognized that building and sustainability are relevant linked concepts.[14] There is convincing evidence that climate changes now underway are due primarily to human activity in releasing CO_2 into the atmosphere. Buildings are particularly implicated in this process, and so it is appropriate that the design and construction process should be a prime target in the war against climate change. Indeed, ecological sustainability is now on the curriculum for all the RIBA recognized courses in architecture.[15] In 1977 Steadman pointed out that the

(a)

(b)

Plate 1.3 Expressions of the energy urban landscape. (a) The universal turf groundcover and scattered buildings. (b) Loss of visual connections to the countryside.

Source: Steve Frost.

external containing envelope of buildings becomes the consequence of their internal organization and material structure, in contrast to traditional architecture, where the design of the exterior shell was a response to the problem of keeping out the weather and protecting the interior from heat and cold.[16] In the twenty-first century, architecture is beginning to respond to environmental limits in their form and function. Natural ventilation is being incorporated into building design, using traditional approaches in new ways; for example, through the use of wind towers, a technology similar to the nineteenth-century wind towers that were used to dry hops in the English county of Kent. Solar photovoltaic cells are in use not only in housing, but also for downtown office buildings, which requires façades to be oriented to the sun. Environmentally effective design involving interior lighting and ventilation has been shown to lead to significant increases in productive and substantial energy savings.

The integration of urbanism and ecology achieved through the design and planning process is our concern here. It establishes links between a local and a larger bio-regional view, and makes connections between disparate elements to reveal possibilities that may not otherwise be apparent. The insights that urban ecology provides, when put together with social and economic objectives, creates a rational basis for shaping the city's landscape.

We should now review the principles that seem the most applicable to this view, since they form a frame of reference for the discussion of city form in the chapters to come. They include *Process*, *Economy of means*, *Diversity*, *Connectness*, *Environmental education begins at home*, *Making the most of opportunties*, and *Making visible the processes that sustain life*.

Some design principles

Process

Processes are dynamic. The patterns of the landscape are the consequence of the forces that give rise to them: geological uplift and erosion of mountains, the hydrological cycle and forces of water that are continually shaping the land, the successional stages of woodlands and the different species of animals and birds that inhabit them at different phases of their evolution. The form of the place reveals its natural history and the continuing cycle of natural processes. So the tendency to view natural phenomena, such as mountains or waterfalls, as static events, frozen in time, is a root cause of the aesthetic dilemmas that we face, in communicating, for instance, the processes that underlie the tourist attraction for 'wild scenery'. When nature is seen as a continuum, the argument of what is beautiful or what is less so in the landscape becomes, if not meaningless, then of a very different order of meaning. The rock falls at Niagara, that caused such a furore in the press in the 1960s with the collapse of the cap, and the pile of 'unsightly' rock that marred the view of the falls from bridal hotel bedrooms, are simply visible evidence of nature at work. Places such as this cannot be regarded as some gigantic engineering toy that has somehow gone wrong and must, within the limits of technological wizardry and irrelevant aesthetic standards, be put right.

The same analogy may be applied to human communities. The form of the city is the consequence of a constant evolutionary process fuelled by economic, political, demographic and social change; of new buildings replacing old and old buildings being adapted to new uses, of urban decay and renewal. The natural sciences can teach us about unpredictability in nature and how this applies to the design and planning disciplines. There is a growing realization among resource scientists and managers that nature is capricious and that a great deal of uncertainty underpins theory about the dynamics of populations and communities.[17] The recognition in the scientific community of the unpredictability of natural systems has close parallels with the evolutionary processes of the city whose form and growth changes and adapts to unforeseen economic and social conditions over time, and over which designers and planners have little direct control. Cities are continually changing and adapting to new conditions and will never be 'completed' as one might finish a painting or a piece of furniture. The role of planning approval processes, such as by-laws, building codes and regulations, are often in conflict with this organic process – the need for people to modify and shape their own living environments (see Chapter 7, pp. 238–9). As social entities, neighbourhoods evolve by themselves when permitted, often in unpredictable ways. We can see parallel examples in the relationship between evolving plant communities in abandoned urban areas, and the pristine horticultural displays that are the pride of urban design in the cared-for parts of town. The former represents nature's processes at work, the latter represents control over these spontaneous natural forces. In effect, the creation and upkeep of urban spaces has traditionally been seen as a static endeavour; once created the object is to maintain the status quo. Design and *maintenance*, when based on the concept of process, become an integrated and continuing *management* function, rather than separate and distinct activities guiding the development of the human-made landscape over time.

One of the unavoidable conclusions about urban natural processes is how intricately they are interwoven with economic, industrial and cultural activity. The inherent conflicts between the human and non-human processes at work in the urban landscape are well demonstrated in an industrial site that was photographed over a period of more than twenty years and how it changed during that time. The issue here is less the loss of an urban wetland, regrettable though this may be. Rather, it is a demonstration of how the natural colonization of an industrial site and social redevelopment forces are constantly at work in the city, particularly in areas of urban renewal.

There is a prevailing view among conservation-minded people that human influences on the land are inherently destructive. The disappearance of forests, wetlands and a host of life forms, both at home and around the world, leaves one in no doubt that this opinion is, in large measure, well founded. The blunt statement of the Manager for State Parks in New York that open space is like virginity – once lost it can never be regained[18] – rings true when we are confronted with the destruction of priceless landscapes and cultural heritage in the face of urban development. The preservation and protection of nature can be argued for, in the context of today's values, on the basis of moral and aesthetic values, of the absolute necessity of maintaining genetic and biological diversity and of keeping options open for the future. Design, however, is also directly concerned with the notion of change, and the constructive opportunities that change provides. David Loenthal makes the point that the Manager for New York State Parks may not be altogether correct in his assertion, since only non-virgins can produce more virgins.[19] This remark contains an important truth when humankind is seen as part of these natural processes. Landscapes may be created that are different from the original, but may result, none the less, in diverse and healthy environments. Human beings, as agents of change, have historically been concerned with modifying the land for survival – draining land to create productive fields, exploiting the earth for fuel and raw materials – but are often unconscious of the effects of their activities on the original landscape. While the world today exhibits countless examples of destructive change, it is important to remember that there are also many that have been environmentally beneficial. W. H. Hoskins has shown that the origin of the Norfolk Broads, a landscape of water and marshlands in south-east England of great diversity and beauty, was for many years a subject of speculation, one theory being that they had resulted from a marine transgression in fairly

(a)

(b)

(c)

(d)

(e)

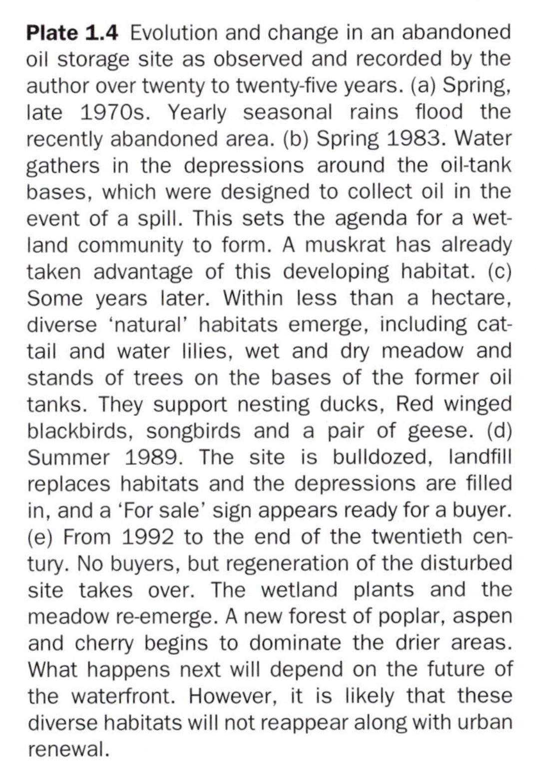

Plate 1.4 Evolution and change in an abandoned oil storage site as observed and recorded by the author over twenty to twenty-five years. (a) Spring, late 1970s. Yearly seasonal rains flood the recently abandoned area. (b) Spring 1983. Water gathers in the depressions around the oil-tank bases, which were designed to collect oil in the event of a spill. This sets the agenda for a wetland community to form. A muskrat has already taken advantage of this developing habitat. (c) Some years later. Within less than a hectare, diverse 'natural' habitats emerge, including cattail and water lilies, wet and dry meadow and stands of trees on the bases of the former oil tanks. They support nesting ducks, Red winged blackbirds, songbirds and a pair of geese. (d) Summer 1989. The site is bulldozed, landfill replaces habitats and the depressions are filled in, and a 'For sale' sign appears ready for a buyer. (e) From 1992 to the end of the twentieth century. No buyers, but regeneration of the disturbed site takes over. The wetland plants and the meadow re-emerge. A new forest of poplar, aspen and cherry begins to dominate the drier areas. What happens next will depend on the future of the waterfront. However, it is likely that these diverse habitats will not reappear along with urban renewal.

Source: These photographs originally appeared in Roots et al. *Special Places: The Changing Ecosystems of the Toronto Region*. Vancouver: UBC Press, 1999 and are reproduced by permission of the publisher.

recent times. In the 1950s, it was shown conclusively that they were the result of deep peat-cutting in medieval times, some four hundred years ago. Since the region was naturally treeless, peat was a valuable fuel. Water seepage into excavated areas eventually caused the abandonment of peat-cutting. Marshes developed and finally enough water filtered in to create the 'artificial' lakes that form the present landscape.[20] In an urban context, one may find flourishing natural landscapes that have evolved from old quarries abandoned long ago. The restored landscapes of Britain's industrial Midlands, being brought back into productive use, are examples of the purposeful modification of natural process to bring formally ravaged places back to health. Human or natural processes are constantly at work modifying the land. The nature of design is one of initiating purposeful and beneficial change, with ecology and people as its indispensable foundation.

Economy of means

From an ecological perspective this could be called the principle of least effort. The greatest or the most significant results that spring from an undertaking usually come from the least amount of effort and energy expended rather than the most. It involves the idea that maximum environmental, economic and social benefits are available from minimum resources and energy. It also involves the idea of doing things small, since it suggests that making small mistakes is infinitely preferable to making very large ones. Over time small mistakes can be adapted to social and environmental conditions; large ones may last indefinitely. Many cities today play an important role as suppliers as well as consumers of materials. Leaves and unwanted organic products are being transformed into compost. Paper, metals, plastics and glass are being put to new uses and products.

In developing countries, where poverty and necessity play a major role in the way many people earn a living, economy of means becomes crucial to survival. Scavenger groups in Jakarta, Indonesia, for instance, collect and sell for recycling much of the rubbish that is thrown away. There is a good market for recyclable materials such as paper, plastics, bottles and metals. Rubbish pickers take their finds to nearby junk dealers who buy their materials and sell them to a variety of industries. For example, used cans are in high demand by kerosene stove-makers. Soy ketchup bottles can be sold back to the factories for refilling. Furniture shops are grateful for new supplies of crates since it is cheaper to use parts of crates than to buy new materials. Thus the recycling process demonstrates the principle of economy of means. Not only does it reduce the amount of waste material that must be disposed of in landfill sites, it also provides a whole sector of the population with a living, and the process provides environmental, social and energy-saving benefits. The *Jakarta Post* summarized these facts in its report on recycling: 'We have finally arrived at the finding that garbage is decomposable while those components that are not can be used as a resource.'[21] In an agricultural context, Dutch fruit-growers at one time maintained the grass under their trees with sheep. The animals kept the grass mown which inhibited competition for nutrients and moisture. This benefited the farmer by giving him two sources of income. Similarly, a policy in Britain of reforesting the verges of the country's motorways has reduced grass-cutting costs while creating wildlife corridors. This has been found to have benefits in increasing the diversity and movement of plants and animals, and an altogether more interesting visual character at less cost in money and energy.

Diversity

If health can be described as the ability to withstand stress, then ecological diversity also implies health. Odum has commented: 'the most pleasant and certainly the safest landscape to live in, is one containing a variety of crops, forests, lakes, streams, roadsides, marshes, seashores and waste places – in other words, a mixture of communities of different ecological ages.'[22] Diversity makes social as well as biological sense in the urban setting since the requirements of an infinitely diverse urban society implies choice. The quality of life implies, among other things, being able to choose between one place and another, between one lifestyle and another. It implies interest, pleasure, stimulated senses and varied landscapes. The city that has places for foxes and owls, natural woodlands, trout lilies, marshes and fields, busy plazas, markets, noisy as well as quiet places, playing fields and formal gardens is more interesting and pleasant to live in than one that does not have such places. The city also needs well-defined, identifiable districts that determine many different kinds of place throughout the urban area and that reinforce the changing social, business, commercial and environmental character of the city on a larger scale. This is particularly significant in urban renewal projects such as waterfronts that have developed a unique sense of place through years of neglect. Outdoor storage areas, rusting cargo ships that once plied the waterways and oceans, lift bridges, rail spurs, storage silos, regenerating vegetation and naturalized road verges all speak to their unique history and character. The temptation to renew such areas without recognizing their inherent personalities is a recipe for missed opportunities to celebrate the city's diversity.

Connectedness

Barry Commoner's well-known principle that 'everything is connected to everything else'[23] has become, in the twenty-first century, the embodiment of a larger regional and global view as well as a local one. In the late 1980s the Royal Commission on the Future of the Toronto Waterfront (a commission set up to examine issues along Toronto's lakeshore) recognized the implications of Commoner's principle when it realized that the waterfront could not be viewed as simply a narrow band along the Lake Ontario shore. It is linked by Lake Ontario to the Great Lakes system, by rivers and creeks to the watersheds, by water mains, storm and sanitary sewers and roads to homes and businesses throughout the Metropolitan area.[24] What goes down the sewer in the local residential area has an effect on the watershed, its rivers and lakes many hundreds of kilometres away. Air quality is influenced by local and regional sources. Beaches, wetlands, cliffs, woodland and meadow along the waterfront are habitat for both resident and migratory wildlife and linked to the hinterland via the river valleys. Human uses of the land – transportation, housing, industry, business and recreation – tie the waterfront to the larger region. To properly understand a local place therefore requires an understanding of its larger context – the watershed and bio-region in which it lies. At the same time, understanding of the bio-region begins with its local places.

Environmental education begins at home

Environmental literacy strikes at the heart of urban life, and consequently at the way we think about and shape our cities. The perception of the city as separated from the natural processes that support life has long been a central problem in environmental thinking. The urban experience of 'nature' is to a large extent a 'disneyfied' experience: too often relegated to the visit to the zoo, where elephants and tigers, safely behind bars,

Plate 1.5 Diversity of city places. Places for people, activity and social contact. Places for wildlife, solitude and education (see Plate 1.4c).

are on display; too often associated with domesticated pets – poodles, tabby cats, rose gardens and floral clocks. It has been said that children know more about nature in distant lands than they do about the natural things in their own backyards, neighbourhoods and cities. The media reinforce this perception in their treatment of nature. The threatened tropical forests in Brazil, the massive hydroelectric projects that flood thousands of hectares of the Canadian north, the disappearance of countless species of animals and plants from all parts of the globe, remain for many, out there somewhere, beyond the cities, remote from the immediate concerns of ordinary people pursuing the often precarious business of day-to-day living. Environmental education is more than the biology lesson in the classroom or the yearly trip to the nature centre. These are no substitute for constant and direct experience assimilated through daily exposure to, and interaction with, the places people live in.

It may also be said that literacy about how the world works is inhibited by the way we have been taught to think about the environment around us and our relationship to it. This was clearly revealed some years ago when some university colleagues and I conducted a workshop for junior school science teachers on the advantages of environmental education in the city rather than in the 'unspoiled' landscapes beyond the urban areas. The discussion focused on the places close to school where children could learn about nature – the nooks and crannies and abandoned places where naturalized plants and animals could be found. At one point a clearly sceptical teacher asked: 'What if there are no natural places near the school, how then does one teach the kids about nature?' One of my colleagues immediately replied, 'Stand them in the middle of the asphalt schoolyard and ask them why they are alive; that would be a good beginning.'

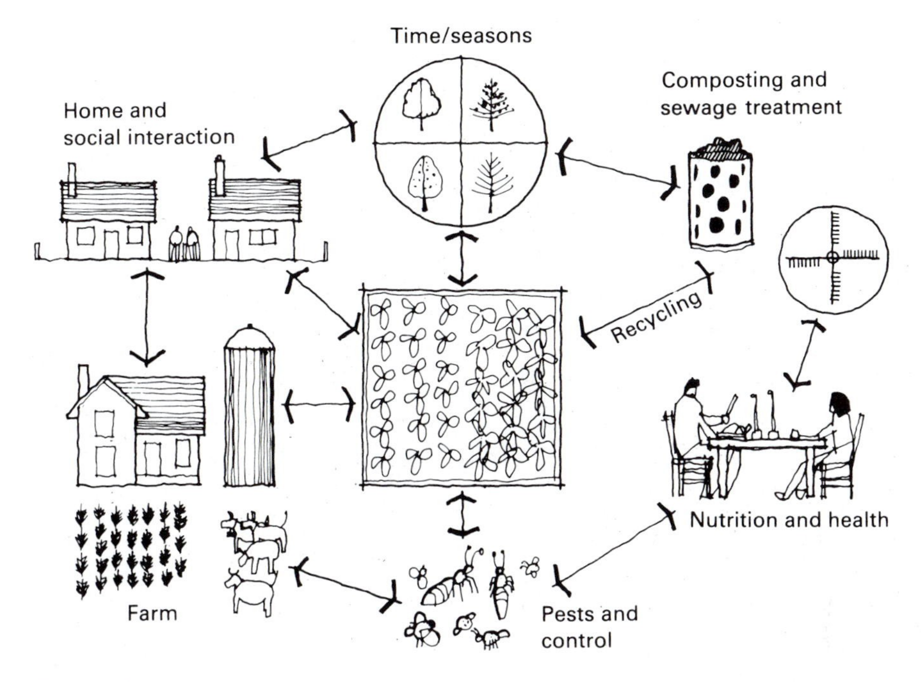

Figure 1.1 Environmental education begins at home. The everyday learning experience in the allotment garden that also makes connections to larger regional and international issues.

The teacher's question highlights a basic problem about how most people think about natural processes. His underlying assumption was that nature is an externality, set apart from human affairs, which can only be studied in non-urban surroundings. In his mind environmental education had little to do with the interdependence of life systems that includes both human and non-human nature. The novelist John Fowles summarizes the problem in his article 'Seeing Nature Whole'.[25] On a visit to Linnaeus' garden in Uppsala, Sweden, he had this to say about its owner, 'the great indexer of nature', who, between 1730 and 1760, classified much of the earth's animate beings:

> Perhaps nothing is more moving at Uppsala than the actual smallness and ordered simplicity of that garden and the immense consequences that sprung from it in terms of the way we see and think about the external world . . . for all its air of gentle peace, it is closer to a nuclear explosion, whose radiations and mutations inside the human brain were incalculable and continue to be so.

Fowles suggests that humankind has evolved into an isolated creature that sees the world anthropocentrically.

> [The] power of detaching an object from its surroundings and making us concentrate on it is an implicit criterion in all our judgments. A great deal of science is devoted to . . . providing labels, explaining specific mechanisms and ecologies – in short, to sorting and tidying what seem in the mass indistinguishable one from another.[26]

Seeing nature whole, understanding interrelationships and connections between human and non-human life, must therefore begin with the places where most people live. The urban allotment garden, through the daily process of growing food, provides a realistic basis for understanding the cycle of the seasons, soil fertility, nutrition and health, the problem of pests and appropriate methods of control. Questions of soil fertility are connected with composting, the source of nutrients in the treatment plant and the recycling of organic materials. The close proximity of food production to home is connected to the energy costs of food production on the farm. The human energy and time invested in urban farming provide economic rewards and social benefits, as leisure time is channelled into productive endeavours. One of the fundamental tasks of reshaping the city is to focus on the human experience of one's home places; to recognize the existence and the latent potential of natural, social and cultural environments to enrich urban places. This provides the best chance of spiritual growth and creative learning since it lies at the heart of environmental education.

Making the most of opportunities

There is a common tendency to regard environmentally sensitive design as that process which minimizes the destruction of physical and life systems. This idea is also reflected in what we have come to know as pollution indexes. The questions normally posed, for instance, suggest an acceptance of negative values. To what extent can an area be urbanized while minimizing unacceptable water pollution or soil erosion? What are the acceptable levels of contaminants for foods, water or air quality? They imply that some loss, wastage or disruption to the environment is inevitable. It is also true that such questions are pragmatic, and based on the realities of current urban conditions, since cities are, after all, imperfect, not utopian, places. They may also be useful tools for constraint mapping where the least number of environmental constraints against a proposed use provide a guide to understanding the limitation of a site or environmental condition. These ways of thinking, however, involve aspects of negative constraint, and inhibit the creative solutions that come from a fully integrated marriage of ecology and human development. Design thinking must go further and ask: 'How can human development processes *contribute* to the environments they change?' Habitat building – creating those conditions that permit a species to survive and flourish – is a basic motivation of all life forms. In nature the by-products of these activities create situations where the altered environment provides opportunities for other species to profit by the change. The action of beavers damming streams, making ponds and cutting forest clearings has extensive impacts on the forest ecosystem. The temperature of the pond may rise above the tolerable level for brook trout, or the dam may impede the migration of fish upstream. On the other hand, drowned trees, while they may cause the end of a food supply for some species, create favourable conditions for others. The eventual wet meadow encourages the growth of aquatic plants necessary to support moose. Over time a new succession of vegetation will invade and cover the area. The by-products of one form of life become useful material for others.

In human terms, the negative consequences of human-made change on the environment occur when the necessary linkages are not made. A house or an entire suburban development is an imposition on the land when the resources necessary to sustain it are funnelled through a one-way system: water supply–bathroom tap–drain–sewer system–river–lake or ocean; or food–supermarket–kitchen–dining-room–dump. The by-products of use serve no useful functions. The concept behind integrated life-support systems is to make these linkages. They actively seek ways in which human development can make a positive contribution to the environment it changes.

The principles of energy and nutrient flows, common to all ecosystems, are applied to the design of the human environment. The unwanted products of the life cycle become the requirements for another. The recycling of organic products restores soil fertility and its capacity for production. Recycling of used water maintains groundwater levels and water purity. The sewage treatment facility, constructed to collect and process sewage and operating as a human-made wetland, provides a new and rich habitat for a wide variety of wading birds that may not have inhabited the area previously. Stormwater conservation improves the quality of water entering streams and rivers. It maintains aquatic life and soil stability, and provides opportunities for restoring impaired habitats. Where change may be seen as a positive force to enhance an environment that has been degraded, rather than simply minimizing its further impact or loss, the chances for a constructive basis for urban design will be enhanced. Thus, this principle suggests that development should sustain and reuse the resources it draws on as a benefit, rather than imposing them on the larger environment as a costly liability. Natural processes become internalized into human activities, rather than being incorporated into human affairs when society thinks it can afford it.

Making the most of opportunities is also the basis for ecological restoration – bringing natural systems back to a state of ecological health and re-establishing bio-diversity and resilience. Bio-diversity is also linked to cultural history and with restoring both human and non-human habitats in larger bio-regional contexts. Thus it is not usually a return to a purely 'natural state', in the absence of human history. It involves the creation of new landscapes – a mix of the natural and the human that may not have existed before, but which recognizes the interdependence of people and nature in the ecological, economic and social realities of the city. At the same time, functioning natural systems cannot be 'improved' upon. Efforts to enhance degraded environments can only begin the process of restoration, after which nature takes over the business of establishing a self-supporting diverse environment.

Making visible the processes that sustain life

Much of our daily existence is spent in surroundings designed to conceal the processes that sustain life and which contribute, possibly more than any other factor, to the acute sensory impoverishment of our living environment. The curb and catchbasin that make rainwater disappear without trace below ground, cut the visible links between the natural water cycle, the storm sewers that dispose of it into streams, and the lakes and rivers that ultimately receive it. We are unaware of the ecological degradation that occurs to aquatic life and to the beaches which have to be closed after a heavy rain. Soft fruits grown in warmer climates and transported thousands of kilometres to cold ones are available in the supermarkets in winter. In Chapter 3 I discuss the significance of community efforts to restore the land, and how the urban forest parks in Zurich are managed to produce saleable forest products in full view of the public. They illustrate the importance of making the natural and human processes that sustain urban life visible. Professor Tjeerd Deelstra, of the International Institute for the Urban Environment at Delft, has suggested that in industrialized countries the management of urban resources is anonymous. The supply of electricity and water, the processing of waste, are not visible to urban people and consequently they do not feel responsible for them. 'You simply turn on the light or open the tap and light and water are there. You buy your food in a shop . . . and put your garbage outdoors and it will be removed.'[27]

Deelstra also makes the point that visibility is essential in economic and political terms. Paying cash to use a toll road is better than paying an annual tax through the bank for the use of one's car. Policies should capitalize on the visibility of the environmental consequences of human actions in the process of daily living. One Dutch city, for instance, calculated how many square kilometres of forest should be planted around it to store the carbon dioxide generated by local household emissions. It showed that there was insufficient available open space in the city to compensate, a fact that

Plate 1.6 Making visible the processes that sustain life. Waterfronts. Redevelopment in Capetown, South Africa. Shipbuilding and repairs within a recreational environment, making visible the industrial and trading functions of the waterfront.

reinforced the need to conserve fuel, gas and electricity at home.'[28] Thus it may be said that making processes visible is an essential component of environmental awareness and a necessary basis for environmental action.

A basis for an alternative design strategy

In the preceding pages we have seen how, in the presence of cheap energy, the urban environment has been shaped by technologies whose goals are strictly economic rather than social or environmental. This has contributed to an alienation of city and country and a misuse of urban and rural resources. We find a preoccupation with leisure as the prime function of urban parks, while other functions which the unbuilt environment of the city as a whole must serve to maintain environmental quality are largely ignored. Health has been understood to mean the promotion of healthy human bodies, not healthy life systems as a whole. We find a preoccupation with aesthetic design conventions that are more concerned with 'pedigreed' landscapes than with the forms that have evolved from the necessity of conservation. The amounts of energy and effort spent creating them do not justify the results when alternatives exist that are cheaper, more effective and more rewarding. Our primary concern is how the city can be made environmentally and socially healthier; how it can become a civilizing place in which to live. As ecology has now become the indispensable basis for environmental planning in the larger, regional landscape, so an understanding of the altered but still functioning natural processes within cities becomes central to urban design. The conventions and rules of aesthetic values have validity only when placed in context with underlying bio-physical determinants. Design principles, responsive to urban ecology and applied to the opportunities the city provides through its inherent resources, form the basis for an alternative design language. They include the concepts of process and change, economy of means that derives the most benefit from the least effort and energy, diversity as the basis for environmental and social health, connectedness that recognizes the interdependence of human and non-human life, an environmental literacy that begins at home and forms the basis for a wider understanding of ecological issues, making the most of opportunities that integrate human with natural processes at a fundamental level, and making visible the processes that sustain life.

We seek a design language whose inspiration derives from making the most of available opportunities; one that re-establishes the concept of multi-functional, productive and working landscapes that integrate ecology, people and economy. As environmental issues gain an increasing sense of urgency for the future of cities and their regions, it is becoming increasingly necessary to meet new goals in the way we shape future landscapes. Urban land as a whole will be required to assume environmental, productive and social roles in the design of cities, far outweighing traditional park functions and civic values. Many of the problems generated by the city, and imposed on its larger regional environment, will have to be resolved within it – a recognition of the impact of the human footprint on nature. All the city's environmental and spatial elements may then be drawn into an integrated framework, to serve according to their capabilities, as producers of food and energy, moderators of micro-climate, conservers of water, plants and animals, and amenity and recreation. The following chapters will examine various opportunities for achieving this strategy in accordance with the principles that have been outlined here.

2 Water

Introduction

In his book *Bellamy's Europe* the botanist, David Bellamy, has this to say about Venice, Italy's city of water without parallel:

> Wherever you walk in Venice, not far beneath your overheated feet is one of over 22 million wooden stakes, the majority of which are as sound as the day they were driven into the soft silts of the lagoon. The reason for their long lasting service is that the lagoon muds are, and presumably always have been, deficient in oxygen . . . tourists and local inhabitants should rejoice in the fact that the waters of Venice are polluted. All the time there is excess organic matter pouring through the canals, the water will be bung full of bacteria thriving on the products of decay and in so doing using up the oxygen dissolved in the water and that helps protect the all important piles.[1]

It may seem to be a contradiction in terms to be suggesting, as does David Bellamy, that 'pollution is a good thing'. Yet in many ways the knee-jerk reaction to pollution inhibits a creative approach to the problem. Conventional wisdom needs to be constantly challenged. While it is ironic that Venice should still be standing thanks to its polluted water, this fact reflects a need to explore the physical and biological properties of water, the way its natural cycles are affected by the city and the implications for urban design when alternatives to its current uses and management are examined.

Natural processes

Hydrological cycles

The vast, never-ending cycle of distillation and circulation known as the hydrological cycle is a well-known phenomenon. The most important feature is its dynamic quality. Water is constantly being replenished. Evaporating off the oceans it circulates over land masses, falls as rain or snow, percolates below the surface and is returned to the ocean via rivers and lakes. At every point in this movement some water is constantly being returned to the atmosphere as water vapour, to circulate around the earth and fall again as rain or snow. As a result the atmospheric water content remains practically constant. Annual precipitation on the earth's land surfaces averages about 69 centimetres (or 98,300 cubic kilometres of water).[2] However, distribution varies enormously. There are vast desert areas where rain falls only rarely and areas where annual rainfall is as much as 1,000 centimetres. The amount of water evaporating from oceans is on average 9 per cent more than what falls back to the oceans as rain. This 9 per cent represents the amount of rain falling over land areas and which produces the flow of all the world's rivers.[3] The water that falls over the land as precipitation may follow a number of directions. Some of it is evaporated back to the atmosphere before it reaches the earth, some is intercepted by vegetation and is either evaporated, or transpired back to the atmosphere, some filters into the soil and underground reservoirs, and some runs off to enter streams, rivers, lakes and marshes on its way back to the ocean.

Forests and watersheds

Forests protect watersheds. They stabilize slopes, minimize erosion, reduce sediment inputs to streams and maintain the quality and temperature of the water. In the hilly uplands of a watershed, where water sources originate, forest vegetation greatly influences the movement of water from the atmosphere to the earth and back again. It performs a vital function of maintaining stream flows; reducing peaks and potential flooding, but sustaining flow in dry periods. About 30 per cent of the rain falling on the forest is intercepted by the canopy and evaporates back to the atmosphere.[4] Some of the water that does reach the ground percolates through the soil into streams. The root activity and decaying matter of the forest floor act as a sponge holding and gradually releasing a great deal of water. Winter snows trapped in the forest are also gradually released to the streams and rivers in the spring because the ground beneath the forest is less deeply frozen than open ground and snow melt takes longer in the shade of trees. Some is returned to the atmosphere by the biological processes of transpiration through the leaves of plants. So the forest has a great effect on the movement of water from the atmosphere to the earth and back to the atmosphere. Together with surface and underground water bodies it performs an important storage role.

Human impacts on watersheds

Measurements of water movement in some watersheds have been made for many years. The Canadian Forestry Service has reported on experiments on the eastern slopes of the Rocky Mountains showing the effect of logging on stream flow. After logging stream flows may increase by as much as 50 per cent, gradually diminishing as the forest regenerates. In one study, thirty-five years elapsed before the stream returned to its natural flow level.[5] The flow regime, or the timing of flows, also changes. Peaks of maximum flow after logging may increase by as much as 21 per cent and low flow by 90 per cent. Erosion increases due to higher run-off, and peak flows remove soil and damage productive land. One study of sedimentation from erosional processes showed a sediment concentration 1,000 times greater in streams running through cultivated lands than in streams from a pine forest. Sediment concentrations were seventeen times the prelogging rate in another study.[6] Thus forests play a vital role in sustaining both the supply and the quality of water.

Lake biology and transitions[7]

Lakes contain practically all the fresh water in existence and maintain the rivers and streams of most watersheds. An understanding of their biological and evolutionary processes is therefore useful to this study of urban ecology.

Lake biology

Lakes can be classified on the basis of biological productivity. Productivity is a measure of the quantity of life in all forms supported by an ecosystem. At the base of an aquatic

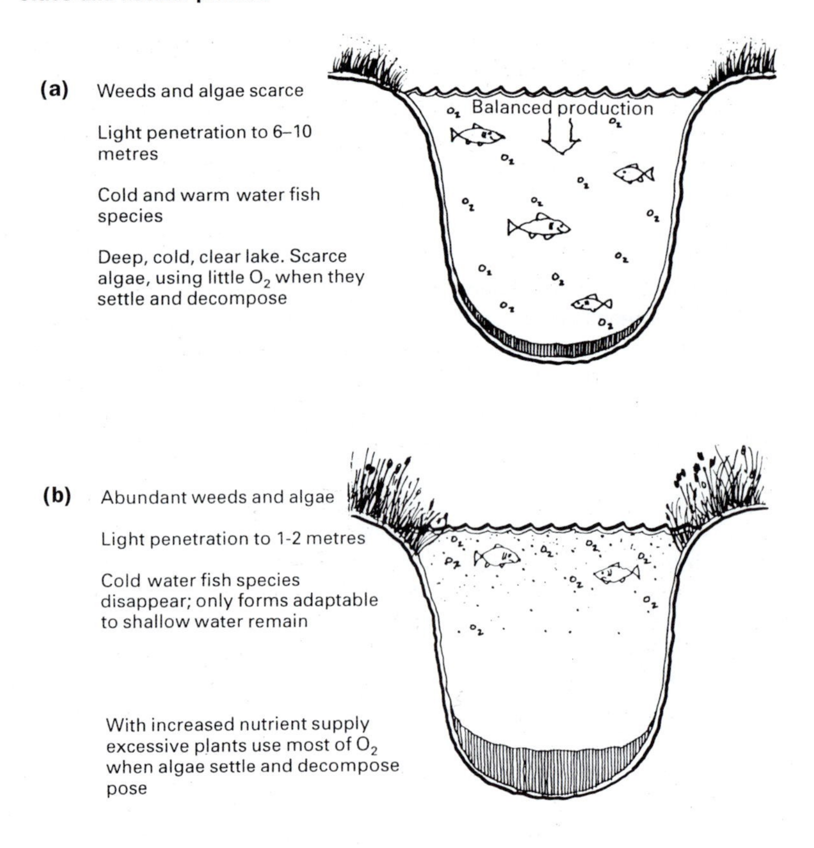

Figure 2.1 Lake productivity. The difference of biological productivity in oligotrophic and eutrophic lakes: (a) Oligotrophic lake (b) Eutrophic lake.

Source: Hough Stansbury and Associates. *Water Quality and Recreational Use of Inland Lakes.* Prepared for the Ontario Ministry of the Environment, SE Region, May 1977.

food chain green plants convert the energy of sunlight into food calories. The abundance and rate of growth of the algae depend particularly on the supply of dissolved nutrients which ultimately determines productivity at all higher levels of the food chain. Productivity may also be considered to be a measure of the organic matter produced by a system. Each stage of a food chain uses some of the production of the preceding stage, but much of it decomposes. In a lake, decaying organic matter falls to the deeper water where oxygen is consumed in the process of decomposition. Thus increased productivity will lead to increased use, reducing the exchange of oxygen between the surface and bottom layers of the lake. Oxygen depletion may proceed to the point at which there is not enough oxygen for various species of fish to survive.

Oligotrophic lakes are nutrient poor. The supply of plant nutrients to the lake is small in relation to the volume of water to which these are added. Productivity is therefore low. Usually oligotrophic lakes are deep, often with a mean depth of more that 15 metres. The paucity of nutrients limits the amount of algal growth. The water is usually very clear, and the algal production that occurs can take place through much of the water column, since light can penetrate to a considerable depth. As productivity is low, decomposing organic matter may not consume a large proportion of the dissolved oxygen in the bottom layers. Fish such as lake trout, which have a requirement for oxygen-rich, cool, deep-water habitats, are frequently found in oligotrophic lakes.

Eutrophic lakes are those that are rich in plant nutrients. They are frequently shallow, with a mean depth of less than 10 metres, so they may contain a small volume of water in relation to the supply of nutrients. The abundant supply of nutrients leads to a large algal population, which causes turbidity of the water. As algae require light for photosynthesis, the algal population tends to be concentrated near the water surface, preventing the passage of light to deeper water. In eutrophic lakes scums of algae on the surface of the water are not uncommon, and beds of rooted plants or filamentous algae may cover the bottom of shallow-water areas or bays. High productivity greatly increases oxygen consumption in the deeper water, by decomposition of organic material. In periods of minimal water circulation, either during the summer when the water separates into warm and cold layers (thermal stratification) or during winter ice cover, oxygen concentrations in the lower layers may be severely depleted. This creates conditions in which deep-water fish cannot survive, although warm-water species such as bass, sunfish and pike are typically found in eutrophic lakes.

Lake transition

Lakes are but a temporary feature of the landscape. Even the largest and deepest of lakes are transitory, undergoing a gradual process of change from youth, to maturity, to old age. Progressing even further, the death of a lake may be equated as the onset of swamp or marshland conditions. Thus the ultimate fate of a lake is to become filled with sediment and eventually to be supplanted by grass or forested land. Average natural sedimentation rates for lakes have been estimated at 1 millimetre per year. This means that approximately 10.5 to 15 metres of sediments have accumulated in most lakes since the recession of the last Ice Age some 10,000 to 150,000 years ago. It is not well recognized that changes occurring as a result of natural eutrophication are more complex and subtle, and proceed much more slowly than was earlier anticipated. In fact, because many deep lakes, such as Ontario's Lake Superior, have continued to remain in an oligotrophic state since the last continental glacier receded, natural eutrophication is really an immeasurably slow process. In general, then, healthy bodies of water are self-perpetuating biological communities that purify themselves through the interaction of aquatic plants, fish and micro-organisms.

Human impact on lake transition

In contrast to the slow natural evolution of lakes from an oligotrophic to a eutrophic condition, cultural or human-induced eutrophication can create conditions in decades or less which would take tens of thousands of years in the absence of human activities. The deterioration of Lake Erie, discussed below, is an example of this process. Induced fertilization in thermally stratified lakes of the Canadian Precambrian shield leads to an increased level of phytoplankton, and the onset of high levels of blue-green algae in late summer, and to reduced pH and dissolved oxygen in the deeper waters. Consequently, numerous lakes accumulate growths of planktonic blue-green algae along shore lines that create unpleasant odours when they decompose. Cold-water fish die off due to reduced oxygen in the lake's deeper waters.

In shallow, naturally eutrophic lakes, induced enrichment resulting from agricultural run-off, urbanization along the system and inadequate sewage treatment increase stresses on already productive environments. When induced enrichment reaches critical values its self-purification capacity is surpassed and rapid deterioration occurs. However, Vallentyne has pointed out that natural and human-induced eutrophication differ in two respects – rate and reversibility:

> Natural eutrophication is slow and for all practical purposes, irreversible, under a given set of climatic conditions. It is caused by changes in the form and depth of a basin as it gradually fills with sediment. To reverse natural eutrophication in this sense, one would have to scour out the basin again – a rather formidable task under any conditions. Human induced eutrophication, on the other hand, is rapid and reversible. It is caused by an increase in the rate of supply of nutrients to an essentially constant volume of water, without any appreciable change in the depth or form of a basin. As a result, eutrophication of this kind can be reversed by eliminating the sources of supply.[8]

A classic example of the ability of bodies of water to recover, and how interdependent natural processes and human actions are, was the destruction of the Lake Erie fishery in the Great Lakes Basin. During the 1950s, water pollution, due to excessive amounts of nutrients from sewage and farm run-off, caused the disappearance of the mayflies that provide food for many fish and birds. Nutrient enrichment of the water fostered prolific growth of algae and other plants in the lake. This resulted in the breakdown of large quantities of plant matter that used up oxygen and killed other aquatic life. Mayfly predators, such as perch, pickerel and bass, declined dramatically. During the 1970s concerted basin-wide efforts to reduce inputs of phosphorus to the lake gradually improved water quality. As a consequence the mayflies returned, and the fisheries recovered.[9] In 1987 the Government of Canada and the United States made a commitment as part of the Great Lakes Water Quality Agreement to develop a Lakewide management plan for the Great Lakes. The plan for Lake Erie is coordinated by federal, state and provincial government agencies. Under the guidance of these agencies the management plan unites a network of stakeholders in actions to restore and protect the Lake Erie ecosystem.

Urban processes

General

It will be clear from the preceding discussion that the biophysical processes of water, land and forests are an interacting system, profoundly influenced by human activity. Since water is a crucial component of the city's support systems, an understanding of these processes is essential to its wise use and management. This is true not only with respect to the larger context of regional watersheds but to the city itself. Many of the pollution problems that affect the water system as a whole begin in the city, so it is here that we must focus our attention. My primary purpose, using this knowledge as a base, is to examine how aquatic processes are altered in cities and what implications for urban design arise from these changes.

There are a number of issues that deserve our attention in an urban context. One is the problem of water supply; another is its disposal. Supply involves the problem of moving water from where it is plentiful – the rivers, lakes and underground reservoirs – to where it is needed in the cities. The natural hydrological cycle is short-circuited by water diversions, artificial storage in reservoirs and urban piped supply systems. Disposal involves the problem of removing it from where it has been used and polluted back to the rivers, lakes and oceans via urban drainage systems.

The urban hydrological cycle

Urbanization creates a new hydrological environment. Asphalt and concrete replace the soil, buildings replace trees, and the catchbasin and storm sewer replace the streams of

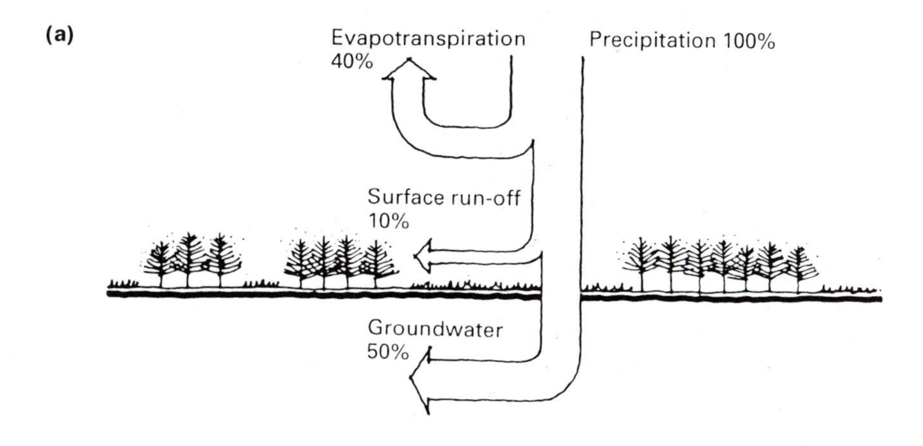

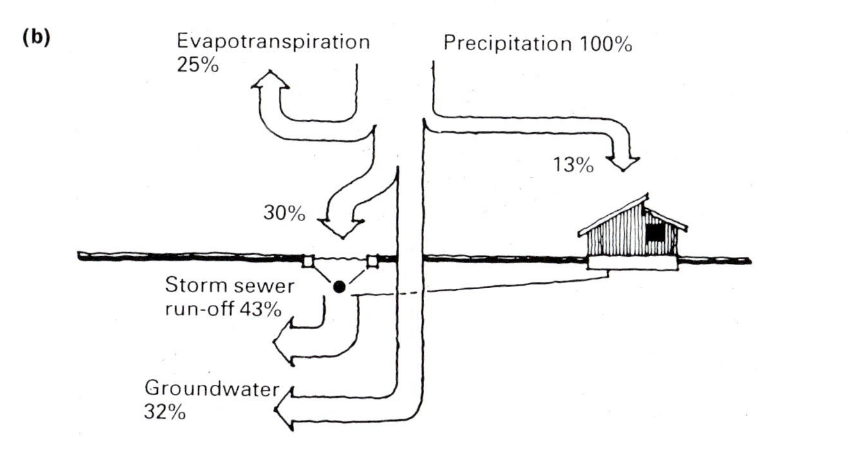

Figure 2.2 Hydrological changes resulting from urbanization: (a) Pre-urban (b) Urban.

Source: Ministry of the Environment. 'Evaluation of the magnitude and significance of pollution loadings from urban stormwater run-off in Ontario'. *Research Report* no. 81, Ontario, 1978.

the natural watershed. The amount of water run-off is governed by the filtration characteristics of the land and is related to slope, soil type and vegetation. It is directly related to the percentage of impervious surfaces. In forested land, run-off is generally absent, as a glance at the undisturbed litter of the forest floor, even on sloping ground, will show. It has been estimated that run-off from urban areas that are completely paved or roofed might constitute 85 per cent of the precipitation. The other 15 per cent is intercepted by streets, buildings, roofs and walls, and other paved and soft surfaces.[10] Piped drainage, designed to carry excess water away from urban surfaces, has two major effects, causing flooding and erosion and impairing water quality, particularly in those climates that suffer from sudden storms.

Flooding and erosion

In urban areas there is a tendency for flash floods and erosion. These are caused by large tracts of impervious paving and the concentration of water flows at specific points. The greater the run-off from a storm, the more swollen are the streams and the size of flood peaks. Conversely, the greater the volume from run-off, the less there is to replenish groundwater and streams. Thus rainfall is accompanied by extremes of flood and low flow. Discharge velocities are also higher than in natural conditions. This is well illustrated in the forested ravine lands that follow a meandering course through densely built-up parts of some cities. In the sudden storms typical of the summer season in Toronto, for instance, valley and ravine streams will rise from a sluggish trickle to a raging torrent in a matter of minutes, carving into already eroded banks and undercutting stone retaining walls. The damaging effects of erosion from water in river courses are greater the more urbanization seals ground surfaces. They increase substantially with the number of urban centres in upstream locations. Downstream solutions to problems caused upstream become more difficult and costly, requiring larger culverts, protection of urbanized floodplains, and the straightening and stabilization of stream banks. There is, consequently, wholesale destruction of natural ponds, marshes, plants and wildlife habitats. The effects of urbanization on the water cycle become clear when we see that on average 72.6 centimetres of water per year leave Ontario cities, of which 31.8 centimetres are from storm run-off. Annual stormwater volumes can exceed sewage flows in low-density urban areas.[11] Calculations show that in an urbanized area that has 50 per cent impervious surfaces and is 50 per cent sewered, the number of stream flows that equal or exceed the capacity of its banks would, over a period of years, be increased nearly fourfold.[12]

There is also a relationship between impervious paved surfaces and development densities. For instance, the amount of impervious cover produced by large lot residential areas may cover 12 to 20 per cent of the land. More intense commercial and industrial land such as shopping centres may produce more than 90 per cent impervious cover.[13]

Water quality

Conventional storm drainage systems impair water quality and disrupt aquatic life. Water pollution of lakes and streams results from combined sanitary and storm sewer systems that still serve many cities. The provision of overflows permits mixed sewage and stormwater to bypass sewage treatment plants in major storms and to enter rivers and streams. The wet-weather loads from combined systems may be many times larger than loads discharged from treatment plants during storms, and can equal or exceed total annual discharges from treatment plants.[14] Water quality is also affected drastically by sedimentation. Surface run-off from land cleared of forest increases overland flow, and consequently the amount of sediments and nutrients carried to streams from agricultural land uses. The result includes loss of soil, turbidity, accumulation of sediments and higher temperatures in water bodies which makes them highly eutrophic. The rate at which urbanization can change rates of soil loss (in kilos per hectare per year) from previously undeveloped land may be seen in the following list:[15]

Forest land	5.5–110
Cultivated agricultural land	110–4,360
Exposed construction sites	552–92,800
Developing urban areas	552–2,208
Developed urban areas	32–160

It has been estimated that in the state of New Jersey the sediment yields that can be expected for forested areas in flat land are 10.2 to 41.5 tonnes per 2.6 square kilometres annually, while those in moderately heavily urbanized areas are 25.5 to 101 tonnes.[16] In addition to sediments, a wide variety of chemical pollutants, salts, heavy metals and debris from roads and other paved surfaces add other contaminants to the water. They are carried through storm sewer systems and contribute to higher temperatures, BOD (biological oxygen demand) and a marked depletion of aquatic life in rivers, streams and lakes.

Some problems and perceptions

The development of a reliable water supply has been a primary determinant in the growth of large and densely populated cities. It has provided the means for controlling disease, raising public health standards and effectively fighting fire. At the same time, abundance and security of supply lead to the perception of water as a free commodity, and result in misuse, wastage and environmental pollution – a fact that is particularly true of cities well supplied with water. A review of how water systems in the city have evolved and how they have influenced urban life is therefore appropriate here.

Keeping clean

Whole cultures have evolved around the ritual of washing. Superb monuments in engineering, architecture and art have been its product. The best known historically are, of course, the Roman baths that were built in many of the cities the Romans occupied. As the great engineers of antiquity, they were concerned with hygiene and public health. Indeed, life in large cities such as Rome would have been impossible without a water supply and sewers. Although the first recorded water-supply system was built in 591 BC to water the fields and palace gardens near the city of Nineveh,[17] it was the Romans who recognized the need for clean water and secured their supplies from distant mountain streams via aqueducts, rather than from the River Tiber. The first aqueduct was

Figure 2.3 The ancient well in Piazza Cavour, San Gimignano, Italy.

Figure 2.4 The fountains of Versailles, France. The traditional role of water in cities – for drinking and social focus – became subjugated to its purpose as art, serving extravagant tastes of the nobility. It is said that when Louis XIV walked in the gardens, each sector of the Versailles waterworks was turned on just before the king arrived. There was never enough water to keep all the fountains running at the same time.

built in 312 BC by the censor Appius Claudius. These supplied the baths and city water. The Cloaca Maxima, Rome's famous sewer, was built about 600 BC and emptied into the Tiber. It is interesting to note that from that time no one bathed in the river or drank from it. By AD 226 no less than eleven aqueducts carried water to the city.[18]

In the centuries that followed, towns in the medieval and Renaissance eras were notorious for their lack of sanitary facilities. The streets acted as a dumping ground for the refuse of the town. However, as Mumford has observed, in spite of these practices the smallness of medieval towns, their accessibility to fresh air and open countryside and the importance of the ritual bath maintained greater health than might normally be expected – a situation that could not be maintained as cities grew larger.[19] Water was supplied via the public fountain which served three purposes: for drinking, as a centre of social life and as a work of art. The tradition of the ritual bath declined in the sixteenth century, which was marked by poor personal hygiene, and was reintroduced into England in the seventeenth century as a luxury. Bathing was seen as a curative rather than a cleansing process.[20] With this in mind, one may surmise that the extensive use of perfumes by royalty in the Court of Versailles (which became known as 'Le Cour Parfumé')[21] undoubtedly had a social and functional role to play. While important suites in the palace were equipped with bathrooms, these were dismantled in the reign of Louis XV – the Age of Reason.[22] Most of the palace's water supply was directed to the fountains, whose waterworks are said to have developed 100 horsepower with a capability of raising 4.5 million litres of water a day 150 metres.[23] In effect, the traditional role of the fountain as a supplier of water for drinking, and as a centre of social life, was replaced by its sole use as a work of art in the pedigreed landscapes created by nobility.

The private bathroom and the health standard of a WC for every family were made possible by the introduction of sanitary sewers in the nineteenth century. The storm sewer, introduced into many cities in the eighteenth century, was built to carry away rainwater, so it antedated the development of the sanitary sewer. Until this time human wastes were emptied by 'night soil men' hired for this purpose. The relationship of this practice to farming has been described by Tarr.[24] The waste, collected from households, restaurants and markets, was sold to neighbouring farmers for use on their land. The law often stipulated that cesspools could be emptied only at night, hence the name 'night soil'. The practice was widely followed in the New England and mid-Atlantic states of America, where wastes were collected in seventy-four cities. Baltimore fertilized garden crops with urban night soil as late as 1910. Tarr reports that 70,000 cesspools and privy vaults were emptied and sold to a contractor for 25 cents per load of 900 litres. Virginia and Maryland farmers bought over 54.5 million litres a year to grow crops such as potatoes, cabbage and tomatoes, which were subsequently sold in the Baltimore market.[25]

As cities became more densely populated, the cesspools proved incapable of handling the increased load of human waste. The introduction of piped water vastly increased consumption. The estimated consumption of 13.5 litres a day before piped supplies were introduced rose to between 180 and 270 litres. In addition, the consequent health hazards from cholera and yellow fever epidemics that periodically swept the nineteenth-century cities brought about the crusade that forced the cities to build the sewerage systems.[26] When connected to every household, they removed raw wastes directly to the rivers and lakes. Thus public health in the cities improved. In London prior to 1850, the old tributaries to the Thames were used only for surface drainage. By the mid-nineteenth century, largely as a result of the widespread use of WCs and a rapid expansion of the population, the River Thames received the untreated sewage of 4 million people. It is reported that in 1858, known as 'the year of the great stink', it became necessary to hang sheets soaked in disinfectant at the windows of the Houses of Parliament to counteract the smell.[27]

The problem of water use in cities has several facets. The first is the continued growth and improvement of technology. Today, consumption of water for domestic purposes in cities like Toronto is estimated, on average, to be 250 litres per person per day, with the bathroom being the largest consumer of water. The requirement that water must be of drinking quality regardless of its use increases pressure on the natural system for its continued supply, and on water filtration technologies for its standards of potability. The same quality of water services fire-fighting, car-washing, irrigation, domestic and industrial uses.

With this kind of use come the immense physical and biological problems associated with the return of used water to the natural system. The sewage treatment plant has provided a partial technological solution to the immediate problem of contaminated urban water. The increasing quantities of pure water taken out of the natural system and returned contaminated have been perceived as a problem requiring engineering, rather than biological solutions. The effect of removing the problem of disposal away from the city, while undoubtedly improving health and eradicating epidemics in Western cities, has also delayed solutions to the ultimate, larger problem of wasted resources.

Keeping your shoes dry

The storm sewer and catchbasin have for decades remained the conventional method of solving the problem of urban drainage and water disposal and have, until recently, been unquestioned. As the established dictum of engineering design, the rules have been simple – water drains to the catchbasin. It is here that, perceptually, the problem stops and connections with larger environmental issues of watersheds are not made. In commenting on the fragmentation of environmental concerns, human activities and policies that beset cities, the Royal Commission on the Future of the Toronto Waterfront began its first report this way:[28]

> At five o'clock in the morning in early July, the rain began, slowly at first and then with increasing intensity. It struck rooftops and trickled down gutters, gathered on driveways, parking lots and roads. Along its way the swirling stormwater picked up animal feces and herbicides from parks and yards, as well as asbestos, oil, and grease from roads. Before the rainfall ended, 4.5 billion litres of rainwater had gushed into the labyrinth of storm sewers under the metropolis.
>
> Between seven and nine o'clock, people began to rise, taking showers, brushing teeth, and flushing toilets in 1.5 million households. By eight o'clock, when most had left for work or school, 770 million litres of wastewater had gone down household drains and into the sanitary sewer system. Combined storm and sanitary sewers were overflowing, and a noxious brew of stormwater and untreated sewage was flowing into local rivers or surging towards the sewage treatment plants. By nine o'clock, the hopelessly overburdened treatment plants began to bypass partially treated effluent directly into the nearshore of Lake Ontario. Unseen by commuters, the brown and swollen rivers in the area disgorged their loads of sediments and toxic chemicals into Lake Ontario. At the river mouths, fishermen tossed their catches back into the lake, mindful of the signs that warned against eating fish. 'Just a reminder to stay out of the water at area beaches for two days after this rainfall' the radio voices continued. By mid-morning, public health officials would be testing water at the beaches lining the waterfront; in less than a week, many would be closed to swimmers. 'Cloudy this morning, sunny later with highs of 25 degrees.' Along with the afternoon sunshine would come high levels of eye-stinging smog. 'And cooler temperatures tonight, especially near the lake. All in all,' said the news readers, 'a pretty average day in Greater Toronto.'[29]

It is clear from this commentary that there are serious problems of discontinuity in our perceptions of urban systems and natural processes. The benefits of well-drained streets and civic spaces are paid for by the costs of eroded stream banks, flooding, impaired water quality and the disappearance of aquatic life. The quite understandable human penchant to keep pets, for instance, is one of the many factors responsible for degraded water quality. The products of animal defecation in public spaces, washed untreated into urban storm sewers, create health hazards for people, fish and aquatic life in general. There has been a failure to grasp fully the hidden environmental and economic costs of local water management practice, such as connecting downspouts to the sewer rather than discharging roof water directly onto the ground. The annual costs in erosion control, channelization of streams and underground stormwater systems are the engineering consequences of the need to keep one's shoes dry. Conventional urban design, in fact, contributes to the general deterioration of the environment by shifting an urban problem on to the larger environment, and by the failure to recognize and act on the relationships between human deeds and natural systems.

There are several ways of approaching these problems that differ from current practice. The first is the obvious and well-established conservation measure of using less, and many cities have adopted water economy policies. The second way of approaching water problems is related to the perceptions and values that have evolved from urban life. The traditional role of the fountain as a vernacular expression of water supply, social interaction and art became subjugated to an expression of art alone in the pedigreed gardens created by the rich and powerful. The aesthetic use of water has remained separated from its functional uses. The preoccupation with the expression of water as display or status symbol, defying gravity with extraordinary feats of engineering, is reflected in the modern city. The sparkling fountain gracing the civic square, symbolic of pure mountain streams, cascading falls and unspoiled places, is made possible by engineering technology – the filtration plant, hydraulic equipment and heavy doses of chlorine. While accepting the validity of its aesthetic purpose, one may well question some other assumptions and values that it represents.

The dichotomy between the euphoric image of nature and the realities of the urban hydrological cycle emphasizes in another way the isolation of urban life from natural processes. It is difficult to reconcile the image of sparkling fountains and children's paddling pools with debris-clogged and muddy streams or the blackened snow that piles up along streets in northern cities over the winter. This sense of isolation has also been aggravated by municipal design and practice. The storm sewer and catchbasin ensure that people remain unaware of where the water comes from or where it goes. Water is drained off streets, parking lots, pavements, plazas, school yards, front and back gardens and parks, and disappears from human consciousness, perpetuating environmentally destructive practices. The ways water and other urban life-support systems are used are not apparent: 'You simply open the tap and the water is there.'[30] The principle of visibility discussed in Chapter 1 is therefore crucial to environmentally responsible behaviour. Thus, the third way of approaching the question of water is to consider the opportunities for urban design that arise when the problems of disposing of the city's 'waste' water become opportunities for restoring hydrological and ecological balance, and enriching the experience and complexity of the city as a place. These alternatives will be examined later in this chapter.

Some alternative values and opportunities

The Thames revival[31]

The deterioration of rivers due to urban pollution is a fact that has been attributed to the costs of progress. But one of the most interesting examples of the process of deterioration and the seemingly impossible feat of bringing a river back to health is the story of the rehabilitation of the River Thames. The 40 kilometres of the tidal river within London are subject to daily fluctuations and have undergone two periods of gross pollution: once in the 1850s and again in the 1950s. As an ecological entity the Inner Thames was once a wilderness of marshes and reed beds, harbouring vast populations of birds. It is reported that spoonbills nested in the area of Putney Bridge up to the sixteenth century, and that Montague and Marsh harriers hunted in the marshes of south London. The water supported a thriving fish industry including salmon and sea trout. By the mid-nineteenth century the widespread use of WCs and the rapid expansion of the population created such polluted conditions that all fish were eradicated, together with the birds that fed on them. The first effort to build a sewage system was completed in 1874. Its main feature was the construction of intercepting sewers that discharged into the Thames at the extreme east end of London at Beckton and Crossness, avoiding the central areas.

By 1900, some of the river's quality had been restored by this work, but it gradually deteriorated again during the first half of the twentieth century due to increased discharges of effluent from both domestic and industrial sources. During the 1940s and 1950s the health of the Thames was at its low ebb, little better than an open sewer, containing no oxygen and permitting the survival of only specialized forms of life adapted to anaerobic conditions.

The crisis conditions of the 1950s led to the creation of several government committees which surveyed the Thames and examined the effects of pollutants. As a result of these investigations the problems of large pollution loads in an enclosed tidal system were better understood, and a programme of improvements was drawn up. It was recognized that the effect of pollutants on dissolved oxygen in the river was a critical factor. As we saw earlier in this chapter when discussing the induced eutrophication of lakes, when the oxygen content is entirely removed, oxidation of wastes cannot occur and anaerobic conditions are created. In this situation hydrogen sulphide is formed, giving off the familiar smell of rotten eggs. Fluctuating tides exacerbated the problem on the Thames, since it may take up to eighty days for water to be flushed to the sea in periods of low rainfall. In 1964, greatly enlarged and improved sewage works were begun, and these were completed in 1974. The most up-to-date filtration, treatment and aeration equipment was installed so that the fluid discharged into the Thames would be pure water.

Two criteria for estuarine quality became the basis for control. First, quality must be good enough to allow the passage of migratory fish at all stages of the tide. Second, it must support fauna on the mud bottom, essential for sustaining sea fisheries. The concept of 'pollution budgets' was introduced: these set the maximum quantity of pollution load for treatment plants that can be tolerated if the desirable water quality is to be secured.

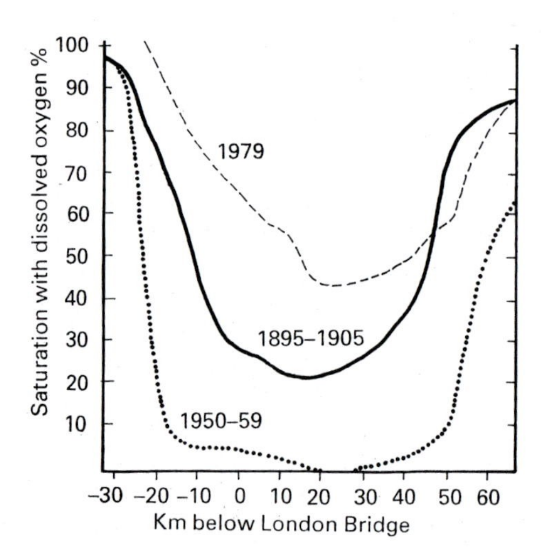

Figure 2.5 Oxygen sag curves, third quarter. The improvement in the quality of the River Thames since the nineteenth century may be seen in this graph showing the percentage of dissolved oxygen for the summer quarters of the three periods 1895–1905, 1950–59 and 1979. The river is now healthier than before records began and represents a return to the quality which was last shown in the mid-eighteenth century.

Source: L. B. Wood. *Rehabilitation of the tidal River Thames.* Unpublished Thames Water Authority paper, n.d.

A goal of 30 per cent dissolved oxygen was set to reach the desired quality. A third criterion was that toxic and non-biodegradable substances such as heavy metals should be excluded from industrial effluent. Early records of oxygen sag curves for the Thames indicated that the objectives had been met. The curve for the third quarter of 1979 showed that despite low freshwater flows, the average minimum dissolved oxygen was above 44 per cent. The Thames that had been devoid of fish for 48 kilometres between 1920 and 1964 now supported aquatic life. By 1975, no less than eighty-six species of freshwater and marine fish had been identified. Large flocks of wildfowl returned to the Inner Thames, and as a wintering area it was attracting 10,000 wildfowl and 12,000 waders.[32]

Impressive as the Thames revival was, however, societal values in the 1980s, and early 1990s, began to reflect broader international concerns for the health of city nature. Vallentyne's comment that 'river ecosystems are integrations of everything that takes place in their drainage basins'[33] reflects the reality that restoring rivers to a state of ecological health involves more than resurrecting downtown waterfronts, or reducing pollution, or making drinking-water less poisonous – a single-minded policy objective of many government agencies concerned with water quality. It involves an environmental view that sees rivers as life-sustaining natural systems, where value cannot be measured in terms of short-term economics, or how much electrical energy can be extracted from river diversions. It involves a view that sees economic objectives linked to social and environmental ones, where urban development processes contribute to, rather than detract from, the environments they change, where economy of means dictates the use of water, and where environmental education can begin by protecting life in the valleys and watersheds. The following case study illustrates how this holistic and ethical view of urban natural systems, combined with a growing determination for political empowerment by local people, can achieve integrated, visionary and pragmatic strategies for the rehabilitation of a river.

The healing of a river: the Don Valley, Toronto[34]

The Don River is one of a system of watersheds in the Greater Toronto bio-region that extend from the Oak Ridges Moraine (the aquifer recharge area that forms their rural head-waters) to Lake Ontario. The Don is significant to the City of Toronto in that it is the most highly urbanized and degraded river in the Greater Toronto bio-region, particularly in its lower reaches. It is a river whose essential natural values have been ignored and despoiled for over two hundred years: its once pristine waters now badly degraded from storm and combined sewers; its lower valley channellized and ransacked by an expressway, a four-lane road, railway tracks, transmission towers and salt dumps; its vegetation and wildlife diversity greatly impaired; its sense of wholeness, beauty and place a fond memory for those who had known it that way.[34] Moves to restore the river became an act of faith by the citizens of Toronto that grew out of the concerns of many people for the natural heritage of their city. Beginning as an informal citizen's organization, the 'Task Force to Bring Back the Don' was formalized and supported by Toronto City Council in 1990.[35] Its purpose was to begin the process of renewal of the most degraded part of the river that flows through the City of Toronto, and ultimately to initiate the restoration of the entire watershed.

The Task Force had four primary goals. First, to restore natural habitats and re-establish ecological diversity in the lower valley in ways that would integrate its cultural history with human and non-human values. Second, to regain public access and restore its sense of place as an urban river. Third, to reconnect the river to the lake and restore estuarine marses; Fourth, to improve water quality. These goals required an understanding of the valley as a whole, and the interconnectedness of its parts as a 'natural' and 'human' system. The concept of restoration, therefore, involved taking more dramatic environmental action where a return to a 'natural state', in a literal sense, was not practically feasible, or even desirable. It required both major and small-scale interventions, continuing over many years, to return the Don River to a state of health.

Historical background

Anyone walking through the Lower Don today would be hard-pressed to recognize it as a natural valley. Yet to bring it back to health it was necessary to understand its past – what the river was and how its present plight came about. 'We have to peel back the rubber mask of today's altered river landscape – the structural changes that have been imposed over its original form – to reveal its original underlying processes and patterns.'[36]

The river's glacial past

At various times, the Toronto area was covered with shallow seas, glaciers hundreds of metres thick, and freshwater lakes and rivers that had basins larger than those of today. Different plants and animals have inhabited the area, responding to changes in climate and land migration routes. Each left its own signature of sedimentary deposits and fossils

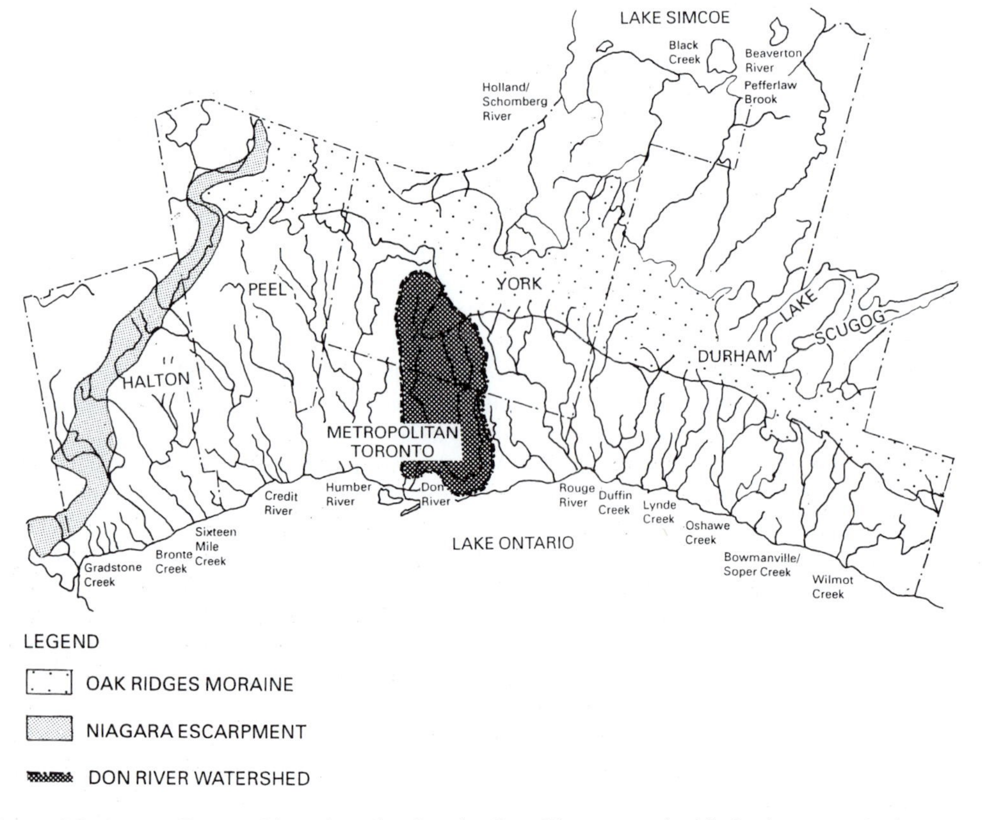

Figure 2.6 Greater Toronto bio-region showing the Don River watershed in its larger context.

Source: Royal Commission on the Future of the Toronto Waterfront. *Watershed.* Interim Report, Toronto, August 1990.

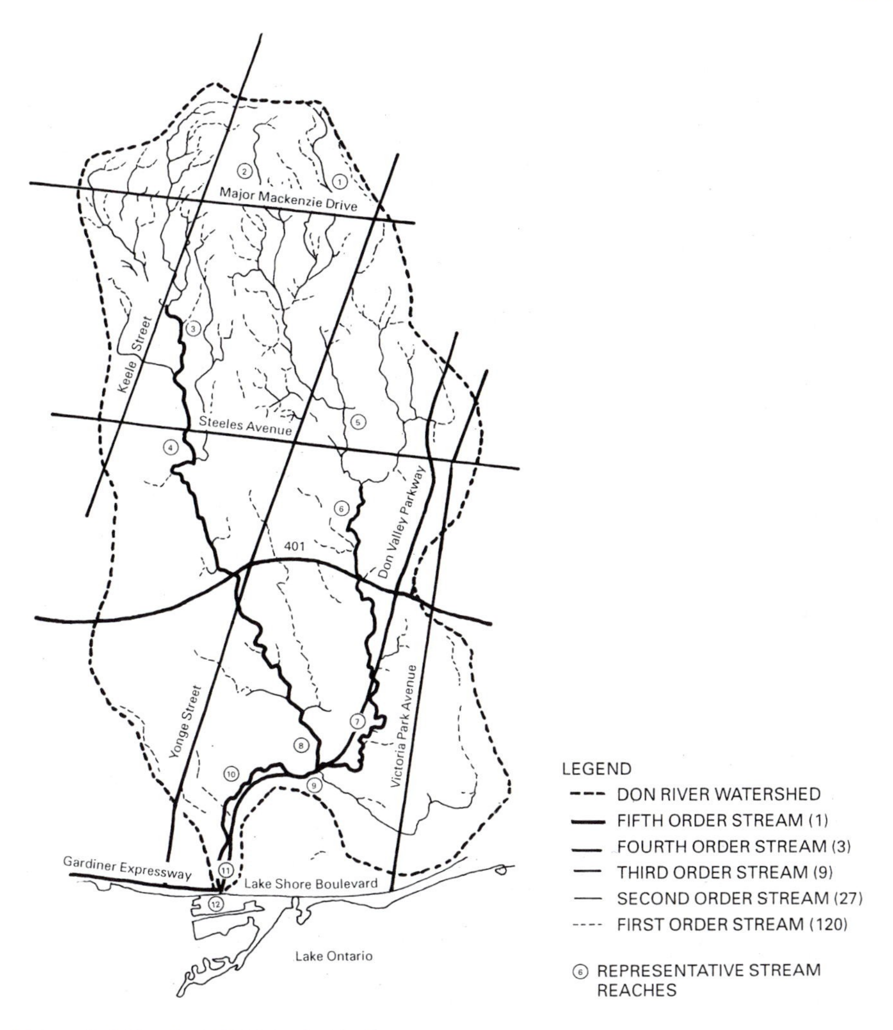

Figure 2.7 The Don River watershed.

Source: Task Force to Bring Back the Don. *Bringing Back the Don.* Toronto: Hugh Stansbury Woodland, Prime Consultants, in association with Gore and Storrie Ltd, Dr Robert Newbury, The Kirkland Partnership, 1991.

in the geological record. During the Pleistocene epoch which began a million years ago, three successive waves of glaciation buried the bedrock beneath thick glacial till. The Don was born at the end of that time, only 13,000 years ago. As the glaciers retreated, streams began flowing south from its source, a porous, water-filled ridge of glacial debris named the Oak Ridges Moraine. In the river's early days it flowed as two streams into a lagoon formed by a sandy baymouth bar that had been created by wave action and shore-line currents, and thence into the early glacial Lake Iroquois. As the glaciers continued melting, the land began slowly to lift up and Lake Iroquois shrank to become the present Lake Ontario. Now the Don flowed as one river out of its old lagoon and south across the flat sediments of what had been Lake Iroquois. As it entered Lake Ontario as one stream, the process of building a baymouth bar and backshore lagoon was repeated, forming the harbour islands spit and a protected lagoon known as Ashbridges Marsh. This was a vast fertile wetland habitat for fish and wildlife, and an important food source for the early Native inhabitants of the land.

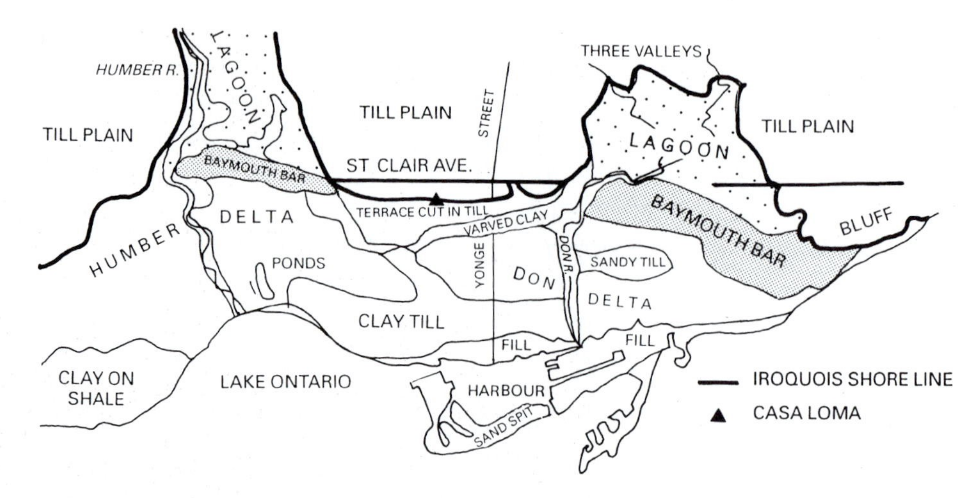

Figure 2.8 Physiography of the Toronto area. Note the shore of early Lake Iroquois in relation to the present shore line.

Source: L. J. Y. Chapman and D. F. Y Putnam. *The Physiography of Southern Ontario.* Ontario: University of Toronto Press, 1966.

Early settlers

Toronto was first settled in 1787 when surveyors laid out a city plan for the future capital of Upper Canada, and in 1793 John Graves Simcoe became its first Lieutenant Governor. The settlers harnessed the river's energy, built mills for lumber, flour, wool and paper, and mined the valley's clay and shale for brick-making, from which much of the early city was built. In less than 150 years, they cleared the lower valley of merchantable trees. The river was also perceived as a threat and an obstacle. Floods swept away mills and bridges, the river was an obstacle to the eastward expansion of the city and the great wetland, its mouth reviled as an unhealthy swamp, lent credibility to the argument that straightening out the river and filling in the marshes would 'secure the sanitary condition . . . to the said river'.[37] By the end of the century, engineers had turned the last 5 kilometres of the river's meanders, where it dropped its sediments, into a canal. The railways were built in the valley, and the Ashbridges Bay marshes were filled in to create the port lands, the most massive engineering project on the continent in its time, forcing the Don into a right-angle turn into the harbour. By the mid-twentieth century, the city had turned its back on the river, a gap between places rather than a place in itself. As a sensory experience it has become a forgotten place; unloved and unused. The roads and expressway, a legacy of the 1950s, have made the valley a transportation corridor, inhibiting access for walkers and cyclists. Other facets of present-day conditions provide an overall image of the river's environmental conditions.

Water quality issues

The Don watershed is 70 per cent urbanized. Consequently the river and its tributaries are subject to various sources of pollution from discharges from some 1,185 storm and combined sewer overflows emanating from municipalities north of the City of Toronto within the Don's 36,000 hectare watershed. Ninety-five per cent of the Don's pollution was estimated to originate from these sources.[38] In addition, there were flows from an existing water pollution control plant, snow dumps, landfill sites, agricultural and rural

Plate 2.1 The original pre-development river meanders imposed on the present channel.

Source: Steve Frost.

sources. Estimates indicate that given an approximate removal efficiency of pollutants of 50 per cent through the elimination of combined sewer overflows and treatment of stormwater, the City of Toronto on its own could expect to achieve a 2.5 per cent reduction in annual loadings to the Don River. This showed that the long-term health of the river was heavily dependent on the cooperation of jurisdictions within the entire watershed (Figure 2.7). Policies to improve water quality and to control peaks and lows of river flow depend, therefore, on mutual cooperation of municipalities throughout the watershed.

Vegetation and habitat

Once, forest cover played a vital role in maintaining the health of the drainage basin and the quality of its waters. It aided the infiltration of rainwater into the water-table and was important in controlling the rate of flow of the river. It kept the streams cool and maintained the diverse fish life and yearly salmon runs about which Lady Simcoe and others marvelled in the early days of settlement. Wetlands provided an immense diversity of aquatic flora and fauna as reservoirs of freshwater, and helped to even out the extremes of stream flows after heavy rain.

Most of the native vegetation of sugar maple, white pine, beech and oak in the lower valley has long gone, cut for farmsteads and settlements for export to England. Similarly, the wetlands that were associated with the bottomlands and river delta have also

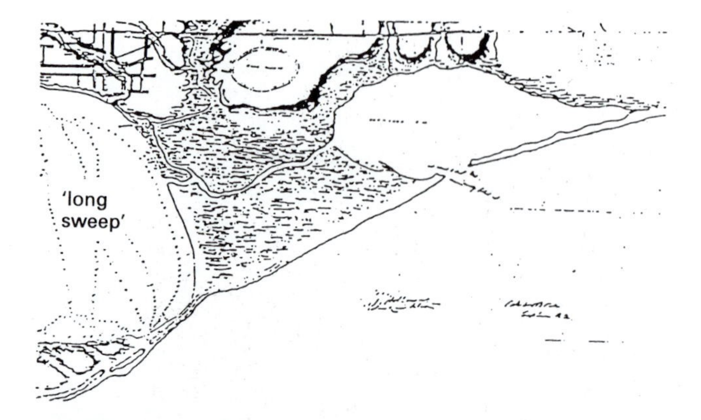

Figure 2.9 Early sketches of Toronto showing the relative positions of its present and proposed defences in 1846 are roughly drawn but purport to show conditions within the Ashbridges Marsh. 'Ashbridges Bay' is named for the first time. The Don in this map has two outlets, that in the north-east corner of the harbour being shown for the first time, although it was said to have been opened for tactical purposes during the war of 1812.

Source: T. H. Willans. 'Changes in the Marsh Area Along the Canadian Shore of Lake Ontario'. *Journal of the Great Lakes Research*, vol. 8, no. 3, 1982, p. 573.

Figure 2.10 A copy of the Toronto Harbour Commissioners plan of 1912 not only gives soundings throughout the area but differentiates between water and marsh. Cottage settlements on the peninsula and the sand-pit are shown, as are early industrial enterprises. The north outlet of the Don has been closed and the river diverted south to the Keating channel, making a right-angle turn before entering the inner harbour of Lake Ontario. Work on the channel has straightened the irregular edges of the mainland and landfilling has already consolidated the marsh to the north and west of the river/channel intersection.

Source: T. H. Willans. 'Changes in the Marsh Area Along the Canadian Shore of Lake Ontario'. *Journal of the Great Lakes Research*, vol. 8, no. 3, 1982.

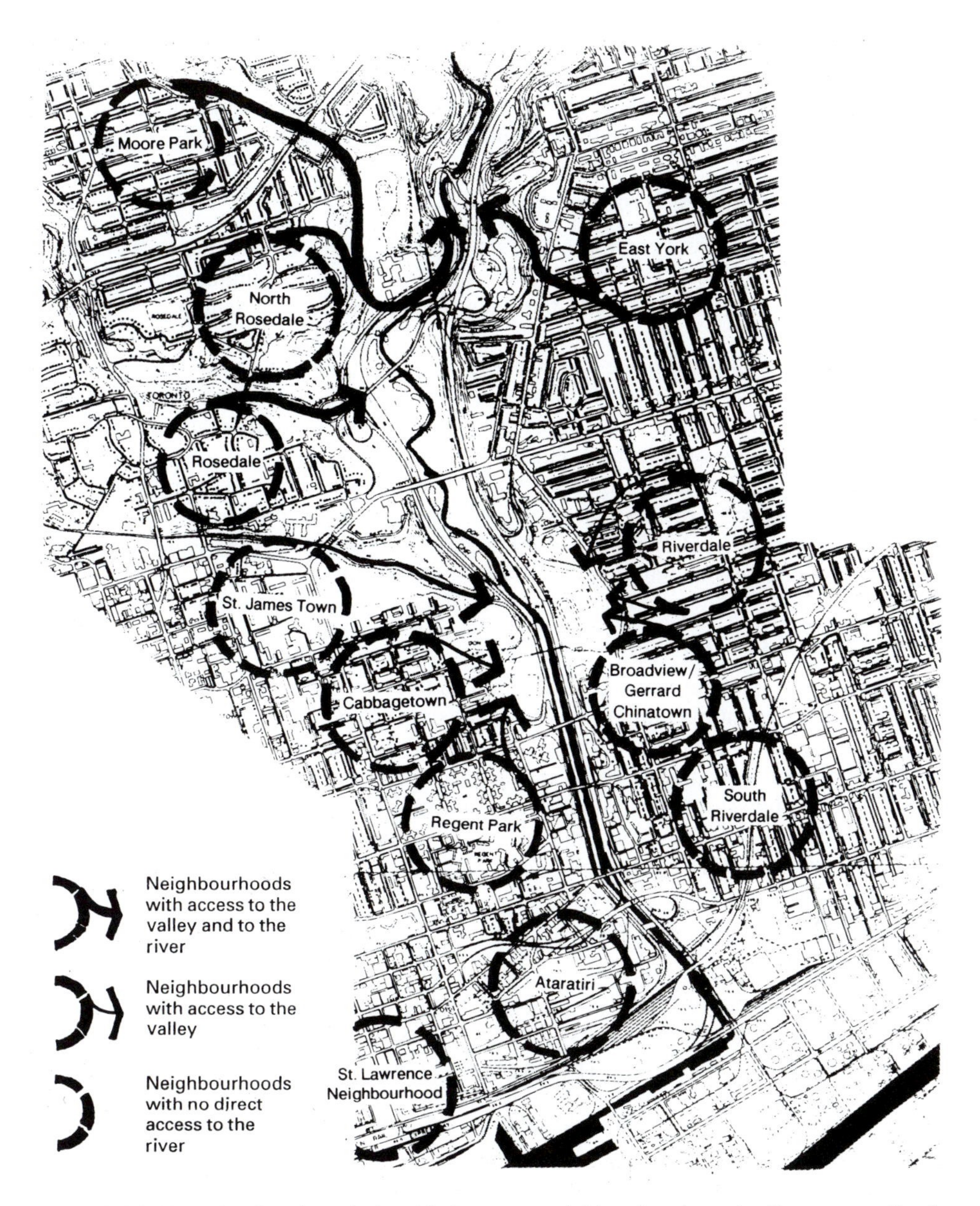

Figure 2.11 Map showing the relationship between neighbourhoods and valley access. It reinforces the Task Force's concerns that the river is a gap between places, not a place in itself. Much of the surrounding development has its back to the river, inhibiting access to the valley.

disappeared. Fish and wildlife populations are no more than a mere shadow of what they once were. Yet a record of the original forest and wetland communities that once dominated the area still exists in the ravine tributaries and places isolated by the valley expressway. The valley continues to serve as a migratory corridor for wildlife. Foxes, coyotes and other mammals are an indication that the river, while impaired, is far from dead and, is therefore a viable candidate for restoration.

The valley as sensory experience

The Task Force spent countless hours in the Don Valley documenting its cultural, industrial and natural features, evidence of its destruction and naturalization and its existence as sensory experience. Among the images documented as planning proceeded were:

- it was a difficult place to get into but, once in, almost impossible to get out of;
- the high levels of noise from the expressway where one had to shout to be heard;
- the magic of places made fortuitously inaccessible by road alignments;
- the presence of industrial objects: fences, dumps, PCB storage, oil drums, switching stations, advertising signs for the benefit of drivers on the expressway;
- the discovery of unusual plants and the sighting of a heron;
- the magnificent engineering of an early bridge crossing the valley;
- the coolness of the valley on a hot summer's day.

Land use and urban policies

It was also apparent that the Lower Don had never been perceived as a single connected valley in current planning policies. It was seen as an edge to a wide variety of planning districts, a transportation corridor unprotected by Toronto's Official Plan. It had, therefore, no status as a place in its own right, worthy of its own policies as a river valley. These circumstances, and a complex ownership pattern of public and private lands and often conflicting land uses within and surrounding the lower valley, influenced the long-range strategies for its restoration.

Strategies for the Lower Don

Restoration strategies were based on a number of interrelated courses of action. These were intended to re-establish its health and diversity, and bring the valley and its river back to the city, so that it could again be treasured and experienced as a valued and essential part of urban life. They were based on establishing a process of design strategies and policies that would allow remedial action to continue into the future.

Observation revealed the existence of three clearly defined landscape types or units, each intimately connected to the whole, yet each with a distinctive natural and cultural character of its own:

- the mouth of the Don where it makes a right-angle turn into the lake, and meets the elevated expressway and the Portlands;
- the channellized and physically restricted section of the river;
- the upper section where the river maintains its original meanders and the valley broadens out into a major floodplain.

These became the basis for future planning strategies. But first the river's hydrology had to be understood as a whole if the goal of reconnecting the river to the lake was to be realized.

(a)

(b)

Plate 2.2 Perceptions of the Don. (a) The urbanized channel: a forgotten place of industrial buildings, advertising signage and expressway interchanges. Early in Toronto's growth the river was seen as a barrier to eastward expansion of the city. It was straightened and encased in a channel by the end of the nineteenth century to facilitate the development of industrial land and to create another railway entrance to Toronto. (b) Heavily forested upper reaches of the Don begin to develop a sylvan character reminiscent of the natural river.

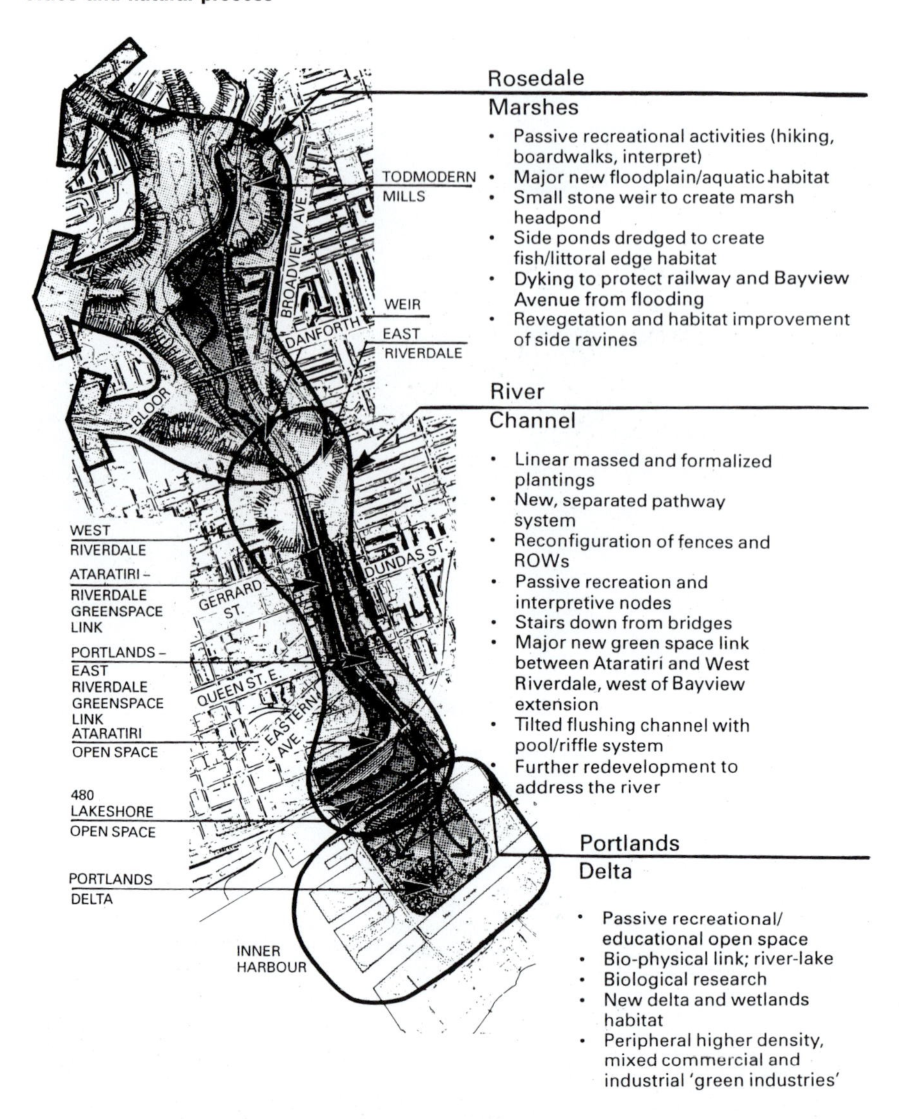

Figure 2.12 Lower Don strategy plan. Three distinct landscape units and the peculiar hydrology of the river formed the basis of the long-range restoration strategy.

Source: Task Force to Bring Back the Don, 'Bringing Back the Don', Toronto, Hugh Stansbury Woodland, Prime Consultants, in association with Gore and Storrie Ltd, Dr Robert Newbury, The Kirkland Partnership, 1991.

Hydrology concepts

The overall strategy for restoring habitat, open space and a delta/marsh in the lower valley depended on modifying the hydrology of the river using natural principles of river behaviour. As a consequence of glaciation and land history that left an almost flat gradient in the lower river, however, its natural tendency has been to fill with sediments carried down from its upper reaches. These sediments have been dredged continuously from the Keating Channel since the river was channellized in the 1880s to protect adjoining development from flooding. Preliminary studies suggested that to establish the new estuarine marshes at the mouth of the Don required an increase in the gradient of

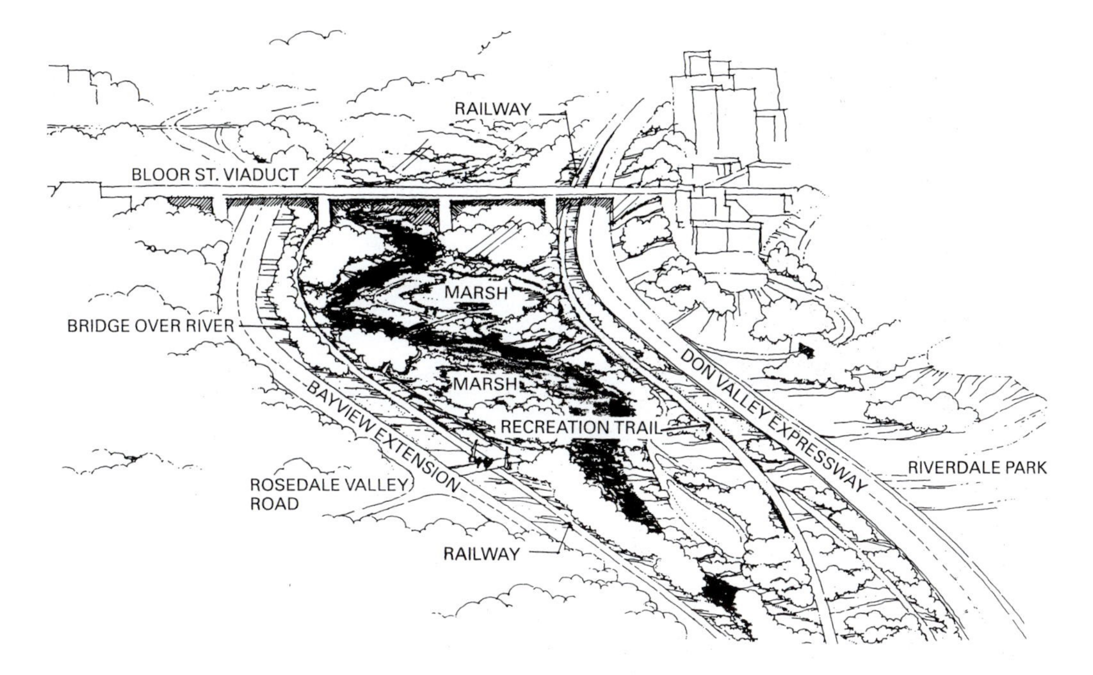

Figure 2.13 Conceptual sketch from the same location showing the proposed marshed, meadows, walkways and passive recreation.

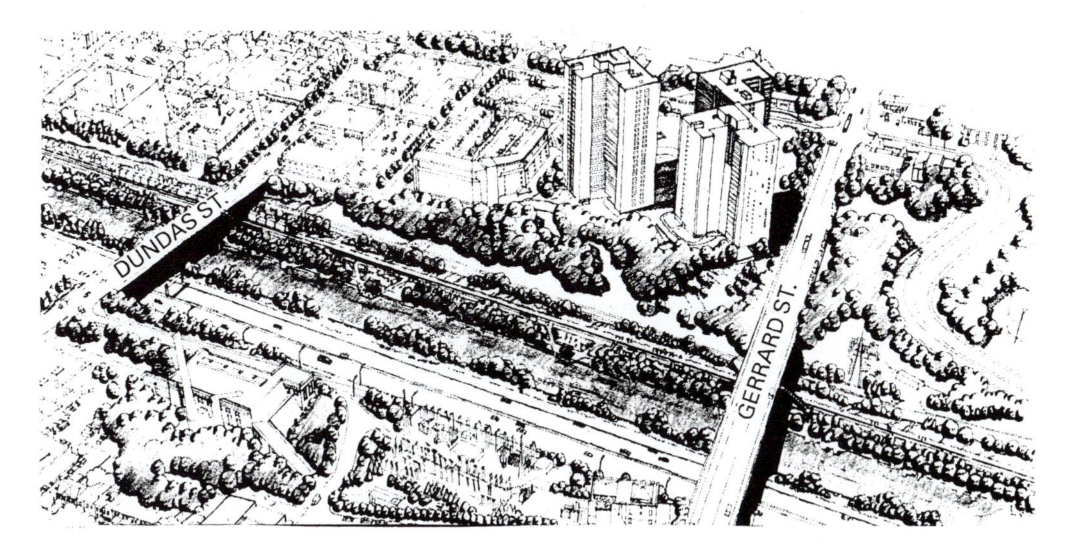

Figure 2.14 Conceptual sketch showing tree planting, bike and pedestrian ways, and stopping places for watching the water. Even with the flat gradient of the river a slight meander is noticeable and can be reinforced with new aquatic planting within the channel, since its width is much wider than that of the natural river channel.

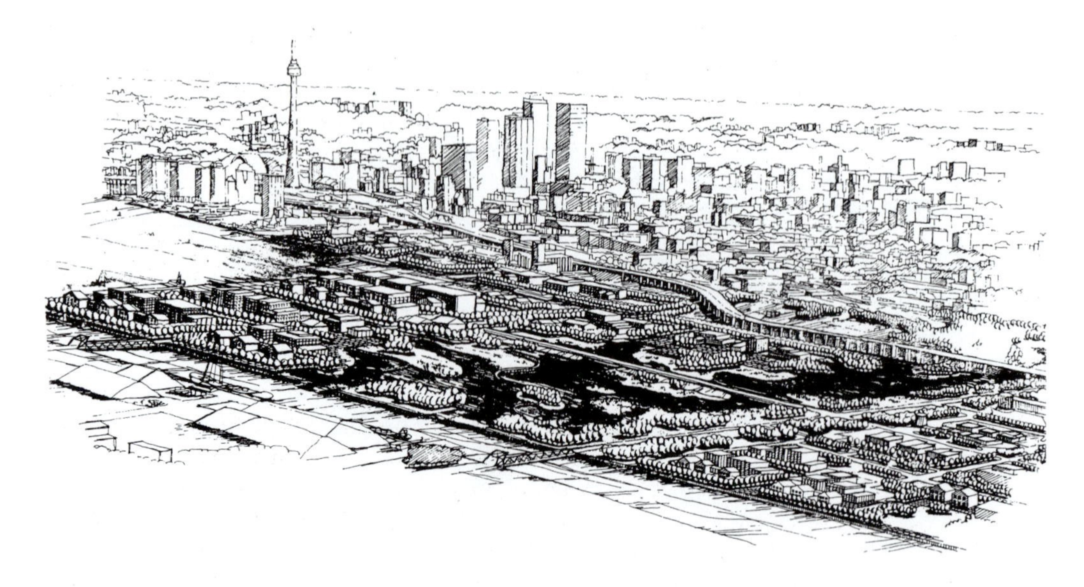

Figure 2.15 Conceptual sketch showing a potential marsh at the river mouth associated with future urban renewal.

the channel to allow the natural delta-building process in the lower valley to occur at the mouth. However, later hydrological examination showed that a more direct realignment of the river over a shorter distance to the mouth would allow a delta to form within the existing gradient. Thus, over time, the delta could link the river biologically and physically with Lake Ontario, and reintroduce a portion of the original marshes previously destroyed. To realize this objective required an integrated three-part restoration strategy up the river. This included:

- the re-creation of a delta and marsh habitats over time from accumulated sediments, where the river meets the lake;
- the re-creation of minor meanders within the confines of the existing channel to create fish habitats and naturally spaced pools and riffles (rapids) along its length;
- the creation of upland marshes, ponds and meadows, whose biological function would be to assist in improving water quality, and to help balance highs and lows in water flows and flooding.

Thus, the journey up the valley would pass through three distinct kinds of place. The river mouth would be integrated with future urban development proposals for the Portlands area. Surrounded by mixed 'green' industrial, residential and commercial development the visitor would look over a large open space – a delta and marsh with the river flowing through it to the lake. A research station would monitor the delta's evolution and provide information and exhibits on its research. People would learn about how the river works throughout the watershed, and about the restoration strategies in the Don River. They would also learn how changes to river hydrology can create new and more complex habitats for wildlife, places to spend quiet leisure time and reinstate beauty and dignity to a once degraded valley.

Walking north, the river's character would change from the open and marshy environment of the delta to a somewhat more formal, linear, urban character, with trees lining the water's edge and improving aquatic habitat, pathways for walkers and cyclists, places to stop, picnic, fish and listen to the sounds of running water, and new points of access along the length of the channel.

At the upper reaches of the Don people would enter a third landscape, a place that broadens out into a large floodplain, with marshes, meadows and forested slopes, walks, interpretive stops and picnic places. The floodplain experience would, in turn, change into a fourth landscape – the narrow forested ravine tributaries that make connections with the local communities and parks of the city. Throughout the walks up the Don there would be evidence of Toronto's industrial history: Todmorden Mills, one of the first water-driven sawmills in the valley; the Don Valley Brickworks, where the bricks that built much of old Toronto were once made.

Implementation: realizing the vision

The picture presented in this restoration strategy was intended to be bold, imaginative, yet pragmatic in its vision. It also needed to be incremental – staged over many years. The Task Force recognized that implementation was connected, first, with political agendas and funding that involved the formation of partnership agreements between various levels of government and private interests; second, with the adoption of the strategy into the city's official plans and policies; and, third, through small-scale incremental action by local communities. Among many implementation- and management-related strategies were:

- cooperative ventures between the city and other watershed municipalities, senior levels of government, the railways and other private interests and community groups

Plate 2.3 The beginnings of implementation of the Don. Staircase from the pedestrian bridge linking the two parks on either side of the river provides the first access to the bikeway in some 8 kilometres.

to improve water quality, create habitats, reforest valley slopes and provide access into the valley;

- establishing major open space links in the lower valley through acquisition, easements and related planning instruments available under Ontario's Planning Act;
- development of approaches to soil remediation that combine new technologies, policies and issues of legal liability;
- planning strategies that will establish a long-needed identity and value for the valley as a distinct place in the city, having its own planning area and official plan designation that protects valley property in private as well as public ownership;
- the creation of partnership agreements between various levels of government to resolve fractured landownership in the valley necessary to implement restoration strategies.

Developments since the early 1990s

Much restoration work was implemented on the ground in the final decade of the twentieth century. With the cooperation of the city, considerable reforestation of the valley's denuded parklands was begun. Stairs were built from the bridges crossing the valley linking the two parks on either side of the river, and providing access to the bikeway along some 8 kilometres of river corridor. A demonstration wetland was created in the river floodplain and smaller ones in the ravines. A major project was initiated in the mid-1990s by the regional Conservation Authority to restore the Don Valley Brickworks and its associated quarry – a heritage site of great significance to Toronto's natural and cultural history.

In its valley location the Don Valley Brickworks, abandoned since 1985, became a vital link in a chain of natural, cultural, industrial and historic places and events taking place in the Don River and its watershed. It is linked to the city's historic growth: the manufacturing of bricks that were made here for nearly a hundred years built much of the early city. Today, it is linked to the citizens of the city engaged in restoring the Don, and who ensured that it remain in perpetuity – a public place for passive recreation and environmental learning. The plan was based on interconnected themes: the internationally significant geology of the quarry's north slope that reveals the changing climates which the region has experienced over geological time and the animals that once lived here; its wetland gardens created to provide life for insects, fish and birds, and to restore water quality to a diverted, once buried stream that flows into the Don River, thus contributing to the renewal of the Don River. Its industrial heritage is based on the premise that history is a dynamic process of change and adaptation – learning from the past to create a relevance to present needs. The Brickworks buildings themselves have changed constantly over their history in response to brick-making technologies. Today they symbolize cultural continuity with the future introduction of pottery and ceramic crafts, with interpretation of brick manufacturing and the machinery that made them.

These themes together establish a central theme – linking heritage and environmental restoration in the Don Valley. It has significance because the restoration of urban places is inevitably tied to human activity and natural processes. The Brickworks also represents an inherent irony. An industry that once contributed to the degradation of the valley has now become part of its restoration – an appropriate way of understanding heritage as a continual dynamic process of renewal.

Plate 2.4 The Don Valley Brickworks – a new life for an old industry. The main access through the conglomerate of buildings that once manufactured bricks to the quarry that supplied the raw material. Arrival at the quarry from the brickworks entrance is a sensory experience of change from noise to quiet; from traffic to a created wetland environment within a vast space enclosed by quarry walls. The contaminated Mud Creek, diverted into the quarry, is purified as it flows through the marsh on the way to the Don River. Boardwalks take the visitor through different wetland areas and different views, habitats and educational experiences against a background of Brickworks buildings.

Plate 2.5 Transformation of a defunct water filtration plant to multi-purpose functions that include an outdoor theatre, experimental water purification with different species of aquatic plants, children's water play and gardens that feature Korea's native plants. Located on an island on the Han River in Seoul, South Korea, the entire design reinforces the experience of its original function adapted to new uses.

Source: Kim Jae Kyung.
Landscape Architects: Seo Ahn.
Architect: Sungyang Zoh.

Reflections on this grass-roots movement in 2003

The Don Task Force's continuing work over twelve years is a remarkable achievement, which leads to the question: What is it that has made this initiative successful? Why has it lasted so long? First, there is political acumen. The Task Force from the beginning aligned itself with the City that provided financial support for its day-to-day operations. It allocated a planner from the City Planning Department to act as a coordinator for the Task Force's ongoing operations, while maintaining control of its mandate. From a social perspective, this greatly reduced its workload and allowed the Task Force to make decisions on community projects, seek funding from a variety of private and government sources and return to their families in the evenings. Second, the Task Force, by its very nature as a grass-roots group, has gained massive support from the Toronto community and politicians, thereby ensuring that the Don's restoration is enshrined in political decisions. It has influenced the city's wet weather flow master plans. For instance, the Department of Public Works initiated a downpipe disconnection programme to encourage home owners to allow rainwater to drain on to their front gardens and lawns. In situations where they were approved for parking purposes porous paving was required. This has recognized that the city itself is part of the Don's larger floodplain and should therefore be required to contribute to water storage and infiltration. Third, it is an integral part of the Toronto and Region Conservation Authority's watershed strategy, and is included in the city's website. In addition, proponents for development projects associated with the Don are required to submit site plans for review by the Task Force. While it does not have the regulatory authority to approve or reject a proposal, its comments are influential since it acts in a similar way to a regulatory body. Fourth, starting as a small group and undertaking small and medium-size community-based projects it has established, over a number of years, a high level of credibility among bureaucrats and decision makers.

This experience strongly suggests that good planning for sustainability is tied to citizen empowerment. It is also important to recognize that the concept for the Don River's renewal is not a master plan, but a *strategy* to guide the work into the future. The long-term vision for achieving the fundamental goal is to heal the Don. The process of getting there will change and adapt in response to changing social, economic and political events over which no one has control. For example, the first design proposal for the mouth of the Don illustrated how the goal of reconnecting the river to the lake might be realized as part of the Portlands area (see Fig 2.15). After further investigation over a ten-year period, it became clear that this approach was technically costly, politically undesirable and unlikely to be implemented. A second proposal, made in 1991, was more economically realistic, hydrologically sound and politically supportable (see Fig. 2.16, p. 55). The plan has now become a key element in long-range plans for Toronto's waterfront and the restoration of the river mouth is one of four priority projects of the Toronto Waterfront Revitalization Corporation. The vision for the Don, therefore, has no finishing post because, like the city around it, it is an organic entity in the process of continuing evolution.

As an ongoing process of renewal and healing, the Don strategy involves key principles, including a fundamental understanding of process as a biological idea that is also integrated with social, economic and political agendas, economy of means where the most benefits are available for minimum input in energy and effort, and environmental education, where the understanding of nature in cities becomes part of a learning experience that begins with community empowerment and action.

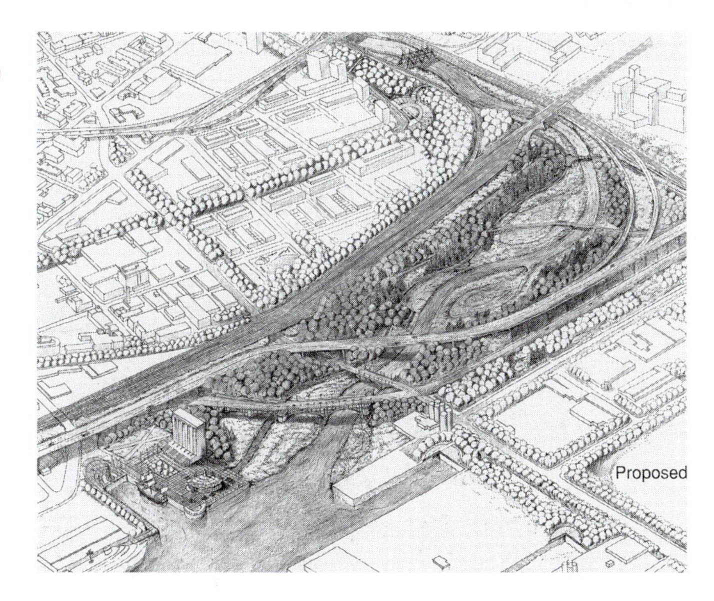

Figure 2.16 Conceptual sketch showing the alternative location to the west of the Keating Channel. Its design and form would create a delta into the inner harbour. This approach has been adopted by the City of Toronto and has become one of the priority projects for the renewal of the waterfront. The restoration integrates hydrological, engineering and ecological functions with art and architecture and the preservation of historic engineering technologies as monuments to the past.

Source: ENVISION the hough group.

Regional watershed developments

After the report's completion in August 1991, the political and community impetus for the Don's restoration began to grow. The Toronto and Region Conservation Authority, whose mandate is to protect and manage the watersheds in the Toronto bio-region, initiated a Watershed Task Force in 1992 – the Don Watershed Regeneration Council – a body made up of citizens, experts in various fields, local politicians and government staff. Its mandate was to develop a coordinated strategy for the watershed and oversee its implementation. This task, to restore a largely urbanized watershed ecosystem, was recognized as a long-term, complex prospect and a daunting effort. A crucial part of the Council's ongoing work required that it oversee the publication of a Report Card every three years to evaluate and measure progress of the river's health. A few of these measures are included here, which, from the perspective of success and failure to meet targets, could be seen to be typical of most urban watersheds undergoing similar management goals.[39]

- *Flow pattern*: Because the watershed is 80 per cent urbanized, the time it takes for rainfall to reach the Don River has continued to decrease, resulting in higher and more rapid peak flows. The river's flow pattern associated with impervious surfaces is the most difficult to change but it is crucial to how the watershed functions. Consequently, there has been no significant change since 1997. Stormwater controls to manage flow required for all new development have been in place since the early 1980s, and some progress has been made in innovative techniques such as the use of porous pipes to enhance infiltration into the ground and home owner programmes to disconnect roof downspouts from sewers. Comprehensive programmes combining lot level, conveyance and end-of-pipe measures have not yet been developed.
- *Water quality* (aquatic habitat) has remained unchanged since 1997. The river has too much sediment especially following rains or snow melt periods, and trace metals, nutrients and ammonia remain a problem. The insecticide diazinon, the herbicide 2,4-D used for lawns and the agricultural herbicide d-ethyl atrazine, were

detected under dry and wet weather conditions. Only diazinon exceeded the International Joint Commission water quality guidelines for the protection of aquatic species.

- *Wetlands*: In addition to creating aquatic habitat, wetlands store water, a crucial function for evening out the rapid peak and low flows of urban rivers. By 1997 only 49.5 hectares of the Don watershed were wetland. By the year 2000 a total of 2.7 hectares of new wetland were created, well short of the target of 12 hectares.
- *Fish*: In 1997, eighteen species of fish were present in the watershed but were completely absent at 12 per cent of the sampling stations. By 1998, twenty-one species were found – an insignificant change from the eighteen species found in 1997. They included two species, rainbow trout and brown trout, that were stocked as fry between 1997 and 1999 in the Upper Don. Five weirs acting as barriers to migration of salmon have been removed or made passable by creating rapids, surpassing the target of three by 2000. Destructive flows, however, are likely to significantly inhibit spawning.

Many of the problems facing the Conservation Authority's efforts had to do with lack of financial support by various levels of government. This has severely limited monitoring, making it difficult to keep track of what is happening in the Don ecosystem. Issues include the inconsistent and inadequate state of stormwater control, the need for large-scale action on the Don's fishery, and the environmental and political issues associated with the creation of an amalgamated city in the early 1990s. The Report Card is, however, frank in its assessment of success and failure to meet targets. Yet it illustrates very well the difficulties inherent in implementing watershed plans on the ground, no matter how imaginative and far-reaching they may be.

Dhaka: a city shaped by rivers

Up until now we have explored community initiatives in returning an urban river to health in a North American environment. There is no question that the movement to restoring degraded natural systems is more prevalent in the developed world where environmental issues are now receiving some attention. The following case study also has to do with rivers and cities, but in the context of a very different scale, political environment and cultural values.* Unlike Toronto, which is a large metropolitan city with a small river flowing through it, Dhaka, the capital of Bangladesh, is a city set within a vast system of floodplains and rivers that empty into the Bay of Bengal. Toronto's community is environmentally concerned and politically active. Dhaka lacks both citizen involvement in government decision-making and environmental priorities. It also has a long history dating back to the seventh century, during which time the city evolved as a unique water-based civilization that grew from its strategic river location to become an Asian centre for trade and commerce. Today, Dhaka is one of the most rapidly growing metropolises in the world, ranking twenty-second among the fifty largest cities, with rich cultural traditions in arts, crafts and fabrics.[40] The way in which new development is currently occurring, however, is diametrically opposed to the environmental and cultural principles from which the old city evolved.

---

* This case study is based on a student major paper by Arifa Hai, completed in November 2001 to fulfil the requirements of the Master of Environmental Studies (MES) degree from York University, Toronto. She also has a degree in architecture from the Bangladesh University of Engineering and Technology (BUET).

The questions that arise, therefore, are these. First, how can contemporary development respect traditional forms of urban design while recognizing the needs and imperatives of contemporary life (the past informing the present)? Second, what are the environmental and cultural approaches to new development that should be incorporated into a planning process aimed at realizing a sustainable future? Third, what are the impediments to a planning and implementation process based on such principles, and how can these be overcome? To answer these questions it is necessary to understand, first, the interdependence between people and the environmental imperatives from which a unique urban form and cultural traditions evolved, and second, the ways in which urban growth is occurring today.

Bio-physical context of Dhaka

Dhaka is at the heart of three major rivers, the Brahmaputra and the Ganges, that originate in the Himalayas, and the Mehgna that begins in the hills of Shillong and Meghalaya of India. This network of rivers empties into the Bay of Bengal, forming a delta of approximately 1.5 million sq.km.[41] The city is surrounded on four sides by tributaries of these rivers which include the Buriganga to the south (where Dhaka originated in the seventh century) within a floodplain environment where the formation of land is constantly evolving from hydrological forces. Historically, settlement occurred on these heights of land that were free from annual floods. The Madhupur terrace, extending from north to south, is one of these heights of land, or 'islands' during floods, and was formed by alluvium and clayey silts. It was on this terrace that Dhaka was originally settled and grew over the centuries. It was this physical limitation that played a vital role in containing urban growth.

Hydrologically, Bangladesh has two distinct seasons: the dry season from October to March, and the monsoon from April to September, which is when the river floods occur. In a larger regional context, the floodplains and wetlands associated with the Madhupur terrace perform a crucial function in storing flood waters, and have long played a central role in the life, culture and economy of the city. The silts and sediments carried by the rivers enrich the lowlands and improve their productivity. Fish and rice are the main sources of food for Bangladesh. The wetland environment, in effect, has provided the indispensable basis for the livelihood of local settlements, uniting the inhabitants into a coherent, subsistence-oriented economy, and forming the basis for Cyprus-gathering for feeding cattle.[42] During the flood season, watercourses become transport routes, and boats are the only form of transportation.

Overall, drainage and hydrological processes have played a key role in controlling and shaping urban growth. Historically, Dhaka was known as the Venice of the East, or the City of Channels.[43] The natural drainage ways, abandoned river channels and depressions that form a network of waterways throughout the island landscape are interconnected with the surrounding rivers and are vital to storing and retaining flood waters during the monsoon. They have also provided old Dhaka with an excellent drainage system, and waterways for communication, trade and commerce.

In addition to functionality, however, the city's waterways established an urban design character based on water from which cultural traditions flowered. Among these is the 'ghat', a typical stepped river wharf which is used to anchor boats and where much of the commerce, entertainment and recreational boating still takes place. The narrow, winding streets of the city provide protection from the sun and are oriented to the river, incidentally using the wind as a natural air-conditioner. The Banyan tree is embedded in the philosophy of Banalee life, symbolizing the unpredictability of life and success, and the tree is where gatherings, festivals and fairs take place beneath its shade.

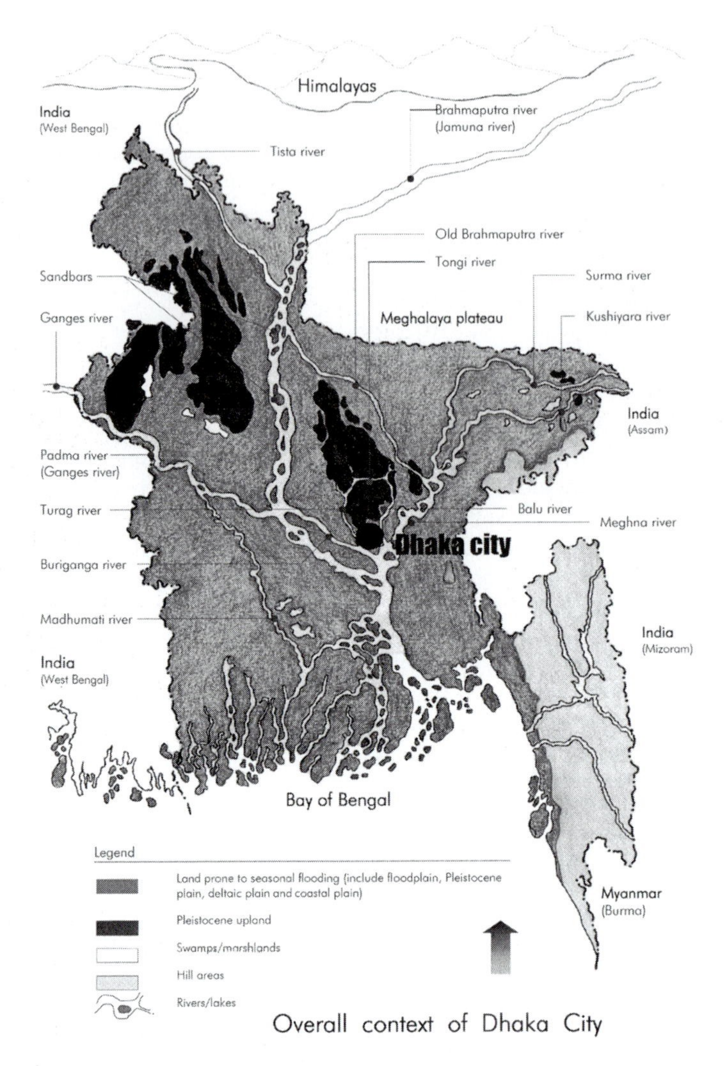

Figure 2.17 Dhaka occupies the central region of the vast floodplain that is the meeting place of the great rivers – the Padma, Brahmaputra and Meghna. They flow south from the Himalayas to the Bay of Bengal and complete the annual hydrological cycle under the influence of the sub-tropical monsoon climate of Bangladesh. Dhaka's land form is an island of Pleistocene upland that is relatively free from yearly flooding – a primary reason why the city was located here. Thus the city is strongly defined by its surrounding rivers. Since it is located north of the sea it is generally well protected from natural calamities, and its high elevation has made it an important regional trading centre. Dhaka has been a centre for economic and cultural activity for hundreds of years.

Source: Arifa Hai.

Plate 2.6 A view of the Buckland Bund road where the city first began. The diverse activities in and around this location and its role in linking land and water have made this a distinctive part of the city.

Source: Arifa Hai.

Plate 2.7
The wetlands and areas subject to flooding have traditionally supplied Dhaka with fish and aquatic vegetation for cattle feed as a major source of livelihood.

Source: Dr. S. M. A. Rashid, Bangladesh Centre for Advanced Studies in association with the Nature Conservation Movement. *Wetlands of Bangladesh* published by BCAS, 1994.

Plate 2.8 A drainage canal in one of Dhaka's neighbourhoods. These drainage areas were once linked to Dhaka's surrounding rivers and have historically functioned both as parks and as water storage areas in times of flood. Almost all of them have now been filled in for urban development (see Plate 2.10c).

Source: Arifa Hai.

Commercially, Dhaka was, throughout its history, famous for its fabrics and as the centre for weaving muslin, closely connected to local climate and the artistic traditions of the region. The muslin industry came to an end under British rule and trading policies.[44] The Buriganga River has influenced the history, culture and economy of the city, from the days when a tiny settlement was founded in the seventh century, to the bustling mega-city that exists today.[45] Its role in the life of the city, it is said, is similar to London's Thames and the Seine of Paris and serves as the major navigation link between Dhaka and other parts of the country.

The evolution of the city may be seen to be a response to natural forces and security from flooding – the product of necessity and limitation. It is Dhaka's unique physical form, cultural traditions and economy that have given the city its regional identity and sense of place. What is relevant today is that the functional connections between climate and land form have, over centuries, become the basis for sustainability.

Dhaka's environment in the twenty-first century

Bangladesh achieved independence in 1971. Since that time and up to 1997, Dhaka's population grew from 2 million to 8 million – one of the highest rates of urban population growth in the world.[46] These population growth pressures, along with the economic and political changes occurring in the country, led to rapid change in the city's landscape. The unusually high flood that occurred in 1988 and the two most severe floods in the country's history in 1998 resulted in a fundamental change in planning policy.[47] This policy, which represents a dramatic change in attitude towards how the

Plate 2.9 Chamery House with its traditional 'Ghat' in the Baldha garden. This was the favourite place of the great Bengali poet, Rabindranath Tagore.

Source: Arifa Hai.

city should grow, has had a disastrous impact on the city's environment. Traditional ways of building in response to river behaviour that had evolved through the centuries were no longer factors to be considered when confronted with the daunting task of accommodating a rapidly growing population. Today, the necessity of working with the processes of the land have been replaced with attempts to control them. Dams and embankments have been built in and around the city to protect it from annual floods, cutting off both the natural and cultural functions that traditionally were the heart of the city. Extreme water pollution from oil-slicks, industrial effluent and sewage, along with construction activities that interfere with normal river flows, is said to be placing the Buriganga River in a state of collapse due to human interference in its hydrological and ecological functioning which, historically, contributed so significantly to the origin and evolution of Dhaka. Yet, as recently as the 1970s, the Buriganga still provided fish from its waters. The streams, canals and natural drainage ways of the city are the essential interconnected system that has long been the water infrastructure within which Dhaka was built. Their functions as rainwater drainage ways, communications networks and flood storage, and their cultural and aesthetic significance to the city, are rapidly being lost as they are filled in for development. In addition, their destruction plays a significant role in the present-day increase in flooding. The lowland marshes in the region have similar functions but also remain unprotected due to a lack of support from the Bangladesh legal system.[48]

The structure plans for Dhaka

One might surmise from this depressing picture of environmental degradation that Dhaka suffers from uncontrolled runaway urban growth that ignores planning constraints. However, two planning reports, the first in 1959 and the second in 1995, outlined the direction in which Dhaka should grow. The 1959 master plan focused in great

(a)

(b)

(c)

Plate 2.10 (a) The Dhaka palace complex showing its grassed site and the Buriganga River. The building and its site are adapted to the annual seasonal floods. (b) Recent development in Dhaka. The walls keep the floods out but also ignore the traditional form of the city that has responded to its surrounding rivers. (c) This street was once a drainage way similar to the channel shown in Plate 2.8. The street follows its original meandering alignment.

Source: Arifa Hai.

detail on the northern extension of the Madhupur terrace, optimizing the higher flood-free land for future expansion of the city. It also gave much attention to the reclamation of the south-west lowlands, considered to be of prime importance for the city's rapid growth in the south. To quote the report: 'In the long run, however, it may well be possible to reclaim selected areas of low-lying land nearer to Dhaka with which to augment the present limited supply of building land.'[49] The 1995 structure plan focused on planning strategies that emphasized new areas for urbanization. It recognized the uncertainty of future events and concentrated on the fundamentals of infrastructure and development, leaving more detailed issues to be resolved nearer to the time when they would occur. Infrastructure, however, did not include internal waterways. The plan also proposed extending urban growth to the lowlands to the Balu River to the east and south-east, and up to the western swamps of the Turag River. It incorporated:

- flood protection walls along the north-western bank of the Turag and north-east bank of the Balu to minimize adverse hydrological effects, risks to life and economic damages;
- building flood-regulated ponds in low areas using excavated materials for landfill;
- raising low (regularly flooded) areas with landfill obtained from dredging the Balu and Turog Rivers.

The hope was expressed that as a result of the implementation plan the city would *get rid of seasonal flooding* (my emphasis). It was the abnormally high floods in 1988 that persuaded policy-makers to protect new developments in these ways. While this is understandable when we look at the city's rapid growth, it represents a total abandonment of the philosophy and practice that made Dhaka a special place. Yet the question remains: What was the cause of these floods? Are they traceable, for instance, to deforestation of the headwaters of the Ganges and Brahmaputra Rivers, or to some other cause?

Flooding: causes and effects

Every year during the monsoon season the Himalayan region receives a great deal of publicity in the press due to disastrous flooding in the vast plains of these two rivers. The peasants of the mountain regions of Nepal are blamed for the events of deforestation and bad farming practices that are said to exacerbate downstream flooding. A 1997 study by Thomas Hofer, a Swiss hydrologist,[50] was undertaken to promote better understanding of the processes that produce annual floods. The study suggested that the extent to which floods in Bangladesh are influenced by climatological and hydrological processes outside its boundaries is a key question to be answered. The study highlighted the fact that it was time to abandon the tradition of blaming mountain farmers for the inundations in the plains. It showed that the floods are a normal process of the Himalayan highland–lowland interactive system, and are poorly correlated with the dimensions of floods in Bangladesh. In addition, flood peaks originating in the highlands were found to be levelled on their way downstream through the plains. With regard to floods in Bangladesh, processes in the Meghna River catchment seemed to be decisive; in some years there was a link between floods of the Brahmaputra in India and floods in Bangladesh. Flow conditions in the Ganges River, however, were important only if their peaks coincided temporarily with those of the Brahmaputra. Finally, the study made the point that the plains have tremendous potential for surplus flood storage, particularly wetlands and ponds. This storage capacity plays a decisive role in the mitigation of flooding further downstream in the watershed.

The Hofer report's conclusions were as follows:

- The Himalayas have a negligible impact on the floods in Bangladesh. Therefore, laying the blame on mountain farmers for flood catastrophes far downstream should be abandoned. At the same time this does not relieve them of their responsibility to farm sustainably.
- Building embankments along large stretches of the major river systems in Bangladesh (as suggested in the Flood Action Plan in 1988, for example), does not take into account the complexity of the flood processes. In combating monsoon floods, a much more important measure is the preservation of natural ponds and swamps for storing surplus water, since the plains have tremendous storage potential. Thus this storage capacity plays a decisive role in the mitigation of flooding further downstream in the watershed.
- Interviews with the people in flood-affected areas revealed that monsoon floods are not perceived as the main problem: the people even count on these inundations, since Bangladesh farmers depend on them and have, over generations, developed sophisticated strategies to deal with, and also take advantage of, the cycle of floods and dry periods.[51] For instance, wetland habitats are retained after the cycle of floods recedes and provide the basis for many commercially important activities that include harvesting of waterfowl and fish. They provide an important source for reed harvesting, and support major rice farming. During the dry season, domestic livestock are grazed on the marshes.

Implications for the sustainability of hydrological processes

Given the conclusions of this hydrological study, it is clear that the direction that was being given to growth strategies for Dhaka is not sustainable. One might conclude that such policies are counter-productive and economically self-defeating, given that large-scale human impacts on the local and regional landscape are themselves the cause of greater flooding. Irrespective of cultural or environmental concerns, the economic implications of attempting to control floods, as opposed to working with them, are increasingly being recognized by governments in the developed world.[52] Much of this discussion may therefore be seen as fairly obvious. Sustainable development is, after all, enshrined in United Nations goals for future growth and development. But the current trend for Dhaka's growth seems to lack this perspective, and this raises a number of questions about why the city took the opposite approach to achieving a sustainable future.

- Given that the Hofer study was written two years after the 1995 structure plan recommending future directions for growth, did it have any influence on how future growth should occur?
- Did Dhaka's long history of environmental and cultural development have any influence on decisions about future growth?
- Was there public discussion about the 1995 plan among the various sectors of the community, particularly farmers and other groups with trades based on traditional flood patterns?

While these questions cannot be answered directly, there are clues that may shed some light on the issues they raise. First, there is the issue of the floods. An interview with a Dhaka government official, who was directly involved with flood protection for Dhaka City, revealed that flood protection walls and embankments were paid for with foreign aid. The design and implementation of the work was directed by foreign consultants.

The official was also unaware of the Hofer study, but he was, at the same time, quite certain that its research and conclusions were both wrong and one-sided, i.e. that deforestation in the Himalayas by Nepalese farmers had a direct impact on Dhaka's flooding problems. In his view, the reasons for building floodwalls and embankments were to protect people, houses and farmers' crops. It is apparent that the long experience of the Dutch, in matters of flood protection and control of rivers, has been persuasive in Dhaka, even though conditions in the two countries are arguably very different. Applying that experience under such circumstances therefore has the potential to end in unexpected and sometimes disastrous consequences. The consequences of eliminating Dhaka's waterway network and filling the low plains have been known, not only to Hofer, but also to the Bangladesh Centre for Advanced Studies since the early 1990s.

Second, there is the issue of Dhaka's long environmental and cultural history and whether it has had any influence on how the city should grow. There are reports in the Dhaka press of Dhaka's poor air quality, its highly polluted rivers, the lack of water treatment infrastructure. There is also the disappearance of many of its significant cultural features – its many lakes and bodies of water, precious open places and the disappearance of culturally significant trees. One can surmise, therefore, that alternative ways of thinking about the form of the city, and Dhaka's adaptation to its unique environment and cultural history, had not been considered priorities, but it also suggests that they are not listening to the concerns of the public.

Third, there is the question of citizen participation in decision-making. Meaningful involvement by citizens in the affairs of the city are largely absent. Developments by the private sector, for instance, are being allowed to occupy open spaces with no input from the community and at considerable environmental cost. The 21 March 2003 edition of the *Dhaka Daily Star* reported on a demonstration by protesters against a housing development that is being built on a much valued open space and public playground, but to no avail. The newspaper article sums up the issue:

> What came as a rude shock to the residents of New Colony is that the construction of so many buildings would have reduced their locality into a highly congested place with no empty space . . . government planners are supposed to be aware of the environmental hazards associated with eliminating parks, water bodies and playgrounds. But the incident is one more example of a great indifference on the part of the housing ministry to the environmental needs of the city.

One might also add, indifference to citizen concerns. This assessment may be typical of many other cities such as Jakarta where development is encroaching on the linear public parks along the Ciliwong River with major environmental consequences. As a colleague reported, 'in this decentralization era, local governments must derive their own revenue. Since Jakarta does not have natural resources, they tend to push the development of trade and industry sectors . . . almost all the green parks are being converted into buildings. It is sad because now we are facing a big natural [flood] disaster. We have never experienced such a major flood before' (a reference to the major flood that inundated much of the city during the summer of 2002).

Linked to the previous discussion is the issue of corruption. As a colleague who has some experience of working in developing countries suggested, in places where everyone talks freely about corruption and is happy to do so, it signals an acceptance of the practice, thus making it legitimate. It is accepted as part of life, as it is in Bangladesh where the priority is making money, not making a better environment. Thus decisions about development agendas such as building on floodplains or creating developable land with embankments and floodwalls have little to do with plans based on environmentally friendly decision-making and cultural history. Corruption does not have

Plate 2.11 Precedent. The Grand Canal, lagoons and islands. Venice was built on a series of islands in the Gulf of Venice, created by river deposits that required wood piles to support buildings. Founded in the early ninth century, its inhabitants depended largely on fish for survival. The city's form and layout, its renowned arts and culture and its centre of trade were shaped by its location. Its canals functioned as streets and transportation routes, with its pedestrian ways, piazzas and urban spaces forming an independent system. Venice has remarkable similarities to Dhaka.

the public interest in mind and, as a Bangladeshi colleague put it, 'talking to well-educated people about the environment, or about the city's future growth, or social and cultural issues is met with derision, and the comment that such things are not a consideration. It is very difficult to make people from the developed countries understand this situation.' In effect, the links between environmental issues and their economic implications have not been made. It is true that many developing countries base their decisions on ideas and advice originating in Western countries, yet the developed countries are beginning to abandon these ideas. Controlling rivers with levees to permit floodplain development and then attempting to deal with the economic consequences of flooding is just one example. Projects such as the damming of rivers for water storage that are shared between countries (the Ganges which flows downstream through India and Bangladesh is an example) are unlikely to be sustainable when uneven distribution results in excessive floods downstream in the wet season and drought in the dry season. Yet, irrespective of the environmental consequences, large dam projects are supported and financed where they are lucrative for the donor countries.

Clearly, another direction is needed if urban growth in Dhaka is to be sustainable. Yet, as the largest city in Bangladesh, Dhaka may be described as a fledgling democracy with elections and women in positions of power. Its communities are interested in what is happening to their city with the courage to make their collective views known in spite of threats from developers. Much of the developed world had, until recently, similar attitudes to environmental and social concerns as Bangladesh has now. The acceptance of sustainability as an idea that links economy, society and environment is only now beginning to emerge.

With this in mind, the following principles and accompanying precedents serve as a theoretical guide to how the city might grow in the future: how its rivers could be

Plate 2.12 Precedent. Harbour City. A proposal to build a major residential community of 60,000 people linked to the Toronto island shore line in Toronto bay. It is a contemporary expression of Venice where water is the dominant landscape. Since Dhaka is politically determined to continue growing in its existing location, Harbour City provides a parallel approach. Major extensions of Dhaka's urban growth, built on piles or fill, could maintain the flood capacity of the lowlands in the monsoon, and maintain their agricultural functions in the dry season.

Source: Zeidler Partnership, Architects.

reintegrated into urban life in ways that reflect its history; and the criteria for sustainability that recognize its river landscape and cultural traditions as its essential infrastructure.

Principle 1. Reinstate the unique character of the Dhaka landscape: its land form and water networks as essential infrastructure and a framework for new urban growth.

- Restore the rivers surrounding Dhaka as the major thoroughfares for commercial, commuter and recreational traffic, replacing cars and trucks with ferries and barges.
- Reinstate the buried channels, canals, natural levees and wetlands into the essential infrastructure of the city as a basis for growth strategies and neighbourhoods.
- Maintain the flood storage potential of the lowlands in urban expansion extending beyond the Madhupur terrace.

Principle 2. Respect cultural traditions in the creation of contemporary urban environments.

- Make the water edges of the river and canals accessible, allowing people to experience their functions, behaviour and beauty through all the seasons.

(a)

Plate 2.13
Precedent. Ontario Place on the Toronto lakeshore. A recreational and cultural aquatic park, built in 1971 on a series of made offshore islands that celebrates the lake and its waterfront. It is connected to the mainland by a pedestrian bridge. There are shops and places to eat and drink, children's water play, outdoor concerts and boating on the canals. There are pastoral landscapes for quiet activities and a marina. Linked buildings, built on pylons and raised above the water, accommodate exhibits and restaurants. The Imax movie theatre also stands free in the water. (a) The marina and the buildings hovering over the water are the main focus of Ontario Place. Its two islands on either side are linked by a pedestrian bridge. (b) The Forum, celebrating Toronto's multicultural communities, was where dance, music and entertainment of all kinds could take place. Four hills surrounding the stage area allowed the audience to sit on the grass and participate. The Forum was later dismantled.

(b)

Source: Zeidler Partnership, Architects. ENVISION the hough group, landscape architects.

- Capitalize on the potential of the water's edge for culture, entertainment, education, tourism and commercial opportunities to bring life and a focus of activities to this part of the city.
- Provide canopy-shaded open-space networks in new city developments for culturally and socially significant traditions to thrive.

Principle 3. Encourage sustainable contemporary urban design strategies for the city that reflects Dhaka's tropical climate, monsoon rains and flood patterns.

- Incorporate new pedestrian streets for walking and cycling routes and local architecture into the system.

Wastes as valued opportunities

As the restoration of the Don illustrates, water pollution issues are best addressed when they become part of an integrated design strategy that combines biology and technology, social and economic concerns. It is a fact that sewage treatment technology becomes progressively more complex as more elements are removed from the water. Primary treatment is a mechanical process of separating used water from the materials that have been added to it. The product is sludge, the total organic and wastewater materials generated by residential and commercial establishments. It is made up of the settled sewage solids combined with varying amounts of water and dissolved material that are removed by screening, sedimentation, chemical precipitation or bacterial digestion. When raw, it is very high in moisture and biologically unstable. When subjected to secondary treatment, it is anaerobically digested and produces methane gas and carbon dioxide. The digested material has a high degree of biological stability, and is rich in phosphates, nitrates, potassium and trace elements. Tertiary treatment aims to remove 95 per cent of all remaining substances and chemicals, leaving the water in a drinkable form which is then ready to be returned to the natural system. The technology involved is, however, very costly and few communities are able to afford conventional tertiary treatment systems.

When the products of the treatment plant are seen to be valuable rather than as wastes to be disposed of, the current practice of dumping, by landfill, dumping at sea or incineration, may be regarded as a misuse of resources. In almost all countries where large-scale technologies are limited or expensive to buy, alternative approaches have to be found. It is simply a question of pragmatic long-term investment in a healthy environment. China, as with other Asian countries, has long demonstrated the importance of human wastes to agricultural development and has provided practical approaches for treating them to minimize health hazards.[53] More will be said in Chapter 5 on the relationship of organic urban wastes to the land, particularly its relevance to urban agriculture and the reclamation of urban land. Given that the presence of industrial contaminants in wastewater is a problem that must be addressed at source, the concept still provides a basis for solutions to the practical realities of waste disposal in urban areas.

Aquatic plants as a filter

Biologists have known for many years that marshes have a very high capacity for recycling wastes. The highly productive nature of marshland ecology promotes the uptake of nitrates and phosphates by aquatic plants. Research in Mississippi has shown that, annually, 0.4 hectares of water hyacinths absorbs 1,600 kilograms of nitrogen, 360 kilograms of phosphorus, 12,300 kilograms of phenols and 43 kilograms of highly toxic

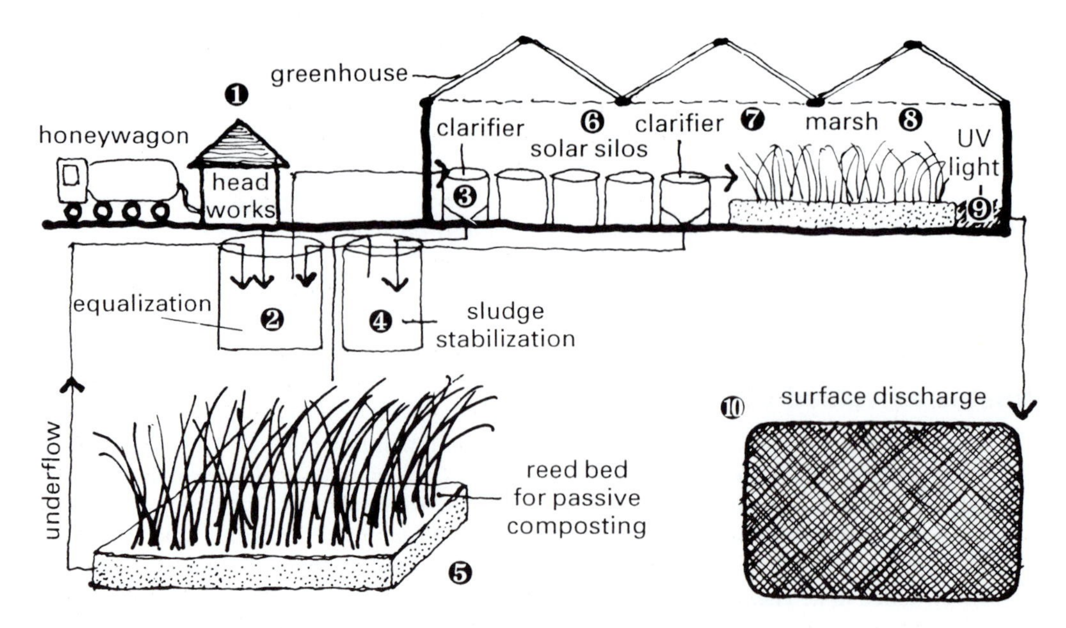

Figure 2.18 Solar aquatics septage treatment system. As a typical SAS septage treatment plant, effluent is discharged into head works (1) consisting of a receiving station and degritter. From there it flows by gravity into in-ground tanks (2) where the effluent is blended to compensate for the variations in loads and is preconditioned. Following equalization and preconditioning, effluent is pumped to a clarifier (3). From here sludge is pumped to the sludge stabilization tank (4) where it is aerobically stabilized. It then flows to the reed bed (5) for passive composting. Underflow from the reed bed is recycled by pumping it back into the equalization tank (2) for further treatment.

Liquid from the clarifier enters a series of solar silos (6) that contain the plants and animals involved in the clean-up operation. After moving through the silos by gravity flow, the liquid enters a second clarifier (7). Solids from this clarifier are recycled back to the equalization tank (2), where they re-enter the system, reseeding microbes as they travel.

Next the liquid moves through a gravel-filled de-nitrifying marsh (8) planted with grasses and is then sterilized with ultraviolet light (9) before being discharged (10). Depending on total suspended solids limits, a sand filter may also be required at this point in the system for final polishing before purified water is discharged.

Source: Ecological Engineering Associates, 13 Marcom Lane, Marion, MA, USA.

trace metals.[54] Conventional wastewater plants release large quantities of excess nutrients and toxic chemicals into city sewers, lakes and oceans. They also create millions of tonnes of sludge each year, much of which is loaded with heavy metals. Aquatic systems, however, suspend solids by aeration, and use solar energy, bacteria, plants and animals in a system designed to enhance natural responses which biochemically change or remove contaminants from sewage. Taking ten days to process, wastewater circulates through a series of large transparent plastic tanks or silos, each of which is a functioning ecosystem. Bacteria and algae use sunlight to digest organic matter suspended in the liquid, converting ammonia to nitrate. Water plants and algae consume nitrates and zooplankton and snails feed on algae. Subsequent steps include engineered marshes with plants that take up heavy metals and other toxic chemicals; other silos contain fish that feed on nitrate-eating phytoplankton, and water hyacinths take up other heavy metals. The final step is 'polishing' marshes from which water flows clear and sludge-free.

A variety of wetland and solar aquatic systems were developed experimentally for commercial purposes in the 1980s and 1990s. Columbia, Missouri, contracted for a 36 hectare sewage-treating wetland and over 150 wetland treatment systems serve cities and towns in the USA.[55] Harwich, Massachusetts, and Providence, Rhode Island, have installed solar aquatic treatment systems, and the Providence facility is intended to process between 75,000 and 113,000 litres per day of domestic and industrial waste.[56] Other, smaller systems have been installed in such places as Muncie, Indiana, and the Toronto Board of Education Nature Interpretive Centre at the Boyne Conservation Area, Ontario. In cold climates these systems must be contained in greenhouses. In warm, sunny regions such as the Middle East, or the south-west USA, they can operate in the open.

Solar aquatic systems require much less land than marshland treatment and are no more land-consuming than conventional treatment plants. The operating cost is also about two-thirds the cost of conventional treatment.[57] In addition, wastewater is seen as a resource rather than a waste to be disposed of in the larger environment, and there is potential revenue to be gained from hydroponically grown flowers and plants. Some problems remain, however. Many heavy metals, such as lead and cadmium, stored in plants, must be periodically gathered and disposed of, a limitation that is, none the less, still a cheaper and safer option than current methods of incineration or burying tonnes of metal-laden sewage sludge in landfill sites from where it is released into the larger environment. Government resistance to adopting such systems is also inevitable since their operation requires a very different type of expertise. There is a need to adapt these natural systems to local conditions and to ensure the control of input, facts that make them 'somewhat suspect in the highly standardized world of water quality engineering'.[58] In summary, however, as Bob Bastian of the US Environmental Protection Agency has indicated, 'much of the future of alternative treatment will depend on whether regulatory agencies push for compliance with strict water quality standards'.[59]

Water purification using vegetated wetlands has also gained recognition in the Netherlands, Germany and other European countries. There are some thirty wetland projects in the Netherlands dating from the 1960s. Some include camp sites where wastewater is purified on a small scale, and others have been built by water boards, municipalities and industry to treat the effluent from treatment plants.[60] In the Dutch government's fourth National Report of Physical Planning (1988, updated 1990) it was suggested that vegetated wetlands should be incorporated in city and land-use planning. It proposes, where space is available, the decentralization of smaller purification systems that can serve multiple functions. Studies have proven, in fact, that the combination of water purification, nature development, recreation, and agriculture is possible.[61]

Rainwater: the principle of storage

The basic lesson that nature provides in the water cycle is one of storage. Natural floodplains and lakes are the storage reservoirs of rivers that reduce the magnitude of peaks downstream, by spreading and equalizing flows over a longer period of time. Vegetated soils and woodlands provide storage by trapping and percolating water through the ground with minimum run-off and maximum benefit to groundwater recharge. Water quality is enhanced by vegetation and storage which in turn will contribute to the diversity of natural and human habitat. Thus storm drainage must be designed to correspond as closely as possible to natural patterns, allowing water to be retained and absorbed into the soil at a similar rate to natural conditions. This principle is now well recognized in Western countries as the realistic alternative to current practice. A discussion of a few of these alternatives is appropriate here.

Groundwater recharge

Where the porosity of the soil permits, rainwater applied direct to the land helps to replenish groundwater reserves. Natural drainage over turfed or vegetated land is a key factor in controlling and managing stormwater. Functionally it aids natural infiltration into the ground and controls the velocity of water flow, which is essential to control erosion and sedimentation. The objective is to achieve a rate of water run-off that is equivalent to pre-development levels, helping to minimize flood and erosion damage. 'Because it turns the hazard of storm flows into the resources of base flows, it is environmentally the most complete solution to the problem of urban stormwater.'[62] Many low- to medium-density housing developments in North America have adopted the practice of overland flow from developed surfaces over turfed areas and channels as an alternative to traditional storm sewers, since it has been shown to be more economical and more beneficial to water quality. Vegetation is the crucial factor which ensures that water is recycled back to the natural system. Natural drainage, woodlands, rough grass and shrubs and small wetland areas provide the functional basis for determining open space patterns. The following two examples, one begun in the early 1930s, the other in the mid-1980s, illustrate this principle of infiltration and groundwater recharge in different ways.

Infiltration: Long Island, New York

The importance of infiltration basins in eliminating run-off near its source while sustaining groundwater is demonstrated in Long Island, New York, since it is groundwater which serves as the area's primary source of municipal water supplies.[63] Begun in the 1930s, 4,000 recharge basins were in operation by 1990 supporting more than a million people within the drainage area. Of the 45 inches of rainfall per year, some 22 inches infiltrates through to the recharge basins. Municipal agencies draw 150 gallons of groundwater per person per day to support residents. Run-off to recharge basins puts back into the aquifer an equivalent of 240 gallons per person per day, more than replacing withdrawals. With respect to contamination of groundwater, soil is an effective filter, protecting streams and aquifers when it contains clay or humus.

> Nitrogen compounds tend to decompose and return to the atmosphere. When infiltrating runoff contains only the common, mostly biodegradable, constituents from residential, commercial and office sites, and not foreign chemicals discharged directly from industry, it is within most soils' treatment capacity.[64]

Infiltration: Kitchener, Ontario

The city of Kitchener has similar conditions to Long Island in that the city relies largely on groundwater for its water supply. In one experimental development, the layout of infiltration basins was integrated with interconnected grassed drainage ways and recreation areas designed to form a focus for housing clusters. In the mid-1990s the Kitchener Parks Department undertook a naturalization study of its overall parks system to determine what approaches should be taken to different categories of public landscape that included stormwater impoundments.[65] The Idlewood Creek subdivision was selected for a naturalization demonstration as part of the larger infiltration project. The fundamental purpose of both the Long Island and Kitchener stormwater systems was similar – to store and infiltrate rainwater into the ground. The Idlewood experimental system, however, incorporated multiple functions in accordance with Parks Department requirements for naturalization. The social recreational needs of the surrounding community

Plate 2.14 Minneola Basin, Long Island. One of the first infiltration basins on Long Island, built in about 1930. It is therefore one of the first in North America. The steep slopes are densely vegetated. In spite of being on the edge of a park, there is no provision for human access other than maintenance.

Source: Caption and photo, Bruce Ferguson.

had to be incorporated into the design, such as children's play facilities and the need for a footbridge to allow access across the basin in wet weather. Ecological functions included diversifying the environment of the greenway system: modifying a mown turf environment to a naturalized retention pond as a wetland habitat for birds, frogs and dragonflies. A key function was the enrichment of a sensorily deprived suburban environment to a changing complex place where children could wake up in the morning and hear bird-song, experience the cycle of sun and rain, the changing seasons, and the natural world in which they lived. Other objectives included naturalizing a small connecting creek that had once been a coldwater stream and which provided a pathway link to other communities and natural habitats.

This example demonstrates two design principles that are necessary to re-establish links between cities and natural processes. The first is the principle of 'least effort for maximum gain'. The most significant results, when attempting to solve a problem, often come from the minimum use of resources and energy. The Long Island installations, effective though they are, have a single function and are generally isolated from their physical and social surroundings. The task for design is to incorporate multiple benefits that here have included social, recreational and learning experiences. The enrichment of a sterile groundwater recharge environment was a primary motive. The creation of a wetland came from increasing the depth of the basin in one area to allow water to collect after rainfall. Wetland plants already present in the damp soil of the basin quickly colonized the pond and brought with them a range of insects, frogs, tadpoles and birds.

(a)

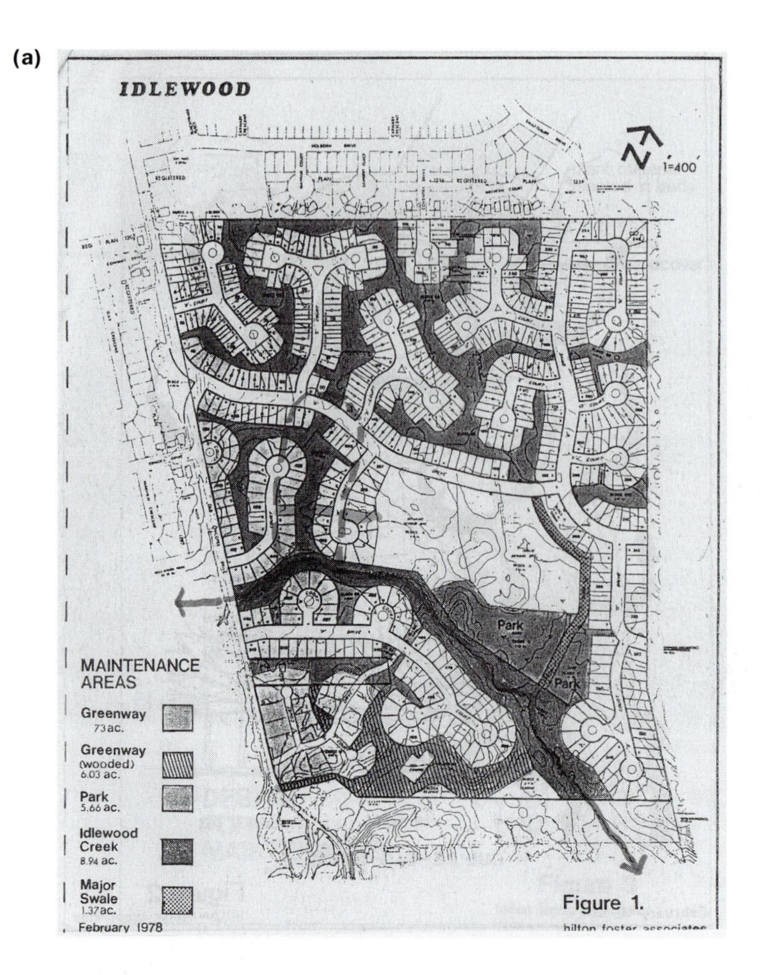

(b)

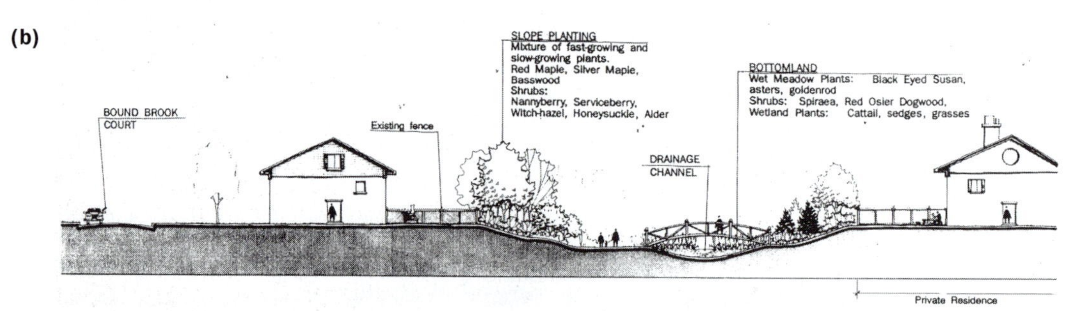

Figure 2.19 (a) Overall site plan for the Idlewood Creek subdivision, Kitchener, Ontario. Organized around a system of greenways that function as flood infiltration basins and intended for active use. Rainwater fills a created permanent pond which then drains to a stream that flows into the Thames river. (b) Section across the basin showing the proposed pedestrian bridge that crosses the depression. Planting on the banks of the basin was part of the naturalization strategy.

Source: ENVISION: the hough group.

(a)

(b)

Plate 2.15 (a) The turf infiltration basin before restoration, a moist site showing signs of succession with aquatic plants. The basin was not used by the surrounding residents. (b) The restored site some four years after construction. The wetland area was dug and allowed to succeed on its own. Planting on the slopes was an active part of the restoration process. The engineer for the greenway/infiltration system came to the conclusion that the permanent pond and wetland did not affect the infiltration properties of the basin. The bridge is a popular gathering place.

Prior to the existence of the subdivision, the creek existed as a coldwater stream, but was degraded by previous agriculture and development. Intensive and frequent discussions with the community during the design phase of the project included the city's maintenance crews who attended every meeting. They answered questions and supported the Parks Department naturalization programme because they were convinced of the benefits it would bring to the park system.

The second relevant principle is that 'Environmental education begins at home'. The residents of Idlewood Park may well have watched nature programmes on TV, but nature was now on their own doorstep. Trading the featureless universal mown turf typical of much of suburbia for an enriching and complex balance between human and non-human landscapes is one of the major functions of the naturalization movement and, significantly, the role of design in the twenty-first century.

Developments since 1991

From the perspective of infiltration, both the Long Island and Idlewood Creek systems have continued to work well, but the primary function of the former, namely to replenish groundwater, has not been enhanced by ecological and cultural values or public access. The latter has also continued to perform well from an infiltration perspective, but has not been repeated, one reason being that the greenway and basin system takes up too much land. Alternative development scenarios, compact, moderate densities in conjunction with greenway systems, have not been seen as a viable alternative to the low density typical of much of Kitchener's suburbs. A concern was raised in the early implementation of this project that allowing water to pond would reduce the efficiency of infiltration. Further investigation revealed, however, that the presence of the pond did not, in fact, change infiltration conditions, but added a great deal to the residential environment. Having participated in its planning the residents living around the basin have, since the mid-1990s, become persuaded that a naturalized environment makes for a more liveable and interesting place than the one that existed when they originally moved into the area.[66]

Retention ponds and lakes

When modelled on nature, retention ponds are a means of controlling water run-off by modifying flows, smoothing out peak loads by releasing water slowly to streams to lessen the danger of downstream flooding. Temporary ponds also contribute to the replenishment of natural groundwater where soil porosity is high and where severely contaminated urban run-off is not an issue. They are effective in reducing flows and enhancing water quality. Pollutants entering the cycle from rain, roads, paved surfaces and rooftops include a range of organic and chemical compounds and heat from paved surfaces, making stormwater hostile to aquatic life. Most pollutants in stormwater are attached to sediment particles, allowing these sediments to settle to the bottom of the pond over a period of days, rather than letting them spread throughout the system. This is therefore an important method for treating and improving water quality. Vegetation is also a significant factor in stormwater management since trees, shrubs and aquatic species contribute to cooling through shading and improve water quality by filtering out pollutants before the water enters streams and lakes.

Storage of water occurs naturally in lakes, ponds and marshes, and often occurs fortuitously in many urban places that do not come under the label of parks and playgrounds. The wealth of abandoned industrial or mining lands, vacant lots, waterfront sites and highway interchanges perform, quite unplanned, a valuable hydrological function by retaining and storing water. This is often lost when redevelopment occurs and sites are 'improved'. Storage of water on low- and medium-density development is becoming an accepted water management alternative in situations where land is available for temporary ponds and as an inexpensive and environmentally appropriate approach to urban drainage. The Ministry of the Environment for Ontario has concluded, for instance, that water-retention facilities to control run-off must be incorporated into urban development.[67] Many alternatives are possible, each depending on the nature of the place, its rainfall, topography, drainage patterns, plant cover and soils and type of development.

Permanent storage

This is appropriate where a continuous supply of water is available, and where inflow and outflow and soils permit stable conditions. Ponds that can be maintained at some minimum water level offer multi-purpose potential for community uses. Balanced ecosystems of plants, animals, fish and other aquatic organisms maintain stability and provide places for nature study, education and fish management. Open water provides places for boating, recreation and visual enhancement. Examples of this concept have been implemented in a variety of housing developments in North America and elsewhere.

Temporary storage

It might well be assumed that densely urbanized areas do not have the capacity to cope with rainfall storage. It was for this reason, in fact, that storm drainage was introduced in the nineteenth century. Where space is at a premium, or permanent ponds are inappropriate, or where an existing storm drainage system is subjected to additional loads, the principle of delayed return to the receiving body may be put into practice. The floodplains of rivers and streams work on this principle, releasing excess water slowly and smoothing out peak loads. In the city, temporary storage is useful in situations where various functions must be accommodated in the same area of ground. It can be designed to accumulate water during rainstorms and drain completely after the storm over a period of time. The land thus serves dual purposes; assisting hydrological functions, but still providing space for various other uses. Golf-courses, playing fields, cemeteries and parks are typical sites where this can occur. In fact, it often does occur by default rather than by design, as the flooded fields and lawns of many urban parks after sudden rain will testify.

Where non-paved areas are insufficient or unable to cope with natural storage, many other types of open space are potentially available that can fulfil hydrological roles. For instance, parking lots take up a significant proportion of most downtown areas. If designed with a hydrological function in mind, such spaces can provide appropriate storage which would not significantly interfere with their parking uses. As an example, it is accepted practice to design parking areas in cold climates for snow storage. A similar practice could be extended into the summer months for rainwater. Streets, which may take up to 27 per cent of downtown space, may also act as temporary storage, as may storm sewers themselves. There are many other realistic ways of reconnecting urban water with the hydrological cycle. For example, the rainwater butt that stores water off roofs and provides for garden irrigation and washing cars can help conserve drinking-water supplies. Downspout disconnection programmes that discharge water on to vegetated ditches, lawns and backyards to collect in shallow depressions reduce stormwater volumes at source and are a strategy that has been implemented with considerable success in the Village Homes community in Davis, California.

Wetlands for stormwater purification

In addition to their function as providers of biological treatment for wastewater, discussed on pp. 69–71, constructed wetlands have become an increasingly important method of enhancing the quality of run-off water in cities in Europe, the UK, the USA and Canada. They also have immense importance in controlling stream flows and downstream erosion, and restoring impaired habitats for aquatic and terrestrial species. While there is considerable variability in the efficiency of wetlands at removing contaminants, it has been shown that they can remove 70 per cent of excess nutrients, and destroy bacteria and viruses. Heavy metals may be accumulated either in sediments, or are associated with organic materials, and many pesticides, oil and grease are broken down by microbes and plants. They have also been developed to treat the acidic water quality of coal-mining and ore-processing plants. For instance, it was found that wetlands created in western Pennsylvania to treat water from old mine sites resulted in improved water quality as well as furnishing habitat for birds, mammals, reptiles and amphibians.[68]

As discussed on pp. 76–7, the chemical and physical mechanisms for treating stormwater include dispersal by evaporation, sedimentation and absorption where dissolved pollutants adhere to suspended solids and settle out, and filtration through vegetation and soils. Since studies have shown that the first 2.5 centimetres of rainfall carry 90 per cent of the pollution load in a rainstorm,[69] the design often includes a pretreatment detention pond to collect and settle out sediments on which much of the pollutants are carried, before the water flows into the wetland. Like all natural systems, each place is unique and requires its own particular adaptations.

Some considerations of design

The basis for design form

In this review of water I have tried to make connections between urban and natural processes. It becomes apparent that if we examine these connections in the context of the urban environment, the city's open spaces become a fundamental factor in re-establishing hydrological balance. City spaces have a significance beyond the transportation, economic or recreational assets that we normally ascribe to them. The notion of investment in the land now begins to acquire conservation and health values. An ecological basis for urban form suggests that when the city's water is recycled back into the system there are reduced costs and increased benefits. Urban development becomes a participant in the workings of natural systems. Urban forests and market gardens acting to purify wastewater and recharge groundwater can also become valued when they provide food, timber and wildlife reserves close to home. The value of urban forests to the city is immense and provides the basis for useful, productive and low-maintenance landscapes. The city's residential parks, open spaces and wastelands, parking lots, playgrounds and roofs can be adapted, where appropriate, to serve a hydrological function by creating temporary or permanent storage areas and wetlands, thereby helping to redress the problems of erosion and pollution.

Design and the new symbolism

Thus far I have dealt with the ecological and functional basis for appropriate design and management that may achieve a measure of environmental sustainability. It will be obvious, however, that water, perhaps more than any other element of the landscape,

(a)

(b)

Plate 2.16 The symbolism of the great gardens of history. As an element of design, water was manipulated to create places of delight and beauty. Operating without electrical energy, their form tended to be an expression and symbol of the place, gravity, and the natural opportunities of the site. (a) Italy. The Villa D'Este gardens. (b) Stourhead, Wiltshire, England.

has deep-rooted spiritual and symbolic meanings to which design must respond. As an element of great experiential power, water has historically been manipulated and shaped to create places of delight and beauty. It has reflected cultural attitudes towards nature. The Romans celebrated water in their engineering and architecture. The medieval gargoyle capitalized on the inherent opportunities to celebrate water falling off cathedral roofs. The exuberant splendour of the Italian water garden, created by the volume, light and sound of its fountains, exploited the hillsides and streams around Florence and Rome by using natural gravity. The Japanese garden symbolized in miniature the oneness of nature and culture. The placid lakes of the English garden expressed the romantic and pastoral qualities of England's rural landscapes.

The task today is to create a new design symbolism for water (and urban natural systems as a whole) that reflects the hydrological processes of the city; an urban design language that re-establishes its identity with life processes. The opportunity for this to occur lies in the establishment of a vernacular landscape whose aesthetic rests on three factors. First, on its ecological and functional basis for form. Second, on the *integration* of design objectives where design becomes multi-faceted and experiential. Single-purpose solutions to problems tend to create other problems. Third, and most important, on the notion of visibility (see some design principles in Chapter 1). The motivating forces that have shaped the city's conventional systems and technologies have been based on the concealment of the very processes that support it. Revealing and enriching the processes of nature and the diversity of the city's cultural landscape, therefore, lies at the heart of the urban experience and artistic form.

Expressions of this notion of visibility from European and North American cities illustrate the opportunities. The engineering functionalism of the sewage treatment plant can be brought into harmony with form and symbolized in unexpected ways, as in the flow form sculptures at Jarne, Sweden.[70] Here sewage water cascades down various sculptured basins and is aerated as it drops. Porous paving materials, that permit water to penetrate through to the soil and ground vegetation to grow while maintaining a hard surface, offer other opportunities for integrating design with hydrology. In Italy, the variety and pattern of ground surfacing, the channels that direct water movement and the beauty these add to the city floorscapes are matched by the hydrological and climatic functions they perform; a welcome relief from the visual tyranny and sensory deprivation of asphalt surfaces. In Canada, art can capture, as part of the city's storm sewer system, a moment in the water cycle.[71] Fish have been painted on city catchbasins to remind people of where the water goes, and on Toronto's Don River, cut-outs of birds and animals that once inhabited the valley were placed on the chain-link fences that now separate walkers and cyclists from the valley's railway corridor.

Hydrology and the inner city

An attempt to give physical form to the inherent potential of rainwater was made in the early 1980s in a housing project in central Ottawa. Called the LeBreton Flats Demonstration Project, it was a joint undertaking by the National Capital Commission, Canada Mortgage and Housing Corporation and the City of Ottawa. The overall project aimed to demonstrate innovation in housing, responding, first, to the social and economic realities of the city and, second, to the need for conserving energy in housing design. Located on the edges of an established neighbourhood in the inner city, the project was concerned with the rehabilitation of derelict urban land within an existing medium-density community. Part of the new housing was grouped around a small 0.5 hectare park, the focus of the entire neighbourhood. It became apparent, after evaluating local community concerns and interests, that the resources of the site and the project's

Plate 2.17 Children and play: sculptured stormwater flows in a German housing development.

Source: Herbert Dreiseitl.

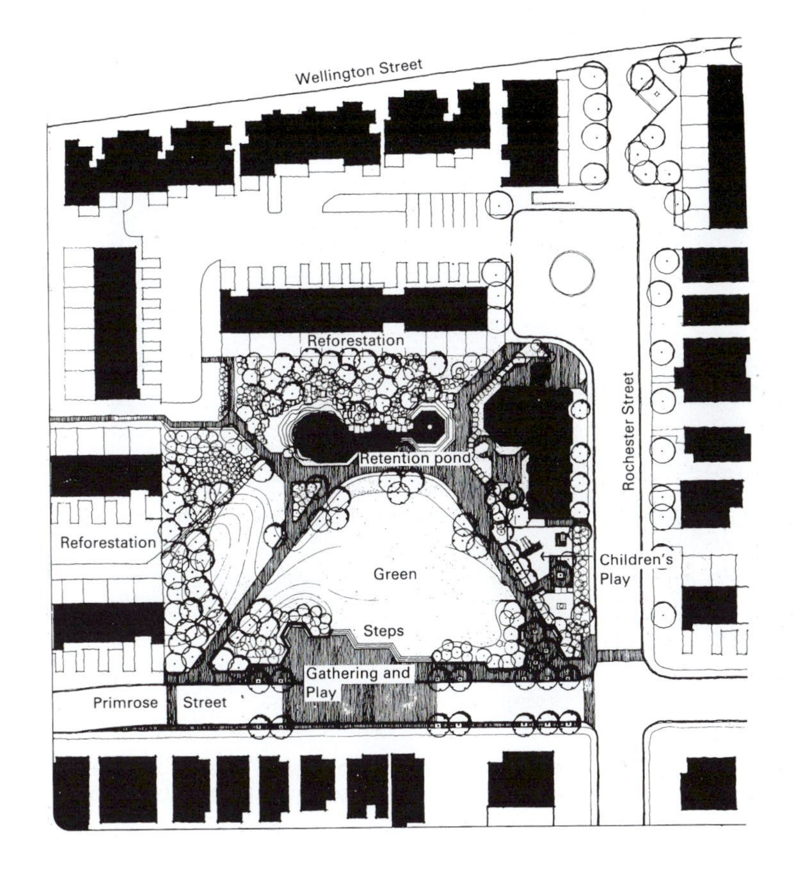

Figure 2.20 Plan of LeBreton Park, Ottowa, and surrounding residential housing.

Source: Hough Stansbury Michalski, Toronto.

environmental goals must extend to the park itself. It was apparent that the concept of stormwater retention should be explored in medium-density residential areas of the inner city. Since the rainfall of the Ottawa region typically falls as sudden storms, the environmental problems of conventional urban drainage are equal to, if not more severe than, those of outlying suburban areas. The potential opportunities were twofold:[72]

- to demonstrate on a small scale the creative and practical alternatives to traditional site development practice in storm drainage in an inner-city park;
- to create a place that would integrate environmental with educational objectives, i.e. storage and slow release of urban drainage to create temporary ponds for play and recreation activities that are dynamically tied to the hydrological cycle.

Several important considerations dictated the design solution. The first was the smallness of the park in relation to the size of the community using it. The second was the volume of water that could be accommodated during a storm while not interfering with other essential park functions and ensuring safety. The third was the community's perception of the kind of park needed. During many intensive meetings with both adults and children it was apparent that there was a strong desire for a 'green' park with grass and flowers, places for sitting and strolling, for meeting and informal games. This was understandable in view of the dearth of parks in the general neighbourhood. In addition, it was evident that most active games requiring hard surfaces were played in local streets, which, like most established neighbourhoods, traditionally have been the social hub for both children and adults.

The way the hydrological functions of the park were expressed as design form was directly affected by these considerations. The need for a green park permitted most of the park to have soft surfaces – grass, trees and planting beds. At the same time the

Plate 2.18 LeBreton Flats, Ottawa: the retention pond flooded. During a rainstorm the pond fills to 46 centimetres and takes on a very different character, inviting unstructured water-related play and visual variety.

Plate 2.19 The retention pond dry. Flexipave surface provides a topcoat for a variety of dry weather activities.

need for socializing, gathering and informal hard-surface games resulted in a new approach to the street design bordering the south side of the park. Children's activities were recognized as a valid function of the local streets. By incorporating safety measures to slow local traffic, the park concept could integrate both the street and the park as one unit, the street providing hard surfaces, the park providing the soft ones. The ability of the park to serve a hydrological function was thus enhanced by absorbing runoff to the fullest extent over its vegetated surfaces and by the inclusion of a stormwater detention pond. This was hard-surfaced, since the notion of a naturalized impoundment with a mud bottom and aquatic plants was rejected as impractical.

The pond was designed to hold 76,000 litres, which is the maximum amount of water that might occur in any one rainstorm over a two-year period. A peak rainfall of 15 minutes, amounting to 2.3 centimetres for the Ottawa region, was assumed for purposes of calculation. Roof water obtained from adjacent housing and overland flow within the park was collected by a series of drainage channels that deliver the water to the storage basin. Water from parking areas and streets in the housing project was not collected to minimize the introduction of pollutants in a general play environment. A small adjustable weir and catchbasin inlet were designed to permit the pond to fill to a depth of 46 centimetres and drain out slowly to the storm sewer over a period of time. An average of two to three days was regarded as a reasonable period for a temporary storage situation.

The character of the pond would therefore be constantly changing: becoming a potential ice surface in winter, and filling periodically after rain during the summer. Its hard surface was designed to take intensive use and was contoured to permit many different activities, depending on whether it was wet or dry. Thus children's play could respond to the changing environment: focusing on water play after a rainstorm and dry weather activities when the pond dried out.

The purpose of this experiment was to integrate an urban natural process – water and its management – with social and education values. It involved natural process and design in the familiar and daily routine of city living. A hydrological function was established for park settings that had traditionally been devoted to recreation alone, and hydrological functions became part of the understanding and experience of the park and of the neighbourhood. In these ways it was intended to bring together solutions to environmental problems in the city with social needs and art. It was intended to delight the senses, provide a place for creative play, function practically and bring people closer to the continuum of natural events. As I remarked in Chapter 1, children learn about life and their environment less by the occasional visit to the nature centre or the museum than by constant and direct experience in their daily surroundings and activities. A pond in the park after a rainstorm that one can splash through is the focus for play and learning while it lasts. It provides the best opportunity for understanding hydrology in cities. Through play children are brought closer to the cycles of rain and sunshine. Water as the ageless material of design can assume a new relevance and it is on this basis that an alternative urban form must emerge.

In hindsight: what was learned from this experience?

During the ten years following completion of the project in 1980, it became apparent that the realities of its evolution were very different from those imagined during its conceptual development. It is here that hindsight provides lessons for what works and what does not in terms of the role of government, community change and park policies.

Things that succeeded

Shortly after completion of the park the detention pond began to be used in interesting ways. Children flocked to the park after rain and used the water-filled pond as a paddling pool, for sailing small boats and for those creative water activities known only to children. During dry periods its surface became a place for other happenings, including biking, roller-skating and tag. One small girl said in conversation that she did not live in the neighbourhood but came there to skateboard every day because it was the best shaped surface she knew of. Adults sunned themselves on the south-facing grassy slopes, flowers were planted, mothers brought their small children to the playground and people talked about the different things they could do in the park.

Things that failed

A number of interrelated events contributed to the failure of the park's original intent. First, upon its completion responsibility for the project was relinquished by the initiating federal authority, and ongoing management became the responsibility of the City Parks Department. The Department had had some involvement in the project's development, but its park management policies, based on limited budgets, were geared to conventional turf maintenance and occasional litter clean-up. Maintenance staff had no experience in the management of the detention pond which quickly began to fill with debris and broken glass, and which became a hazard for children. Second, people in the community who had worked with the design team to create the park and who had decided to live in the new housing complex began to move out for a variety of personal reasons, while others moved in. Over a period of years, therefore, fewer and fewer people who knew the reasons behind the park's design and who were able to communicate their knowledge to newcomers remained. Acts of vandalism, often by older children, began to occur, such as destruction of trees, lighting and benches. Few of the plants were replaced and a sense that the park lacked ownership began to grow. Third, the channels supplying water to the pond were located inside backyard boundary fences rather than on public property. As people altered their gardens to suit their own needs drainage courses were dammed up and the pond began to lose its retention functions during rains, and has ceased to flood. A visit I paid to the project in June 1994 confirmed this situation. While some of its main features remained intact, people had no idea that the pond had hydrological or educational purposes.

Interesting lessons may be learned from this experience both from the point of view of policy and design. First, building a sense of community in a new neighbourhood is an ongoing process that must be nurtured through appropriate social mechanisms. These might include a continuing involvement in the project between the development agency and the community that lives there through local community leaders. Neither the federal authority nor the city had the mandate or the available budgets to finance such initiatives. At the same time, a rough calculation revealed that the cost of replacing damaged property caused by vandalism in the park would have made a substantial contribution to the cost of hiring a resident community leader, thus helping to avoid vandalism in the first place. The situation illustrates the importance of making social and environmental links in the evolutionary processes that dictate the development of human and non-human communities. It also raises related issues, discussed in Chapter 3, of how the sense of ownership and control of space can be developed in the design of human communities and the role of the designer in that process.

Thus far, we have explored the role of water in nature under natural and urban conditions. We have seen how developments in supply and disposal technology, by improving human health, have been a central factor in urban growth. But the benefits of health and urban growth have been achieved at the expense of disturbed natural cycles and the creation of a general environmental deterioration with respect to a worsening climate, water pollution and diminished wildlife habitats. By themselves, the technological remedies designed to mitigate some of these problems have only shifted them from one place to another – from the city where they began to the larger environment.

It is evident that the opportunities for alternative solutions lie in a better understanding of the nature of the places we live in. Urban ecology provides the conceptual vehicle for city design, for which the principles invoke a basic shift in values. We begin to see wastes as opportunities that can contribute to environmental health and diversity, drawing maximum benefits from the means available. Wastewater returned to the land with appropriate treatment can enrich the soil, provide nutrients for produce, crops and urban forests, giving biological, social and civic value to currently wasted urban land. Stormwater retained in the city's open spaces contributes to the restoration of the hydrological balance and can help ameliorate urban climate. It enriches the potential for integrating environmental, social and aesthetic benefits within parks and urban spaces, and brings nature's processes closer to everyday life. And while many specific issues remain to be solved in this process, the fundamental basis for this alternative view remains true.

Our discussions thus far have shown that water, being a part of the whole interconnected system of natural process, affects every aspect of the subject at hand. It is central to the maintenance of biological communities – plants and animals – which are themselves vital to the city's environmental health. This chapter has shown how plants are directly connected to the hydrological cycle. They have from time immemorial shared pride of place in civic design. It is now time to turn to plants themselves, since they reveal other issues that must be examined in our search for an expression of the city as a place in balance with nature.

3 Plants and plant communities

Introduction

The enquiring observer of plants in the city may be struck by the extent to which they depend on horticultural and technical props for their survival and health. That individual may wonder what motivates people to use plants the way they do and what purposes they serve; pollarded lindens at the base of thirty-storey tower blocks, tropical plants, dusty with age and neglect, crammed into the dark and unused recesses of an office interior, the unbroken turf and isolated trees of every city park. Why do people put so much energy and effort into the nurture of cultivated and fragile landscapes that are usually far less diverse, vigorous and interesting than the 'weedy' landscapes that flourish in every unattended corner of the city? Why indeed is the one tended with such care and attention and the other ignored or vigorously suppressed?

My task here is not to provide another guide to design with plants in the conventional sense of discussing principles of form, space, colour, texture and so on. This has been, and continues to be, done very successfully. My intent is to consider plants and plant communities from other perspectives, to seek a valid basis for aesthetics that has its roots in urban ecology, to explore functions and opportunities for urban plants that are consistent with the ideals of an ecologically sustainable philosophy, and to examine city spaces in relation to their functions and patterns of human behaviour. To this end we must first review briefly some aspects of natural processes, how these are altered in the city environment, how attitudes and perceptions have shaped the city landscape and what opportunities exist for alternative ways of using plants in cities.

Natural processes

Plants are the basis for life on earth. They produce all the oxygen in the earth's atmosphere; they provide the food and habitat through photosynthesis that supports all living creatures. Plants are environment specific, evolving different forms and communities that have adapted to specific climates, rainfall, soils and physiographic types. These major plant communities are grouped into general regions which range from Arctic tundra, northern coniferous forests, temperate deciduous forests, tropical regions, savanna, grassland and desert. Our concern here is primarily with the temperate-zone deciduous forests of eastern North America and Europe. It is in these regions that the majority of industrialized cities have evolved.

Succession

Each forest type goes through a period of infancy, youth, maturity, old age and rebirth. In some forests such as the northern boreal regions, this process of succession is dependent on fire that reduces them to ashes, after which rebirth starts again immediately. In the deciduous regions the process is continuous under natural conditions. Parts of the community die and regenerate, but the forest as a whole remains. Odum describes this constantly evolving pattern of forest growth, starting with a field from which the forest has been cleared and abandoned.[1] The original forest that occupied the field will return, but only after a series of temporary plant communities have prepared the way. The successive stages of the new forest will each be different from that which ultimately develops. While subject to much discussion by ecologists today, Odum describes succession as being based on three parameters:

- it is the 'orderly' process of community changes which are 'predictable';
- it results from the modification of the physical environment by the community;
- it culminates in the establishment of as 'stable' an ecosystem as is biologically possible on the site in question.

With each successive stage the insects, birds and animals change, giving way to other groups of creatures. Succession is rapid to begin with in early stages, and slows down in later ones. The diversity of animal and plant species tends to be the greatest in early stages and subsequently becomes stable or declines.

Succession starts with the invasion of grasses and other colonizing plants. Over time and depending on the degree of soil disturbance, these are replaced by shrubs and fast-growing pioneer tree species. These in turn are replaced by climax species, such as hemlock, oak, maple and beech. Once mature, a plant community is said to have reached a 'steady state'. The climax vegetation perpetuates and reproduces itself at the expense of other species that have been crowded out. Most ecologists consider this steady-state situation to be more of a theoretical concept than actual, however. Odum describes it in terms of bio-energetics; 'energy fixed tends to be balanced by the energy cost of maintenance'.[2]

Structure

Forest communities grow in a series of layers. The highest, forming the canopy, are the dominant species that control the environment for the rest of the forest. Beneath these are smaller under-storey trees and shrubs that are adapted to living in partial shade. At the lowest level on the forest floor are the ferns, mosses and herbaceous plants. Each basic group of plants is structured to take advantage of various conditions. The canopy species are exposed to maximum sunlight. They intercept much of the rain and evaporate a great deal of moisture back into the air. At the same time the canopy is an important modifying climatic agent for the life that exists beneath. The under-storey plants growing in the shade of the canopy flower in the spring when it is bare and sunlight can reach the forest floor. At the level of the soil are the fungi, moulds, bacterial and other decomposers that recycle nutrients from rotting leaves and fallen trees and branches back into the system via the soil and roots. The number of layers varies depending on the forest type. Highly developed, northern deciduous forests usually consist of four strata. In southern communities the layers are more complex and some typical rain forests may be arranged in as many as twenty-seven groups.[3]

Urban processes

Influences

As we have seen in Chapter 2, forests regulate the flow of water in streams and rivers, and water storage underground maintains its purity and health. Forests have a great effect on the movement of water from the atmosphere to the earth and back again. They are thus interwoven with the physical and biological processes on which life forms depend. Cities have created modified environments to which natural plant communities have generally not had time to adapt. In terms of biological time, cities first appeared less than 10,000 years ago. Flowering plants are the product of an evolutionary process that began in the Mesozoic era some 200 million years ago. Trees have been exposed to more than 100 million years of selective pressures to adapt to natural environments.[4] However, their survival in the city is subject to many environmental pressures to which they have not previously been subjected.

The city climate is warmer, which has had a marked effect on plant distribution and survival. The atmosphere of industrial cities, as we shall see in Chapter 6, contains chemical pollutants such as sulphur dioxide from residential and industrial combustion, ozone from the photochemical breakdown of automobile exhausts, nitrogen oxides and fluorides and fine particles emitted from industrial processes. These pollutants interfere with the normal transpiration and respiration processes of plants. Their root systems, adapted to forest soil conditions, must cope with disturbed and compacted soils and paved surfaces. Such conditions reduce water penetration and supply of nutrients, lower groundwater levels and interfere with the transfer of air and gases. In northern regions, the soil is contaminated with salts that are applied to the streets of many cities to keep them clear of snow and ice. There are also other physical problems that plants must contend with in the city environment: for instance, confined soils and exposure to cold from restricted or raised planting areas, exposure to heating and cooling vents on buildings, exposure to high winds or extreme heat, waterlogging from old basement structures or excessive dryness, the altered structure of urban soils, and continuous disturbance from construction and maintenance activities.

Urban plant communities

The impact of the city has been far reaching on plant communities. There are three general groups that deserve attention – cultivated, native and naturalized.

The cultivated plant group

This is the group of plants that is the product of horticultural science – the cultivation and selective breeding of plants to satisfy the environmental and cultural demands of urban conditions. Through cloning and grafting techniques a range of plant species has been, and continues to be, developed in response to increasingly stringent requirements. For example, a large portion of the seed and stock of Canadian shade trees has been imported from the USA and western Europe.[5] These must be resistant to a growing number of diseases, leaf rusts and insects. They must withstand drought, restricted soil conditions and doses of road salts. Their branches must be resistant to breakage from high winds or snow loads. Their leaves must tolerate poisonous gases and particulates in the urban atmosphere. They must also respond to rapidly changing mechanical constraints. For instance, roots cannot interfere with underground services, branching

Plate 3.1 The ideal cultivated street tree. Perfectly formed, even aged and resistant to disease, similar species and age, are used by the million for planting along streets. The American chestnut and American elm were both wiped off city maps when blight struck. This photo from Seoul, South Korea, where massive planting programmes are underway, in this case Gingko biloba, illustrates the problem repeated – a disaster in waiting.
Source: Jae Cheon.

habit and height must not compete with overhead wires. In addition to a multitude of environmental and physical limitations imposed on trees, their selection and breeding has been dictated historically by prevailing aesthetic values and conventions. How should trees behave? What is their ideal form? How profusely can they be made to flower?

The 'ideal' city tree must be fast growing, but long-lived. It must be symmetrical and perfectly formed. Characteristics such as messy fruit or slippery leaves, thorns, peeling bark and other inconveniences are unacceptable. These requirements are built into every designer's specifications for planting and have helped set the aesthetic and cultural standards for plant nurseries. The physical conditions under which plants are grown in nurseries are geared to uniform standards of soil, resistance to disease and form. They are grown to be transplanted to equally uniform sites. The unwary designer attempting to select moisture-loving plants for a wet, poorly drained site may well find they succumb to these conditions. The nursery plant, grown on well-drained loams, is no longer adapted to its original environment. The procedures necessary to ensure the survival of plants involve complex engineering and horticultural requirements for transplantation, moving, soil preparation, guying, irrigation systems and protection.

The native plant community

In many cities one may still find native plant communities that have remained relatively unaltered. These remnants of natural forests or wetland have been surrounded by the

Plate 3.2 The native plant community. Remnants of old forests that have somehow survived, even though they have been drastically altered in many cases. They provide irreplaceable natural history and educational links with nature in cities in all seasons.

Source: Steve Frost.

advancing city, but still retain elements of the original ecosystems that once prevailed. In some North American cities they survive more by good fortune than by design. Topographical obstructions or, more recently, planning policies have forced the city to move around them. Some have since been incorporated into urban park systems. The Tinicum Marsh and Fairmount Park in Philadelphia and the ravine lands in Toronto are examples. Others, such as the common lands of forest and heathland in British and European cities, have been kept open for public use since the first days of village settlement and have never been enclosed or cultivated.[6]

Many natural areas have been encroached on by transportation links and development. Changes in drainage patterns, erosion and human use have severely altered their original character. But in spite of these disturbances something of the natural diversity of the original natural community may still be found. Many wooded remnants still retain plants and animals that have become locally rare due to their isolation in the urban region. It is here in the middle of the city that one can still find trout lilies flourishing on the forest floor, or observe the annual migration of birds. Such places are one of the irreplaceable links between natural and urban processes. They are a small but vitally important historic and educational opportunity for nature in the city.

The naturalized urban plant community

These are the plants that have adapted to city conditions without human assistance. The evolution of civilization has fundamentally altered the original ecosystems that once flourished in the absence of humankind. Pollen analysis has shown that from its earliest beginnings the disturbance of land has been associated with plants of the open ground such as the dandelion, chickweed and plantain. These plants colonized the land following the Ice Age. Their habitat and distribution were subsequently restricted by the afforestation that followed, but expanded again when agriculture and cities again laid the soil bare.[7] Modern industrialized cities have had a profound influence on plant communities. Many have been lost or become extinct through the disappearance or changes to habitats in which they are able to grow. At the same time new associations have become established in humanized habitats as a result of species migrating from other climatic zones. While urbanization reduces the amount of vegetation, it has been shown that there is a comparatively high number of species present in European cities compared to the surrounding agricultural countryside.[8] The types of plant associations that have adopted the city are different from the native associations that the city replaced. Continual disturbance of the urban environment creates ecologically unstable conditions that are favourable to the invasion of numerous pioneer species typical of early succession stages of natural ecosystems. Ragweed, for example, migrated across Canada eastwards from the prairies where it originated as the railways, highways and settlements prepared the way.[9] The increasing number of alien species colonizing the European city has been shown to have originated from warmer areas of the world.[10] Their success is due to a warmer climatic environment; their expansion and naturalization have been greatly influenced, not only by direct introductions and environmental changes, but also by spontaneous hybridization with cultivated plants. From these, new species are continually evolving, adapted to the special soils and climatic conditions of the city.

Plate 3.3 Naturalized plants downtown. Plant diversity in a regenerating residential lot. In many parts of the city, dense stands of trees, regenerating on their own, provide essential summer shade, where it is needed, to buildings and paved surfaces. The compound leaves of the Tree of Heaven (*Ailanthus altissima*), a tree widespread in North American cities, sprout late and drop early, minimizing obstructions to the spring and winter sun.

The city offers a wide range of sites for these naturalized urban communities. They include wastelands that have been created by the demolition of old buildings, abandoned waterfronts and industrial lands, railway embankments, road rights of way and similar places. It is here that associations of white poplar, Manitoba maple, dandelion, chickweed, yarrow and other species grow in vigorous profusion. Even in the most paved-over parts of the city, mosses and grasses colonize rooftops and walls of buildings, and dandelions, thistles and docks push their way through corners and cracks in the pavement. Ailanthus finds a foothold in the foundations of buildings and appears through basement gratings, where the warmth, moisture and alkalinity of the soil give it places to flourish. Plants, in fact, are everywhere in the city, a testament to their extraordinary tenacity and ability to evolve and adapt to the new conditions and environmental niches that the city provides.

Venice is an excellent example of the phenomenon of naturalized plants in an intense urban setting that has, over time, become almost devoid of 'green' space in the conventional sense. Within the city limits 147 vascular plants have been recorded growing wild, including seven ferns and twenty grass species. The delicate maidenhair fern may be found in shaded cool corners of the city's footbridges, a rare plant originally from the warmer, wetter coasts of Britain. Rock samphire grows along the edges of the canals, ivy-leaved toadflax on the Bridge of Sighs and many other plants take advantage of the special micro-climates created by the city's nooks and crannies.[11]

Perceptions and cultural values

Horticultural science and the aesthetic that has evolved with it is ingrained in our urban tradition. The standard dictionary definition of horticulture – the art or science of cultivating or managing gardens – embodies the ideal of nature under control. Each tree, shrub and flower is a symbol of human ingenuity, an artefact in a humanized landscape. Like sculpture, plants are moulded and shaped, admired for their form, flower, leaves, unusual character or uniform repetition, as individual specimens not as part of a community. Set in an unrelenting carpet of mown turf, the ornamental trees and shrubs are as unchanging as the architectural setting in which they are placed. There is an unending human struggle to maintain order and control. It is evident in the formidable array of machines, fertilizers, herbicides and manpower, marshalled to maintain a landscape as close as possible to the form in which it was conceived. The dynamics of plant succession are subjugated in our expectations of how plants should perform and behave in the city.

The urban landscape is a product of conflicting values. It expresses a deep-seated affinity with natural things. The spring bulbs displayed in every civic space clearly demonstrate this emotion. Yet these expressions of nature take place only on our own terms, subject to standards of order and tidiness imposed by official public values. The diverse community of plants that flourish in profusion in the adjoining abandoned lot, in every crack in the pavement and invade every well-kept shrub border and lawn, represent, in the public mind, disorder, untidiness, neglect. The too frequent landscape improvement, intended to 'rehabilitate' a neglected area of the city, replaces the natural diversity of regenerating nature with the uniform and technology-dependent landscape of established design tradition. Few would question the value of the formal as part of the city's civic spaces. Street trees survive in the hostile habitat of downtown largely through the technical science of horticulture. The problem is not with science but with its application and the assumptions that go with it. The preoccupations of research in developing plants that are more and more resistant to air pollution is a case in point.

The picture of ultimate technological success, where the air we breathe chokes us to death but our trees flourish unscathed, is one of misplaced priorities. This frame of reference is concerned more with a scientific interest in plants as individual phenomena than with their place in the economy of nature.

At the same time urban trees, native or non-native, and plants in general, have a crucial role to play in the urban environment besides an aesthetic one. Their role in modifying cold and warm climates, reducing air pollution, improving water quality, serving cultural traditions and having economic value are all functions that I shall be examining in this chapter.

Some alternative values and opportunities

A functional framework

Our task here is to explore the alternative functional frameworks for urban plants. It is from this that a new aesthetic and environmental purpose can emerge, a process that has begun in many Western cities. It involves two basic shifts in perceptual thinking. The first has to do with a frame of reference that embraces cooperation rather than confrontation with nature's processes. The second involves an approach to land that derives its inspiration from old traditions of organic farming practices and in permaculture.[12] The sensory appeal of such landscapes is derived from their functional role as working environments and arises from the integration of many ecologically sound objectives – protecting soils, producing timber, conserving water, natural features and wildlife and providing places for recreation. To the established principles of design and aesthetic use of plants in cities are added multi-functional objectives inherent in ecologically sound management. 'The New Forestry', an approach developed in Washington State that encompasses the entire forest ecosystem of plants, soils, animals and micro-organisms,[13] can provide the inspiration for managing city landscapes. We stand to gain from this approach in economy, educational values and overall benefits. A range of opportunities are made available for a richer, more diverse and more useful environment, and as the basis for an investment in the land. These are explored in more detail in the following sections of this chapter.

The naturalized plant community: fortuitous succession

One of the characteristics of modern cities is the considerable amount of land that lies idle or underused along transportation routes and service corridors, empty building lots, old industrial workings and waterfronts awaiting renewal. Many areas have been sealed from public use by ownership rights, or simply left unreclaimed. These places, once destroyed and subsequently abandoned or neglected, have, over time, evolved as naturalized urban plant communities.

W. G. Teagle has described the conditions in the industrial Midlands of Britain, where over two and a half centuries of industrial activity has left an indelible mark on the region. The original native heath landscape was substituted for one of hills and valleys, created by coal-mining and quarrying, which greatly influenced plant distribution. The proliferation of canals in the nineteenth century encouraged the spread of aquatic plants and invertebrates which previously had been unable to establish themselves in the fast-flowing streams that drained the surrounding uplands. The railways that followed played a significant role in the dispersal of plants such as the Oxford ragwort.[14] Their embankments and cuttings also provided a sanctuary for plants in the immediate surroundings that were being destroyed. Strategies for rehabilitating derelict urban areas, however, also involve other problems that must be addressed. They have to do with industrial soil contamination, the cycling of toxic materials through the ecosystem and their impacts on the health of life systems and industrial heritage. These will be addressed later in this chapter and in Chapter 7.

Such issues notwithstanding, naturally reclaimed industrial landscapes often have an ecological, historic and topographic diversity that is far richer than those created by reclamation and redevelopment. The 'green desert' monocultures of grass and trees are no match for the complex relationships of plants, soils, water, topography, microclimate, wildlife and the complex industrial history that are found there. In many cities the sheer cost of effecting such engineered 'improvements' has saved many significant habitats and led to urban parks naturalization programmes across North America, Britain, Europe and Australia.

These initiatives, many originally born of economic necessity and subsequently inspired by public awareness of environmental issues, are the most telling argument for the thesis under discussion, that the diversity, vigour, beauty and wonder of natural process is available at little cost to enrich the city. These opportunities are being increasingly recognized in North American and European cities. Look behind the petrol station, the forgotten space used as a junk yard, or the alley tucked away behind the city's main thoroughfares and you will find magical places of dense forests and varied groundcovers that have appeared on their own. So-called 'weed trees', such as the Manitoba maple and Tree of Heaven, often provide the only shade over streets and sidewalks, creating, in many cases, an environmental quality that is not often matched by purposeful design. A comment by the botanist Herbert Sukopp is relevant here: 'Ecosystems which have developed in urban conditions may be the prevailing ecosystems of the future. Many of the most resistant plants in our industrial areas and in cities . . . are non natives.'[15] As urban society is beginning to realize, we cannot afford to rely entirely on technology to create urban landscapes while ignoring the real nature of the places in which we live. Changing perceptions, however, have begun to accept the value and economy of naturalized planting. Complex native or naturalized gardens are increasingly being planted in back and front gardens in many North American cities, and old city by-laws that required all front gardens to be cut to 5 centimetres are being replaced by a greater tolerance of public expressions of social diversity.

Some alternative strategies

General concerns

Since the 1980s the grass-roots environmental movement has clearly recognized that the ecological processes operating in the city form an indispensable basis for restoring the urban landscape. The interrelationships of climate, geology and geomorphology, water, soils, plants and animals provide the fundamental ecological information on which environmental planning and management of land are based. At the broad level of spatial design the remnant natural and naturalized plant communities that occur all over the city require cataloguing and evaluation. Places that have habitat potential for plants and animals – patches of woodland or meadow, valleys and other linkages – should be integrated into the planning and social network of city places. They provide, among other things, alternative opportunities for diverse and rich recreational and educational experiences. A biological classification and management approach to open space with respect to its ecological sensitivity to human intrusion is an essential component of urban

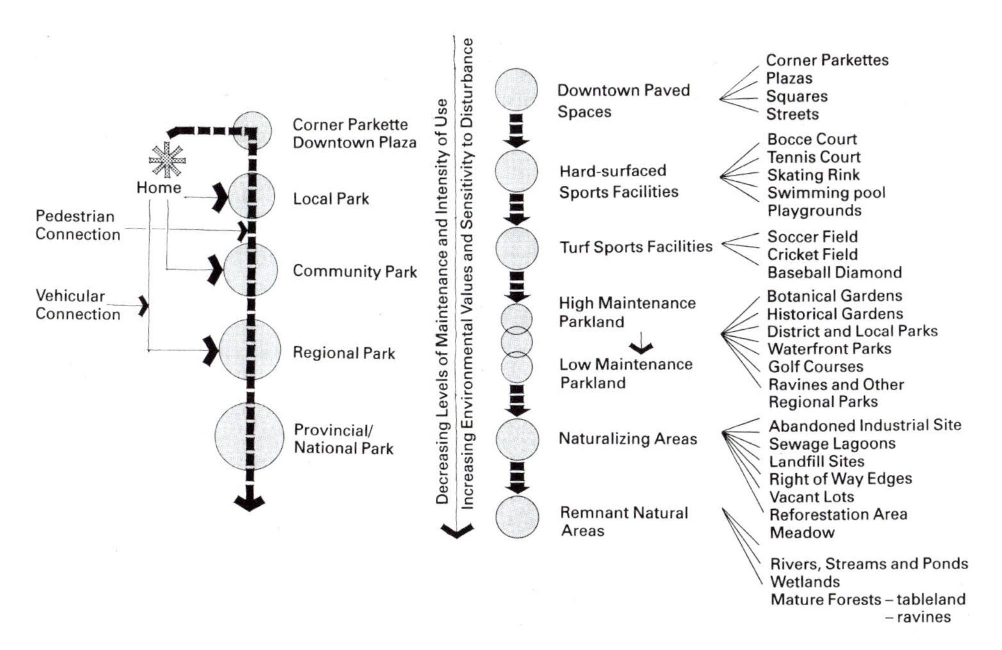

Figure 3.1 An environmental classification system for urban spaces to supplement or replace the conventional hierarchy based on activities and size. An ecological hierarchy that can classify spaces according to levels of sensitivity to human presence.

Source: M. Hough and Suzanne Barrett, *People and City Landscapes*, Toronto: Conservation Council of Ontario, 1987.

landscape planning. Many cities have developed methods for rating environmentally sensitive areas (ESAs) and are derived from an assessment of such factors as the presence of rare or endangered plant or animal associations, the size of the area and species diversity.[16] Parallel examples are the International Biological Program, or various non-urban parks classification systems. A 1987 system proposed by the author for classifying urban spaces was based on ecological criteria and involved a sliding scale, going from nature reserves of high sensitivity to human intrusion, and requiring restricted access, to downtown spaces that can support intense human activity and permit unrestricted access. In addition, a matrix was developed that relates different types of landscapes to various types of human activity, and provides a basis for deciding what activities should be permitted where.[17]

Many alternatives become available once such a framework is integrated into the planning process. In addition to protecting remnant native communities, the naturalized plant community becomes a valued resource for establishing vegetation on poor sites and sterile soils. Plants may be seen as components of constantly evolving communities rather than static individual phenomena. Colonizing plants adapted to urban soils can enhance and modify them and provide alternatives to importing fertility to the city. Landscape maintenance can become a process of integrated management, based on ecological parameters, and gives us the practical tools for maintaining productive and self-sustaining landscapes.

Ecological restoration and plant succession

There is an aesthetic appeal to old trees rising clear from an open grassed sward uninterrupted by obstructions. It is a landscape that is achieved all at once, which is necessary and appropriate where the use of the ground plane is needed for human activity, but in the absence of ongoing tree replacement policies, the ultimate death of the old trees leaves nothing in their place. The countless streets that were changed overnight by Dutch Elm disease are evidence of this fact. It calls into question the long tradition of planting similar species of equal age throughout the city, and the questionable urban design convention that sees beauty in similarity rather than in diversity. Such practices need to be challenged.

An approach to planting that reinstates natural processes provides a basis for a self-perpetuating adaptive environment in areas where traditional horticultural approaches are inappropriate. This may be defined as the process of naturalization that brings an ecological view to the design and maintenance of the urban landscape. It involves the introduction of natural landscape elements into the city that include the re-establishment of woodlands through the reforestation of some lands, the creation of wetlands where hydrological conditions are appropriate, the development of meadow communities through modified turf management and the establishment of varied wildlife habitats. These, outlined in the restoration of the Don River (see Chapter 2), are inherently productive self-regulating communities achieved through ecologically sensitive management rather than total maintenance control. When this approach is placed in context with areas requiring higher levels of upkeep, it achieves benefits in environmental, social and aesthetic diversity and in overall economy in energy, materials and manpower.

Restoring urban woodland

Urban forestry, an increasingly well-understood concept in Britain, Europe and North America, involves the transfer of ecologically sound forest management practice from the rural to the urban setting. Its objectives are based on the premise that forests, existing or introduced into cities, function to create low-cost and self-sustaining landscapes. At a regional scale, there are major forest projects ongoing in Britain and Germany that include a national forest in the Midlands, twelve community forests near urban areas[18] and the major forests naturalizing the Emscher area in the industrial Ruhr valley of Germany.

Urban forestry requires a management philosophy that integrates aspects of horticulture with ecology and provides environmental social benefits. An integrated management policy of this kind may also produce direct economic benefits in the form of forest products. Urban reforestation also involves land that has often not supported trees for a long period of time. Its intent is to create diverse plant associations that are in harmony with the nature of the site, i.e. with its soils, topography, climate and related environmental conditions. In addition, its long-term objective is to rehabilitate sites that have degenerated over time through soil compaction, removal of topsoil, reduction of productivity and nutrients. The philosophy of woodland establishment under discussion is based on the principle of natural succession speeded up and assisted by management. It follows three general phases:

- an initial planting of fast-growing, light-demanding pioneer species that quickly provide vegetative cover, ameliorate soil drainage, fix nitrogen and stimulate soil micro-organisms, and create favourable micro-climatic conditions for more long-lived species;
- an intermediate phase of plants that ultimately replace the pioneers;
- a climax phase of slow-growing, shade-tolerant species that are the long-lived plants.

In practice, the planting of these phases may be done all at one time, or introduced at intervals in the development of the woodland. (In the author's experience based on evaluation of research plots over eight to ten years, however, the latter practice has proven to be more successful than the former.)[19] The initial and final composition, character and uses of woodlands will be quite different as they evolve. They provide varying and useful places from the beginning. Initially they become socially useful and durable in a short time, something conventional, climax-oriented planting does not do. Later the landscape will evolve to a different character, responding to changing uses and environmental conditions. The approach obviates the current problems that occur when whole sections of urban plant materials die from disease. It also maintains a dynamic sense of process and diversity that gives us a realistic view of nature and a valid basis for design with plants.

The successional method for establishing urban woodland can also be applied on city streets. This involves planting fast-growing trees in combination with slow-growing, climax ones. As the former grow larger they are pruned upward to allow the latter to develop. Over time the climax species take over. This management technique can be constantly repeated by replanting other species in gaps left by dying trees. This approach has the advantage of maintaining a permanent street cover. It involves, however, a change from established parks planting and maintenance regimes, to a successional approach that requires an alternative urban forestry practice, with implications for cost, personnel training and the availability of appropriate tree species.

A naturalization programme for an urban parks system

The parks, parkways and waterways that form a comprehensive open space network in Canada's capital city are a part of its long-range development plan initiated in 1950 by the French planner Greber. Both in its planning and design quality it compares favourably with the best urban tradition to be found in North America and Europe. Its overall image as a landscape of beautifully manicured lawns, well-maintained shrub borders, banks of flowers and cultivated trees has, over the years, come to be regarded as the ideal of civic planning. This image represents a period of history when this kind of intensive maintenance was considered an appropriate and proper expression of national pride and commitment to good design. By the 1980s, however, serious questions were being posed for the National Capital Commission – the agency responsible for the implementation of the plan. How could a programme of development and ongoing maintenance be sustained in the face of escalating costs in energy, equipment and manpower? With over thirty landscape corridors comprising an area of some 2,748 hectares to be developed and 3,000 hectares of turf requiring weekly maintenance, the financial burden had become increasingly difficult to bear.[20] The question arose: 'Are the accepted quality standards for landscape development and upkeep appropriate for all city lands?' Apart from being costly, these standards had resulted in a large-scale landscape that provided little variety or diversity from one place to another – an image of monotony denying the city's unique natural setting. When examined together, issues such as this make a cogent argument for seeking other ways of developing and maintaining the city's landscape.

With these considerations in mind, in 1981 the National Capital Commission initiated a naturalization programme for its parkway corridors that includes the creation of meadow communities through a modified mowing regime and an experimental reforestation study programme designed to gain long-term knowledge of methods for establishing new woodlands. This radical departure from conventional practice on the part of a major public organization represents the beginnings of an alternative approach to the urban landscape that over time will become low-maintenance, economical and self-sustaining.

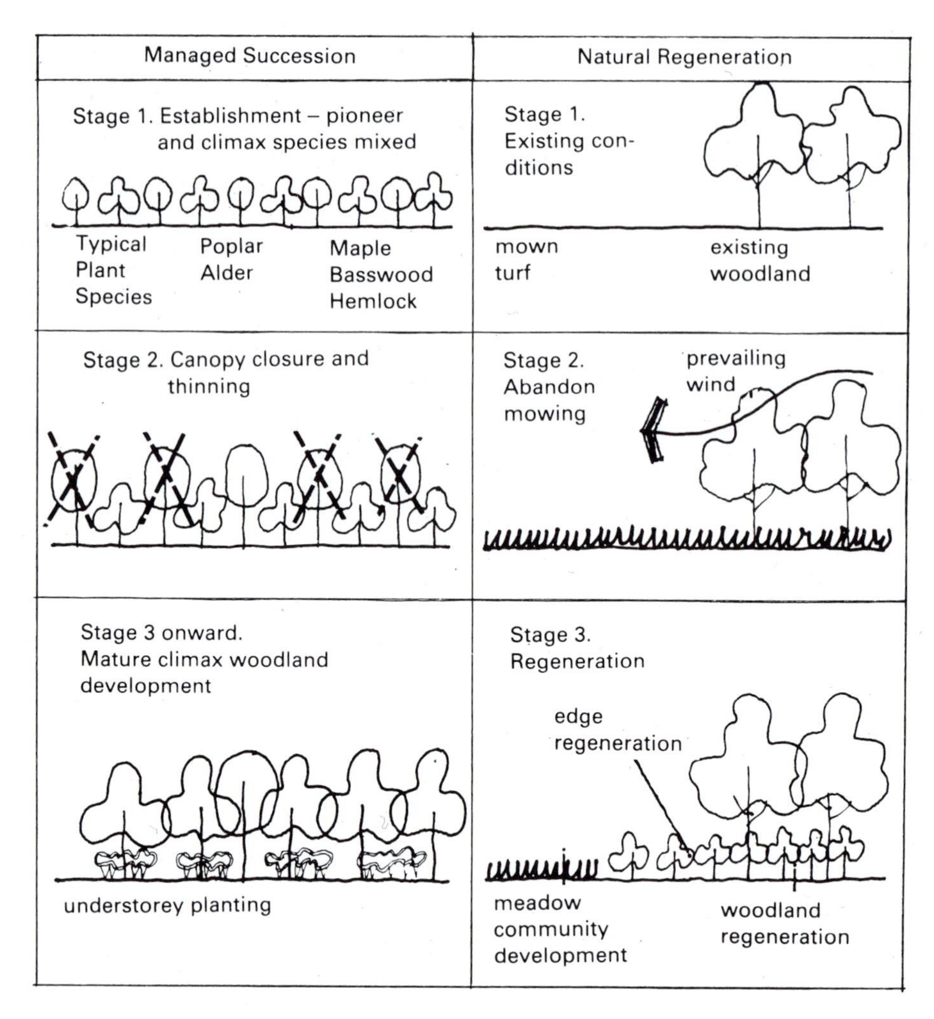

Figure 3.2 General reforestation categories. Plantation involves the planting of predominantly similar species where the final woodland composition is determined by the initial planting. This is the normal procedure of forestry practice and is based primarily on commerce objectives.

Managed succession developed in the Netherlands and Britain is based on the principle of natural succession and assisted through management. The initial and final composition, character and uses of the woodland will be quite different as it evolves. The nurse crop functions to ameliorage soil drainage, fix nitrogen, stimulate soil micro-organisms and create a microclimatic environment suited to the development of climax species. This approach is therefore concerned primarily with the rehabilitation of derelict landscapes, rather than with commercial objectives. Arguments on the advantages and disadvantages of native versus non-native plant species may be less important than considerations of structure, wildlife habitat, adaptability to soils, local climate, air pollution, drainage and so on.

Natural regeneration involves discontinuing mowing regimes in areas where a woodland seed source is available. In the absence of disturbance a woodland landscape is re-established naturally over time.

Source: Hough Stansbury Michalski Ltd, *Naturalization Project*, report, Ottawa: National Capital Commission, 1982.

Prescription	Planting Procedure	Comments
Alternative 1 Mechanical/manual cultivation – cultivation of planting area prior to planting (fall) to kill ground vegetation – manual cultivation regularly during growing season (monthly)	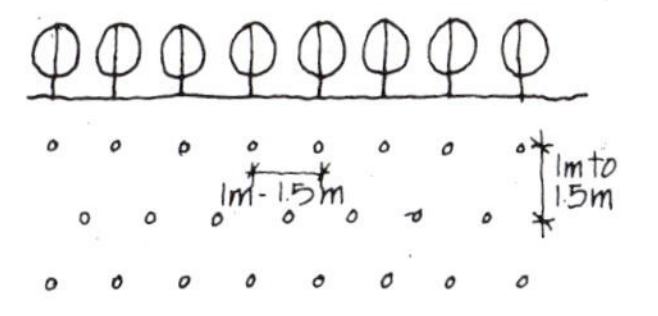	– labour-intensive – application to small or awkwardly sheltered areas – application to closely spaced planting where fast canopy closure is a high priority – constant maintenance required during growing seasons
Alternative 2 Chemical treatment – application of Round-up or equivalent herbicide to kill ground vegetation (fall) – mechanical/manual cultivation of area 7-10 days following chemical application – application of Simazine or equivalent herbicide (spring, prior to planting) in doughnut pattern around tree locations	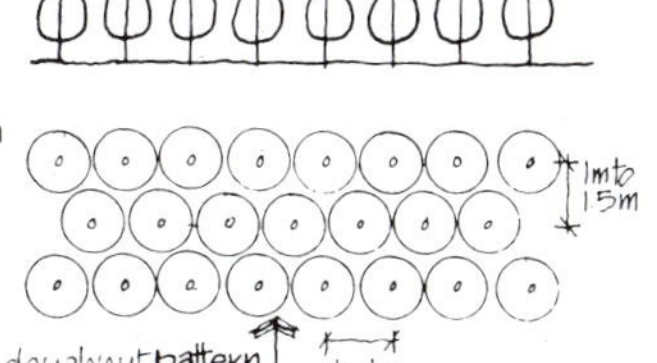	– greatly reduced labour requirements, since only one application required per year (depending on rate of application) – applicable in small or awkwardly shaped areas and to closely spaced planting – chemical treatment in urban areas may present problems of health and public acceptance Note: Simazine is registered for use only for white pine and balsam fir plantations; while Simazine is also applied to hardwoods, many species are sensitive to chemical herbicides
Alternative 3 All mechanical cultivation – mechanical cultivation of area prior to planting (fall) to kill ground vegetation – mechanical cultivation between rows done regularly during growing season	mechanical/hand planting mechanical/chemical maintenance 	– greatly reduced labour requirements – suitable for large areas where machinery may be economically used – growing season cultivation still requires ongoing maintenance (monthly) until canopy closure
Alternative 4 – application of Round-up or equivalent herbicide to kill ground vegetation (fall) – mechanical cultivation of area 7-10 days following chemical application, between rows – application of Simazine or equivalent herbicide (spring, prior to planting) by mechanical spray between rows	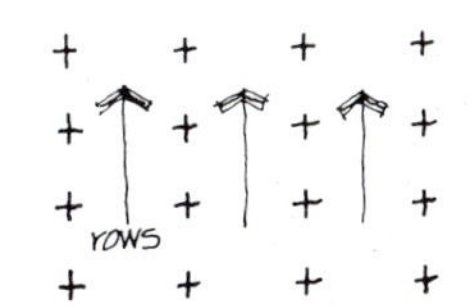	– all mechanical treatment – therefore least manually intensive and least costly – chemical treatment required only once a year (spring) depending on rate of application, until canopy closure – requires large areas where machinery can be economically used – chemical treatment in urban areas may present problems of health and public acceptance

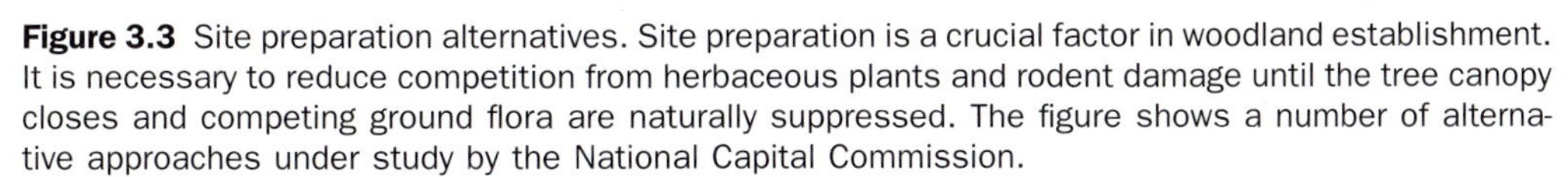

Figure 3.3 Site preparation alternatives. Site preparation is a crucial factor in woodland establishment. It is necessary to reduce competition from herbaceous plants and rodent damage until the tree canopy closes and competing ground flora are naturally suppressed. The figure shows a number of alternative approaches under study by the National Capital Commission.

Source: Hough Stansbury Michalski Ltd, *Naturalization Project*, report, Ottawa: National Capital Commission, 1982.

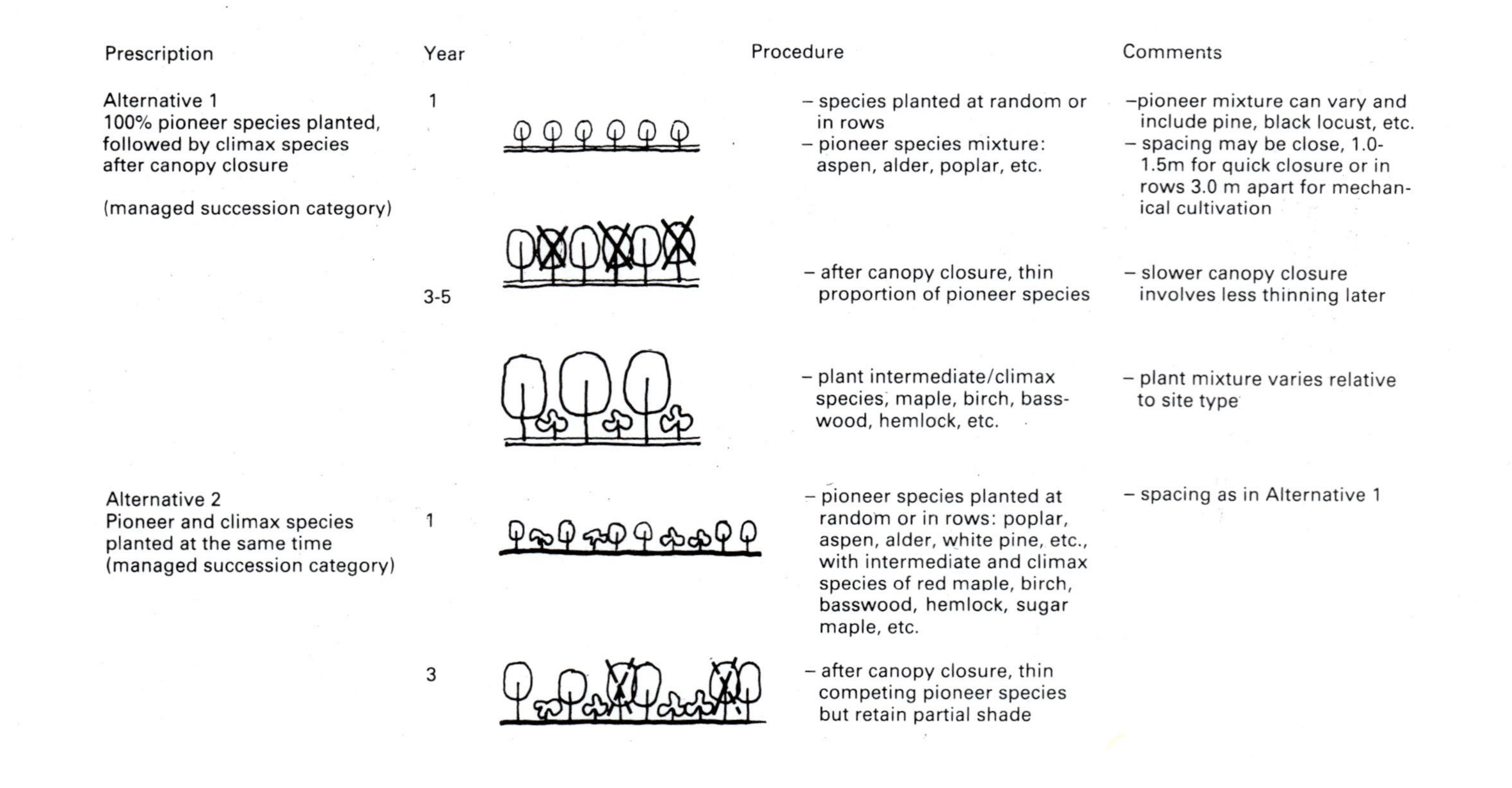

Figure 3.4 Planting techniques. The layout and spacing of plant materials depends on a number of interrelated factors that require investigation; for instance, the balance between closely spaced plants that achieve fast canopy closure but involve costly hand cultivation, versus widely spaced plants that achieve slower closure but involve cheaper mechanical cultivation; the relative merits of an initial 100 per cent pioneer planting versus mixing fast- and slow-growing species together.

Source: Hough Stansbury Michalski Ltd, *Naturalization Project*, report, Ottawa: National Capital Commission, 1982.

The project began with a series of test plots designed to evaluate specific planting techniques, management procedures and public acceptance of the idea. While in general terms the approach taken was based on managed succession, there were many questions that required answers and that dictated the test plot design, i.e.:

- the proportions of various species through the successional range of plants suited to the soils and climate of the region. Four relatively simple groupings of plants were selected that related to the well and poorly drained sites;
- the most effective types of site preparation techniques relative to cost factors, manpower, competition and speed of plant establishment;
- the best methods of ground treatment to control competing plants such as grasses and damage by rodents;
- the types of management required up to the establishment of woodland (canopy closure) and subsequent management of the evolving woodland (thinning of stands in relation to long-range objectives).

Test plots were initiated in 1983. Monitoring and evaluation of the results took place in 1987 to record what had been successful and what had not, and to determine the best approach to woodland establishment for future naturalization projects.[21] This led to a number of observations and conclusions:

- An approach involving managed succession is by far the most effective way of initiating woodland.
- Poor soils are not an impediment to growth provided that they are reasonably well drained with a neutral pH. Much of the soil in the test plot area was excavated subsoil from a nearby construction site, and was highly alkaline averaging a pH of 8. This required the addition of sand and sulphur integrated into the top 30 centimetres of soil to neutralize its alkalinity.
- Ground treatment that inhibits competition from grasses and winter damage to saplings by rodents can achieve a closed canopy of pioneer species and the beginnings of a forest floor association in four to five growing seasons, given the climatic conditions and leda clays of the Ottawa region.
- Of the various methods used to keep plots free from grasses and other competing species (manual weeding, rototilling, chemical applications, clover groundcovers), perforated plastic sheeting covered with mulch proved to be the most effective from the point of view of growth rates and management cost. In one plot prepared with this method, poplar pioneer species attained a height of 8 to 10 metres in four growing seasons.
- An initial close spacing of plants (about 1 metre) achieves a much greater survival rate and a faster canopy closure than the wider spacing used in conventional landscaping.[22]

Using the techniques established in these experimental plots, a length of parkway was planted to native forest, and proved to be successful and economical as an alternative to the manicured approach that had traditionally been taken by the Commission's Maintenance Department. Figures 3.2–3.4 and Plate 3.4 illustrate some of the design and management principles on which the experiment was based and some of the results.[23]

Developments since 1984

Following the completion of the second phase of this applied research effort, the NCC abandoned the project, further research was terminated, and the different plots were left

(a)

(b)

Plate 3.4 An example of growth rates in one plot in the National Capital Commission's research in woodland establishment. (a) View of plot 3 taken in spring 1985 at the time of establishment, showing perforated plastic and mulch treatment (background) versus tilled ground treatment (foreground). (b) View of the same plot four growing seasons later (autumn 1988). It shows the contrast between the low survival and growth rate of trees in the tilled portion of the plot (foreground) and the excellent tree growth in the plastic and mulch treatment portion (background). No maintenance was done on this plot during this time, barring watering during the first summer. Encouraged by the long grass, mice and voles decimated the foreground plantings.

Plate 3.5 The experimental plots some fourteen years after Plate 3.4b was taken. Undisturbed natural succession after abandonment of the research has resulted in various plots coalescing, creating a low-maintenance uneven-aged woodland environment that replaces much of the maintained greenway turf. The planned 'nucleation' plot, however, failed

Source: Richard Scott.

to develop on their own. An immediate impression, after some fifteen years, was how difficult it was to identify the various plots. They had grown together into one large woodlot, with long-lived species such as Bur Oak appearing in the under-storey, brought in from the adjacent woodland by squirrels. Nature was taking over the job left unfinished. The lack of interest by the NCC in continuing the work was, at the time, a continuation of attitudes about horticultural maintenance that had been the stimulus for this applied research in the first place. Long-held values about the aesthetics of mown turf and what constituted good care of the landscape began to change, however. In the late 1990s there was evidence of new woodland naturalization projects being implemented along the greenways using the techniques that had been developed in the 1980s. The management of grasslands along the parkways had progressed from an exclusive manicured regime to a mix of meadow and mown turf, and there is evidence that the movement to alternative approaches to management was having a visual impact – one that was more diverse, more visually interesting and which reflected the character of Ottawa's natural region, a sense of place. There was a time when the high costs of horticultural turf maintenance were thought to be a persuasive argument for change to ecologically appropriate landscape management. This has not been the case, however. It has been the changes in values, occurring over the past two decades, that has stimulated many municipalities and parks departments to make the change to less costly practices.

The residential landscapes of Delft: the Gilles Estate

The landscape of many housing developments and parks in the Netherlands has, over the years, been brought to a high level of sophistication and design. As early as the late 1960s and 1970s, the Delft Parks Service realized that a radically new approach to urban living was necessary, in contrast to the sterile and tidy environments of previous Dutch housing developments. The intention was to surround the housing with all the informality of the countryside and to create an educational landscape geared to unstructured, natural play environments. In the high-density apartment development of the Gilles Estate, courtyards were planted as urban woodlands in addition to providing open space for nursery schools, play areas and sports facilities. These inner court landscapes were based on certain fundamental and social objectives.

- There must be freedom of movement and play by children and adults. In these woodland landscapes the sheer vigour of early plant associations and their density provide a tough and highly varied environment. They must withstand the pressures of play and other activities, even in high-density housing environments. The Dutch believe that it is unrealistic to attempt to confine children to specific play areas, since this is not the way they behave in real life. While these are provided, the whole environment is available for play with a variety of opportunities and sensory stimulation for children ranging in age from pre-school to teenagers. The complexity and ruggedness of the woodland landscape accommodates pressures that would soon reduce conventional design layouts to ruins.
- Making a natural landscape. Plant material was selected on the basis of several criteria: to create intermediate scrub vegetation that would provide as rich a diversity as possible; and to emphasize the use of local native woodland species to minimize cost, maximize durability and take advantage of their ability to coppice. Plants were planted as whips in a random pattern spaced at 1 metre intervals. Specific site requirements dictated plant groupings (for example, locating willow in wet places), and the planting of larger trees to provide a mix of ages and initial shelter. Management responded to use patterns and circulation over time. For instance, secondary paths, created in response to natural desire lines, were gravelled and left as pedestrian and cycle routes once they had been established as permanent patterns of movement. As the woodland matured and responded to use, a self-perpetuating stable vegetation developed requiring little upkeep. The objectives of ongoing management were intended to produce a usable indigenous vegetation with a character of its own.[24]

To the observer, the quality of these housing landscapes is therefore radically different from conventional ones. They have the natural, informal, somewhat untidy, yet functional character of heavily used places. It is an aesthetic derived from the interaction of vegetation and human activity. A study of one apartment complex by the Institute of Preventive Medicine at the University of Leiden concluded that the landscape design and its use by children resulted in a greater attraction and more efficient use of the available space compared with the open spaces created by more conventional planning.[25]

A 1993 review of the Gilles Estate

The woodland landscape of the Gilles Estate was a revolutionary departure from convention for the Dutch that has had enormous influence in shaping the Green Cities movement in Europe and North America. The housing landscapes created by the Delft Department of Parks were a bold and imaginative response to the need to create more humane, varied and useful environments for the people living there than the previous housing developments of the post-war years. The three main post-war suburbs in Delft, of which the Gilles Estate was one, clearly show the changes in opinion as well as the financial constraints that shaped urban landscapes in the Netherlands.[26] Ton Jacobs, however, in his review of the Gilles Estate, suggests that the attempt to replace traditional horticulture by ecologically sound vegetation management has failed. The reasons have to do with the realities of human as well as natural evolution. He notes that while ecology is very much a science dealing with change and development, it is also logical to consider human behaviour and their organizations in the same way.[27] The successful

(a)

(b)

Plate 3.6 The Gilles Housing Estate, Delft, the Netherlands. Comparison between the courtyard plantings in (a) 1980 and (b) 1993. Indiscriminate thinning in the interests of openness and safety has greatly altered the original character intended by the initiators of the concept.

development of a plant community in an urban setting depends on the life cycles of the human environment. Jacobs describes the first Gilles residents as a homogeneous population, typical of the new suburbs, who moved into more spacious dwellings than they had been used to, and whose children loved the green landscapes of the estate. Over the years, children became teenagers and adolescents, parents got divorced and moved elsewhere, migrants from the Mediterranean moved in, policies changed to favour a wider spectrum of people, and the founders of the experiment left the Department of Parks to be replaced by people with less attachment to its ideals. Expectations and attitudes towards the green landscape, therefore, changed. Vandalism became an issue to be combated at all costs, an attitude that is not particularly favourable to green policies. And as the landscape matured people began to complain about safety,[28] demanding open, well-lit spaces (an issue discussed earlier in this chapter). The response was a thinning programme that removed much of dense under-storey vegetation in many places. The long-lived species, however, were also cut and thus encouraged the continued growth of pioneer vegetation. Given the hierarchical structure of Dutch bureaucracy, the necessary changes in the relationship between park officials and the people from a 'them and us' attitude to a cooperative partnership did not mature, and, according to Allan Ruff, is still in an early stage of development.[29] In addition, environmental priorities among action groups focused more on issues with a more global impact such as acid rain, climate change and the ozone layer, with little concern for local urban ecological issues. Thus, as Jacobs comments, 'the Gilles experiment got international recognition as an experimental landscape. But local people hardly knew about it.'[30]

There are interesting comparisons between the perceived decline of this landscape and the LeBreton Flats experiment discussed on pp. 80–85. They have to do with behavioural and social management issues as well as the landscape itself, and how these must be integrated. In the light of this experience, a further visit to Delft in the spring of 1993 by the author (just thirteen years after the first) was interesting for what it revealed about the place. Recognizing the validity of Jacobs' comments, there is much to be learned from what may be observed there. Much of the under-storey has been cleared but large patches remain. There is an overall sense of tree canopy (even though most remains as fast-maturing pioneer species) and long vistas giving a sense of safety. There are areas of turf for group activities, and a wonderful quality of coolness. Children commandeer the place, using and playing in it as they did in the 1970s. Sunny private gardens for residents have been landscaped by them to their own tastes. Overall the place has a sense of ownership and control that is recognized by children and adults alike.

A further review in 2002

In November 2002 I returned again to the Gilles Estate with my friend and colleague Sybrand Tjallingii to observe what changes had occurred over a nine-year period. There were further surprises. Based on the 1993 experience of entering a heavily thinned treed landscape dominated by turf, one might have expected to see further clearing, or even the complete disappearance of tree cover given complaints of safety that initiated the process in the first place. Instead, a vigorous regeneration of under-storey trees and shrubs in large areas had evolved, enclosing clearly defined turfed spaces. The development felt like a winter woodland with generous glades flowing through it, but with woodland dominating the overall environment.

Other social indicators of change, not previously noticeable, could be seen in the building apartments. There was clear evidence that people lived there. Laundry and Turkish carpets hanging from apartment balconies, bicycles, brightly coloured awnings

Plate 3.7 The same courtyard late in 2002. Vigorous regeneration has taken place since 1993, creating a mix of dense woodland and open glades. The early stages of a mature woodland environment have evolved that satisfy the needs of the culturally diverse inhabitants while achieving a variety of environmental benefits.

providing shade, individuality and character to bland façades, spoke of people of different cultural origins – both Turkish and Dutch – claiming control of their living places. There was also evidence of work: a healing centre in a ground-floor apartment, a centre for the production of orthopaedic instruments in a converted wing of the local school. There was a sense of maturity and strong identity in this living complex and its landscape, which had clearly grown and changed over the twenty years since I had first visited the Gilles Estate.

It is also clear that this remarkable experiment will continue to evolve in accordance with changing political, social and ecological influences and decision-making over which no one has control. Its sense of place is still powerful, and the lessons to be learned continue to be relevant to those who were originally inspired by it.

Green economics and the urban forest

Thus far we have examined plants as communities, focusing on this role as an ecologically based alternative to much current practice. It should be noted that the term 'urban forest' does not reflect a true forest (which may be described as a highly complex, interacting community of naturally associated plants and organisms), and in which trees are dominant. It is commonly used by urban foresters to describe a collection of trees of various unrelated species growing in the city, many or all of which may be alien to the natural region. In addition to its traditional roles for streets, parks and open spaces, however, the urban forest is increasingly being examined for its value in mitigating many of the environmental problems found in urban and suburban environments. They include air quality, stormwater management and energy use. Of particular importance in these functions is the density of the canopy cover. The higher the density the greater the influence on the environment.

Air quality. As mentioned above, the density of the canopy area of trees greatly increases the ability of vegetation to filter dust, reduce temperatures and absorb a variety of chemical pollutants. The benefits to air quality from canopy cover are also achieved by the uptake of carbon dioxide. How much carbon dioxide can be stored depends on forest types. On average urban forests store about half as much as natural forests. A three-year study in Chicago (the Chicago Urban Forest Climate Project 1995) set out to quantify the effects of urban vegetation on local air quality and to help city planning and management organizations increase the net environmental benefits derived from the city's urban forests. Tree canopy density ranged from below 5 per cent (in the most intensely developed downtown areas) to nearly 40 per cent (in lower density residential areas) and averaging 11 per cent overall tree coverage. The results showed that in 1991 trees in the city as a whole removed an estimated 17 tons of carbon monoxide, 93 tons of sulphur dioxide, 98 tons of nitrogen dioxide, 210 tons of ozone and 223 tons of fine particulate matter. The study also showed that these trees sequester approximately 155,000 tons of carbon per year. Other related studies have shown that trees can mitigate ozone pollution by lowering city temperatures and directly absorbing the gas. The economic value of pollution removal in 1991 was estimated at $1 million for trees in Chicago itself, and $9.2 million for trees across the larger study area (which includes Chicago, Cook County and DuPage County).[31]

Stormwater retention. The engineering systems required to control urban stormwater run-off and repair flood damage involve very large financial expenditures. Trees provide an alternative option to engineered solutions by storing water through their leaves, branches and trunks and soil. Some water evaporates back into the atmosphere and some soaks into the ground, reducing the volume of water that must be stored. An urban ecosystem analysis for the Houston Gulf Coast region assessed the loss of tree canopy cover over 27 years (1972–99). Tree loss resulted in an estimated increase of 360 million cubic feet of stormwater during a flow peak storm. Restoring lost tree canopy would provide considerable benefits. For example, a 40 per cent increase would provide $3.5 billion in one-time stormwater benefits (a 163 per cent increase from 1999 conditions).[32] In another study of the Lower Columbia region of northwest Oregon it was found that tree loss between 1972 and 2000 resulted in an estimated increase of 963 million cubic feet of stormwater flow during a peak storm event (based on an average maximum, two-year, 24-hour storm event). Using a local cost estimate of $6 per cubic foot to build stormwater systems in urban areas, and $2 per cubic foot in rural areas, this vegetation loss is equivalent in value to a $2.4 billion system.[33] The total stormwater retention capacity of the region's tree cover in the year 2000 is valued at an estimated $20.2 billion, down from 1972's value of $22.6 billion, based on the avoided cost of having to manage this stormwater.[34]

Energy use. The urban forest influences the use of energy for cooling and heating due to the forest's moderating influence on climate and energy demand (see Chapter 6). A case study conducted to evaluate these influences was applied to seventy-one county subdivisions in Sacramento County. Heating, cooling and changes in peak electrical energy use resulting from modifications of solar radiation, air temperature and wind speeds by the existing urban forest were estimated for both residential and commercial buildings. When combined with tree canopy and density, annual cooling savings approximated US$18.5 million per year, or 12 per cent of total air-conditioning in the county.[35]

Urban studies of change over time. American Forests has, since the 1990s, undertaken regional ecosystem analyses of a number of urban regions in the USA including the Chesapeake Bay Region and the Baltimore–Washington Corridor.[36] This was undertaken to determine how the landscape had changed over a twenty-four-year period (1973 to 1997). Using GIS technology to measure the changing structure of the landscape, the overall study area covered some 11.4 million acres of land. A more detailed study involved 1.5 million acres in the Baltimore–Washington Corridor.[37] The results identified a number of trends with largely similar results and illustrate the economic implications of this tree loss for stormwater and clean air over this time period.

In the Baltimore–Washington Corridor:

- overall tree cover declined from 51 per cent canopy coverage to 37 per cent coverage;
- there was a decline of 32 per cent of heavily treed cover;
- areas with little or no tree cover increased from 31 per cent to 49 per cent.

There were also economic implications of tree loss for Stormwater management and clean air:

- Tree loss between 1973 and 1997 resulted in a 19 per cent increase in storm run-off (from each two-year peak storm event), an estimated 540 million cubic feet of water. To build stormwater retention ponds and other engineered systems to intercept this run-off would cost $1.08 billion ($2 per cubic foot of storage). The total retention capacity of the urban forest in 1973 was worth $5.7 billion. This capacity dropped to $4.68 billion in 1997.
- In 1997 the tree canopy removed approximately 34 million pounds of airborne pollutants valued at some $88 million. Tree cover as it existed in 1973 would have removed 43 million pounds of pollutants, a loss of $45 million.

The study concludes with some key observations that have application beyond US cities. They reinforce the value of urban trees from both environmental and economic perspectives, and have application beyond US cities.

- Maintaining and restoring tree cover is a cost-effective way of enhancing the natural processes of the city's landscape.
- All cities (in the Chesapeake Bay Region) should conduct similar local analyses and incorporate them into their planning and growth management processes (see Chapter 7, pp. 235–8).
- County and city governments should be recruited as partners in creating regional vegetation models.
- They should consider the (monetary) values associated with trees when making land-use decisions.
- Overall tree canopy cover in urban areas should be conserved and increased to 40 per cent.
- Planting trees in urban and suburban areas and buffers along streams throughout the region would contribute to water quality, improve air quality, wildlife habitat, conserve energy, sequester greenhouse gases and enhance quality of life.

These examples and case studies are highly relevant to the enhancement of the urban environment. Useful as they are, however, they need to be part of an ecologically integrated system. For instance, the benefits of tree canopies in reducing stormwater volumes can also be combined with many other benefits such as wetland fish and wildlife habitats, discussed in Chapter 2.[38]

Issues of safety

Many communities with minimal or no experience of naturalized landscapes have questioned the viability of the Dutch approach when these landscapes are proposed for their own particular social and physical environments. Questions about safety and security consistently arise, which in many cases have the effect of restricting the potential to bring diversity and sensory richness to public places. The basic right of the public, particularly women, to feel safe from crime in their neighbourhoods, public streets, parks and workplaces is central to civilized life. It is also important to recognize that *the perception* of safety is more significant than actual statistics since it is perception that lies at the heart of the problem. Studies in the USA and in Britain point to the economic decay of downtown areas as a direct result of fear of crime.[39] The Safe City Committee of the City of Toronto has stated that sexual assault, the crime most feared by women, is also a far more terrifying crime than robbery, which is the crime most feared by men.[40] It is for this reason that a number of cities including London, Amsterdam and Toronto have focused on making public spaces safer through concentrating on violence against women. The City of Toronto Safe Cities Committee has published a guide for planning and designing safer urban environments which includes ways in which the safety of places may be enhanced. The measures include:

- the need for appropriate lighting for pedestrians as well as for motorists in public spaces;
- sightlines – the ability to see what is ahead along a route, the avoidance of blind corners, impermeable landscape screens and so on;
- avoidance of tunnels, pedestrian bridges, narrow passageways that offer no alternative choice for pedestrians;
- avoidance of 'entrapment' spots, such as small, confined areas (for example, recessed entrances);
- the need for places that have visual surveillance, where people can watch others;
- the value of activity generators, such as food stands that maintain informal surveillance of places;
- ensuring a sense of ownership or territoriality in neighbourhoods and public spaces;
- the need for appropriate signage and information about places, such as main roads, exit signs and main pedestrian routes.[41]

These guidelines for safety and security are, among others, a necessary consideration in the planning for public space, depending on location, social environment and the specific characteristics of the place. They are, however, particularly relevant to the alternative strategies under discussion, since the relationship between naturalized landscapes and safety must be addressed in modern cities.

Woodland parks and integrated management

The remnants of natural forest communities that have somehow survived in cities are irreplaceable, though often neglected natural elements. They stand as classic examples of the need for ecologically based management. Many have deteriorated due to urban pressures, including reductions in size, habitat fragmentation, by colonization of remnant native woodland by exotic and cultivated species leading to loss of species diversity, and general environmental degradation. The urbanization of watersheds affects woodlands by altering stream erosion processes, sedimentation patterns, soil chemistry, vegetation and animal communities. The need to preserve both locally rare and common

Plate 3.8 Zurich's forest parks. Integrated urban forest management included forest products, wildlife, small-scale farming and recreation. Sophisticated silviculture maintains the productivity of these woodlands and their aesthetic appeal. It shows how a park system can be, at least partially, economically and ecologically self-sustaining, contributing in ways other than recreation to the public good. As a multi-functional, self-sustaining landscape it provides social, economic and environmental benefits and calls into question the assumption that parks are exclusively for leisure.

features, restore species diversity, and protect wildlife and scenic quality is therefore crucial. Management objectives that integrate the physical and social influences of the surrounding city with the dynamics of a changed but evolving ecosystem under urban conditions provide irreplaceable opportunities for education.

Urban natural areas are the field study centres of the city, where plants and animals can be observed, where community dynamics can be studied and where the interactions of urban and natural processes may be measured. They are outdoor laboratories for teaching reforestation and silvicultural techniques. It is here that different management objectives must be integrated in the light of many and sometimes conflicting demands. They provide the essential links to understanding the nature and functions of forests beyond the city. Understanding the processes of nature and human intervention in familiar local surroundings is, perhaps, the most effective way of ensuring informed public concern for the larger environment.

In countries where significant rural timber is lacking, the city may play a role in producing timber to offset the costs of recreational facilities and park maintenance. Zurich, a city of 360,000 people, has a major proportion of its park space (nearly a quarter of the urban area) in forest and common land. These lands, some 2,200 hectares in extent, are within half an hour's tram ride from the city centre and lie within or on the edges of the city. They have for many years been maintained on an integrated management basis, protecting groundwater, providing timber, recreation and athletic facilities, wildlife, agriculture, visual amenity and education. The forests are a mixture of deciduous beech, oak and maple stands and conifers. Forestry is carried out by the

City Department of Forestry. It involves a variety of silvicultural techniques that include shelterwood, seed tree and patch cutting systems, depending on the forest type. The aim is to produce an unevenly aged forest of young and mature stands with a major emphasis on an aesthetic forest quality, and great care is taken in cutting to ensure that this quality is maintained. Cutting occurs in the winter and the logs are stored along an extensive system of forest roads for shipping to various sawmills outside the city. Forest roads are also designed to be used for walking, nature trails, cross-country skiing and exercise by the citizens of Zurich. Fitness trails and exercise stops are also integrated into the forest setting and water for the park's picnic sites is supplied from the groundwater that the forest protects.

The basic aim of the forestry is to produce commercial timber for sawlogs and pulp. These products bring a return and help support the increasingly extensive and sophisticated recreational facilities that the city provides. In 1979 the balance sheet for the city's operations showed the costs of all recreation and forestry operations to be 4.5 million Swiss francs. This also included financial assistance to private clubs. Income from forest products amounted to nearly 2.5 million Swiss francs. Income therefore amounted to approximately 55 per cent of the total costs of the parks system. An interesting aside on the economics of the operation was demonstrated in the sale of beechwood to Italy for fruit boxes. These would arrive back from Italy with fruit, bought by the Swiss. The cost of the fruit is thus paid for by the sale of the boxes.[42]

It is apparent, therefore, that the concept of bringing rural occupations to the city in the form of urban forest products provides numerous benefits that conventional parks operations are unable to do. As we will see in Chapter 6, vegetation has a marked influence on climate and the urban environment generally. As well as the social benefits of a more varied and useful landscape, the urban forest can provide the basis for education in forest practice. Its presence as a part of the life of the city and under the scrutiny of its citizens ensures that high standards of forestry are maintained. It thus helps close the perceptual gap between urban and rural areas by creating a better understanding of their ultimate interdependence. In addition, it makes potential economic sense, a fact that has relevance in different ways for cities in both developing and developed countries that are trying to meet increasing demands for public amenities with diminishing budgets. Finally, the management of the Zurich forests, under full public scrutiny, makes the process visible and ties issues of forestry into daily life.

Management and the evolving urban landscape

Management and maintenance

The alternative ways of approaching plants that we have been examining offer design opportunities which provide a valid functional and ecological basis for urban form. They are also allied to the concept of continuous evolving management. The implementation of design, from paper planning to layout on the ground, is only the start of the design process. It presents a different picture, condition and usefulness at different stages, guided by a management process that determines its form over time. Conventional maintenance, however, deals with the landscape inorganically, as a static form. Its object is to keep the design as close as possible to the sketches the client has accepted – the fixed picture. It brings with it a formidable arsenal of mowing equipment, leaf vacuums and blowers, fertilizer spreaders, herbicide and pesticide sprayers, to keep plants under control. The maintenance regime is high in energy and resources and aims to achieve standardized results. Obviously some types of urban landscape may require this kind of careful maintenance. Those subjected to intense human pressure, or those with intended gardenesque or historic objectives, are evident examples. But others do not. The role of management in cities must be to provide the greatest diversity possible, fitting a great many situations and needs.

The green carpet

Perhaps the most pervasive element of the cultivated landscape is the lawn. Instant wall-to-wall grass appears everywhere, in hot climates and cold, in prestige projects and in those that are run-of-the-mill, in large landscapes and small. Questions arise about the civic and political symbolism that 'official' landscaping represents, and more importantly the environmental and economic costs that mown turf engenders. An article by Michael Pollan in the *New York Times* in May 1991 is indicative of the emerging values of urban nature that have arisen since the 1980s. It suggested that if the President of the United States truly wished to be remembered as the 'environmental President' the White House lawn was not the appropriate symbol. Pollan argued that:

> [T]he democratic symbolism of the lawn may be appealing, but it carries an absurd and, today, insupportable environmental price tag. In our quest for the perfect lawn, we waste vast quantities of water and energy. . . . Acre for acre, the American lawn receives four times as much chemical pesticide as any US farmland.[43]

The lawn is a symbol, in effect, of our relationship to the land, an expression of human control over a natural diversity that extends worldwide, from Britain to California to the Far East. According to the Lawn Institute in Tennessee, America has more than 80,000 square kilometres of lawn under cultivation, on which is spent US$30 billion a year.[44] Pollan continued with suggestions for more appropriate symbols of power and authority as a public setting for the White House grounds, such as a meadow that includes so-called 'weed' species and a once-a-year mowing; a wetland, expressing one of the richest and most important of habitats; a vegetable garden that could make the White House self-sufficient, or feed Washington's poor; and an apple orchard, productive, beautiful and *the* American fruit.[45]

The tradition of the lawn in North America dates back to the nineteenth century and Frederick Law Olmsted. Olmsted was part of a generation of American landscape architects and reformers who set out to beautify the American landscape, although, as Pollan observes, 'That it needed beautification may seem surprising to us today'.[46] In 1870 Frank J. Scott published *The Art of Beautifying Suburban Home Grounds*. Intended to make Olmsted's ideas accessible to the middle class, the book probably did more than any other to determine the look of the suburban landscape in America.[47]

Today's turf depends on standardized requirements for topsoil, fertilizers, herbicides, watering and cutting heights, a total control that continues to challenge its credibility as a living material. As a high-cost, high-energy floor covering, it produces the least diversity for the most effort.

There are simpler ways of producing a diverse landscape through a broadly based ecological approach to management processes. One alternative is a more intelligent and less intensive use of mowers to permit plant diversity and wildlife habitat to become management objectives. A study for the National Capital Commission in Ottawa on the implications of modified mowing regimes for the city's maintained grasslands showed the value of such an alternative. By cutting only those areas that were necessary for recreation, fire hazard and similar factors, and leaving remaining grasslands as meadow during the summer months, a far greater diversity of bird species was created in a very short space of time. The number of exotic 'nuisance' species such as starlings and house

Plate 3.9 Alternatives to mowing. Cutting where it is needed, around groups of trees and shrubs, and associated with established plant communities, provides visually diverse environments. Other alternatives to mowing can also mean leaving naturally regenerating landscapes to evolve, an increasingly popular management approach for highway verges.

sparrows was reduced and the number of grasslands species such as bobolinks rose dramatically as the appropriate habitat became available.[48] This illustrates the need for grassland management objectives that take into account the recreational, visual and functional aspects of parkland that require areas of short turf, together with the encouragement of habitat diversity. If the maintenance of grassland rather than woodland regeneration is the objective, then cutting unmown areas may be periodically required depending on objectives and the stability of the meadow. In order to preserve optimum grassland for bird habitat, mowing should be limited to late autumn, after breeding is over and after migrating birds have taken advantage of meadow seed sources.

Visually, the combination of rough and fine turf can add a complexity and variety to the landscape of a large grass area that universal mowing can never achieve. The problems that often arise when grass is left unmown are to a great extent perceptual. The image of neglect is nowhere more apparent than where the interface between mown and unmown turf is poorly considered. Turf left to grow long adjacent to human activities has tended to represent neglect and abandonment of responsibility. There are also practical aspects to the perceptual problem. Naturalized turf close to walks, roads or housing collects litter which involves higher maintenance costs to remove; and there is always the risk of fire. The creation of well-designed edges that accommodate mown areas for recreational and functional use while maintaining graceful lines establishes a sense of purpose and design intent. There are numerous opportunities for creating herbaceous groundcovers and these are used extensively in Dutch parks and open spaces. They are also being used to rehabilitate industrial spoil heaps in Britain and elsewhere. A survey of metalliferous mine wastes in Britain included field experiments which suggested that

Plate 3.10 Orchard in the Netherlands: sheep maintain the turf groundcover, reducing competition for the fruit trees and providing another source of income for the farmer – a good example of economy of means.

with the high concentration of toxic metals, rehabilitation could be achieved more effectively with naturally occurring grasses than with commercial varieties. Naturally occurring populations grew faster and persisted longer. They provided excellent stabilizing cover with the application of adequate fertilizer, and have persisted for many years. Three cultivars are commercially available that are tolerant of various pH and metal concentrations.[49]

The options available and their relevance to different situations are discussed in a report for the Nature Conservancy Council in Britain.[50] They include the introduction of native herbaceous flora and meadow communities. The conditions under which these thrive in nutrient-deficient soils are typical of much derelict urban land. This, in fact, is the opposite condition demanded by cultivated turf grasses. The elimination of imported topsoil and fertilizers makes their installation a great deal cheaper. Their floral diversity and low long-term maintenance are ideally suited to large-scale reclamation.

These alternative groundcovers are applicable in such places as roadsides, vacant lots and similar areas where pressures of human activity are low. The high-resilient wild areas of Dutch landscapes are seen as constantly evolving systems that adapt to wear and tear. This kind of ground surface is fortuitously created when turf along pedestrian routes is replaced with colonizing weeds, resistant to trampling and often covering the ground more effectively than the original turf.

Another alternative for managing urban grasslands is to replace machinery with livestock. The uniform green carpet quality of the lawn is a product of modern technology, which could not exist before the age of fossil fuels. Grass meadows were previously maintained by grazing animals. The eighteenth-century English landscape garden was conceived of as an extension of the pastoral, agricultural landscape of a livestock economy and it used agricultural methods for the maintenance of its grounds. Sheep are remarkably effective as mowing machines. They graze close to the surface, nibbling at the grass shoots to produce a low, even turf that is both appropriate to many recreational uses and visually attractive. Sheep also encourage species diversity by selective grazing, something that mowing machines cannot do. They require few physical facilities or personal attention, apart from protection from dogs, access to water and shelter from the wind. Anyone who has visited the upland sheep country in the north of England will attest to their hardiness and ability to keep precipitous slopes in almost immaculate condition.

Sheep and other grazing animals are extensively used as an alternative to mechanical mowing in many country parks in Britain.[51] In some towns and villages, churchyards and meadows are maintained by sheep as a result of arrangements with local farmers. However, this practice, which was once common in cities, has now all but disappeared. Yet its advantages as an economic and practical alternative to modern machinery are considerable, as the occasional example reveals. Laurie points out that many urban commons that are kept in a fairly natural state are maintained at an extraordinarily low cost in relation to their size and the use made of them.[52] Some urban commons are still grazed today: Newcastle in England is an example. A petroleum company in Toronto has used sheep to mow the raised berms surrounding its fenced fuel storage tanks since the 1970s and applied this technique to its holdings in other Ontario cities. Sheep were bought from the local stockyards in the spring and sold back in the autumn. Ten sheep maintained 1.2 to 1.6 hectares of steeply sloping grassed banks in a close-cropped and fertilized condition. The cost from the stockyards of each animal was Canadian $50 per head (in 1977 dollars).[53] Even without taking into account the resale value of the animals, $500 per season for grass cutting was considerably less costly than mowing mechanically; nor do mowers fertilize the grass. In addition, the company had no need for equipment or other maintenance outlays barring an occasional bale of hay in very dry periods and a water supply.

There are various factors to be considered if such a practice is to be applied to the city. The major limitation is domestic dogs. Dogs worry sheep and can cause grave injury, so the need for fencing is necessary. In the absence of basic research into effective means of keeping dogs and sheep separated, or of enforceable laws to keep dogs under control, this strategy has practical applications for those urban open spaces that have to be fenced for other reasons, or can be adequately supervised. Industrial areas, high-security installations, private or quasi public lands and cemeteries are examples. The British country parks use sheep, cattle and roe-deer, as much for the education and enjoyment of visitors as for keeping the grass under control. However, the isolated zoo-like environment of the demonstration city farm that some parks provide for deprived city children has less educational relevance than functioning farm management situations where these are part of the daily life of the city. For example, the Newcastle commons, mentioned above, are grazed continuously by cattle through arrangements with local farmers. The commons are also in daily use for recreation and as a pedestrian link between different parts of the city. The presence of grazing animals is therefore an accepted part of urban life and establishes important links between city and countryside.

Management costs

The rising cost of maintenance is a universal issue for parks departments and other agencies concerned with amenity open spaces. The Dutch have made some interesting calculations of maintenance costs based on their woodland approach to urban design in comparison to more highly maintained traditional parks. Taking the maintenance of woodland landscape as the index against which other forms of maintenance are evaluated, calculations by the Parks Department at Rotterdam in 2002 are shown in Table 3.1 for different types of plantings. It is evident from the table that woodland landscape was the least expensive form of planting with the exception of grassland and vegetation. It is interesting to note that turf maintenance is relatively low in cost, while annuals are more than ninety times as costly as woodland. Also included in the table for comparison are the figures for 1980. These figures relate to the Dutch experience and highly developed traditions in parks management. In addition, many landscape developments incorporate a mixture of all or many of these categories, but they provide an interesting guide for comparison with other approaches, bearing in mind the unique problems that every city must face. They also have lessons for us in the development of low maintenance landscapes. Some comparative costs for woodland creation and naturalization illustrate the economies in establishment as well as continued management. It should also be borne in mind that with conventional landscape development the management costs rise over time, whereas with naturalized landscapes they fall over time.

A comparison between the figures for 1980 and 2002, published twenty-two years apart, is instructive. For instance, the differences in cost for woodland, naturalized groundcovers, hedges and lawns are only marginally different. The Rotterdam Parks Department made several comments. First, that the costs of maintenance remaining almost the same were due to hand labour being replaced by machinery and by further specialization of the equipment used. Second, while the cost of labour has increased, the use of labour over forty years has been reduced by mechanization. Third, maintenance practices have reflected different approaches over the past forty years. During the 1960s these were traditional horticultural practices. In the 1970s the parks system was expanded considerably which involved the use of chemicals for weed control. The 1980s saw public opinion turning increasingly against the use of chemicals, and more natural maintenance techniques were begun which, together with reductions in maintenance budgets, led to an intensification of this natural approach. Today, more and more attention is being given to developing ecologically based management. The Parks Department estimates that approximately 30 per cent of its total park area of 2,000 hectares is being managed in this way.[54]

Table 3.1 ***Plants: comparative costs of maintenance***

Type of open space	*No. of hrs 1980*	*No. of hrs/100 m^2 2002*
Woodland	1.7	1.6
Shrubs	1.8	3.5
Groundcovers	8.0	8.5
Roses	15.2	20.8
Annuals	69.1	150.0
Perennials	29.9	39.0
Hedges	22.0	18.6
Lawns	3.3	2.0
Naturalization		0.5

Source: Aanemerij Plantsoerien vande Gemeete, Rotterdam. Public Open Space Management Department, Rotterdam, 1980 and 2002.

Some design implications for city landscapes

Changing roles of city spaces

The discussion so far has dealt with alternative strategies that have included an ecological classification for city spaces, issues of naturalization and an ongoing management process that may be described as 'design over time'. We need now to examine city spaces as a whole, not only from an ecological and habitat perspective, but also from a social and behavioural point of view, since natural and human processes in the city are inseparably linked. The ability to choose between one place and another, each satisfying the immense social diversity of the city, has to do with quality of life. A recognition of these essential differences is necessary if the inherent cultural vitality and diversity of contemporary urbanism is to be embraced in the twenty-first century. The basis for design therefore lies in the recognition that the city's landscapes have different potentials depending on such factors as who uses them, accessibility, biological and physical character, ownership, zoning, legal constraints, costs and practical applicability. Thus, as in the intelligent application of any integrated management plan, not all uses apply everywhere, or necessarily all at one time. Such an integrated view of urban spaces is, however, emerging, as much the result of changing times and social forces as it is a function of purposeful planning. Parks, like city life, are under a continual process of change.

Social activity has become part of every kind of city space. People jog, cycle, rollerblade and walk the dog on public and private open space, river valleys and abandoned rights of way. They play chess on busy street corners. Residential streets become places for conversation, street parties, hockey, and kids earning pocket money feeding the absent neighbour's cat. The public demonstrate, rally and celebrate public holidays and football games along the city's main thoroughfares, halting traffic. They practise Tai Chi and hold concerts and craft markets in the public squares. Birding and botany take place in abandoned and once ignored naturally regenerating industrial areas. The elderly look for quiet places to rest, socialize and admire the spring flower displays. Different groups seek different places for different kinds of recreation and the peculiar requirements of urban life such as growing food in front yards, on rooftops and in community gardens.

Public parks today are assuming environmental functions; establishing wild places in abandoned waterfronts that recognize the importance of ecological diversity to urban places – the value of stormwater detention, naturalization programmes, and environmental education for local schools. The Asian, Caribbean and South American communities, the elderly, disabled, those with special needs and interests, all have very different ways of perceiving and using city places. Social forces, in effect, have begun to shape multi-functional parks and park connections throughout the city. The city itself has become the park, an idea that has become infinitely more complex and diverse. And while the traditional park has long been planned *for* people, it is people who ultimately both shape and create them according to their needs, in ways that reflect contemporary needs. What we see today in many cities is a radical shift in how they are perceived and used. Beginning as islands of green within the city where recreation has traditionally taken place, social forces have reshaped their original purpose and established new ways of thinking about parks as networks. This has come about in response to a diversity of lifestyles, necessity, societal needs, health and environmental imperatives in city life.

Today the notion of the park as an isolated place surrounded by buildings has given way to an entire city landscape that now performs countless functions in countless ways and forms. The concept of networks, which is inherently part of most cities, will be explored further on a regional scale in Chapter 7, but first we should examine the spatial resources that, to one extent or another, are present in most cities.

Residential parks and public spaces

Streets

Possibly the most important function of the residential street is its role as community space. Yet an examination of space allocations in a typical residential area in Ottawa, Ontario, reveals the contradictions between planning regulations, jurisdictional liability and by-law, and the realities of how the streets are used. With a density of about seventy-one houses per hectare, the amount of space between house frontages taken up by streets is approximately 42 per cent of the total cross-sectional area. Not surprisingly, the streets are a focus of intensive social interaction and play by the adults and children of this French/Italian neighbourhood. The greatest social activity is occurring on space officially designated in the by-laws for vehicles alone. People do not recognize these demarcations of space and ignore them where they interfere with their normal patterns of living.[55] Planning by-laws give way to common usage because the streets are the best places to do many things for which other public spaces are less inappropriate.

The tendency to use streets as social space is spontaneous and natural, as Jane Jacobs showed in the 1960s.[56] Since time immemorial the natural function of the street has been as a focus rather than a separator of social activity. A study of a residential area occupied by lower-income families in the inner city of Baltimore during a four-month summer period showed that half of the people counted, who could be determined as residents, were pursuing recreational activities. Only 3 per cent were using the parks; the remaining 54 per cent were in the streets, alleys, sidewalks and porches.[57] Mark Francis has suggested a number of ingredients necessary for the success of existing streets and for designing new ones.[58] Among these are:

- *Use and user diversity*. Healthy streets are used by different people for a variety of activities. Yet they are designed primarily for one group or a particular function, such as walking or driving. Lively and successful streets are those that recognize this need for variety. The social success of the 'Woonerf' that are widely used in Dutch cities is due to their multi-functional design, as places that foster a diversity of social activities, and where the car is not excluded, but enters on sufferance.
- *Control*. The sense of control that can be exerted over one's immediate environment is central to the social success of streets. Control is real for residents who maintain the sidewalk or street trees; it is symbolic when they feel that their private space, such as their front yard or entrance, extends into the public environment.[59] Control also has to do with such issues as safety and security, and with the environmental values of culturally different groups. Crime may be more controllable where a well-established community has an investment in its own neighbourhood than one without that investment. Observation tells us that people with a long-term investment in their own places shape them in ways that are appropriate to their needs and tastes. While it has long been an article of faith that designers should be able to predict human behaviour, the opposite is usually true, as we see so often in the gathering spaces that are empty of people, or the playgrounds that children sensibly avoid. How many times has the designer created a playground full of originality and the latest design fashion and found that during construction the place is full of dirt and children, but once completed is devoid of them? For the designer,

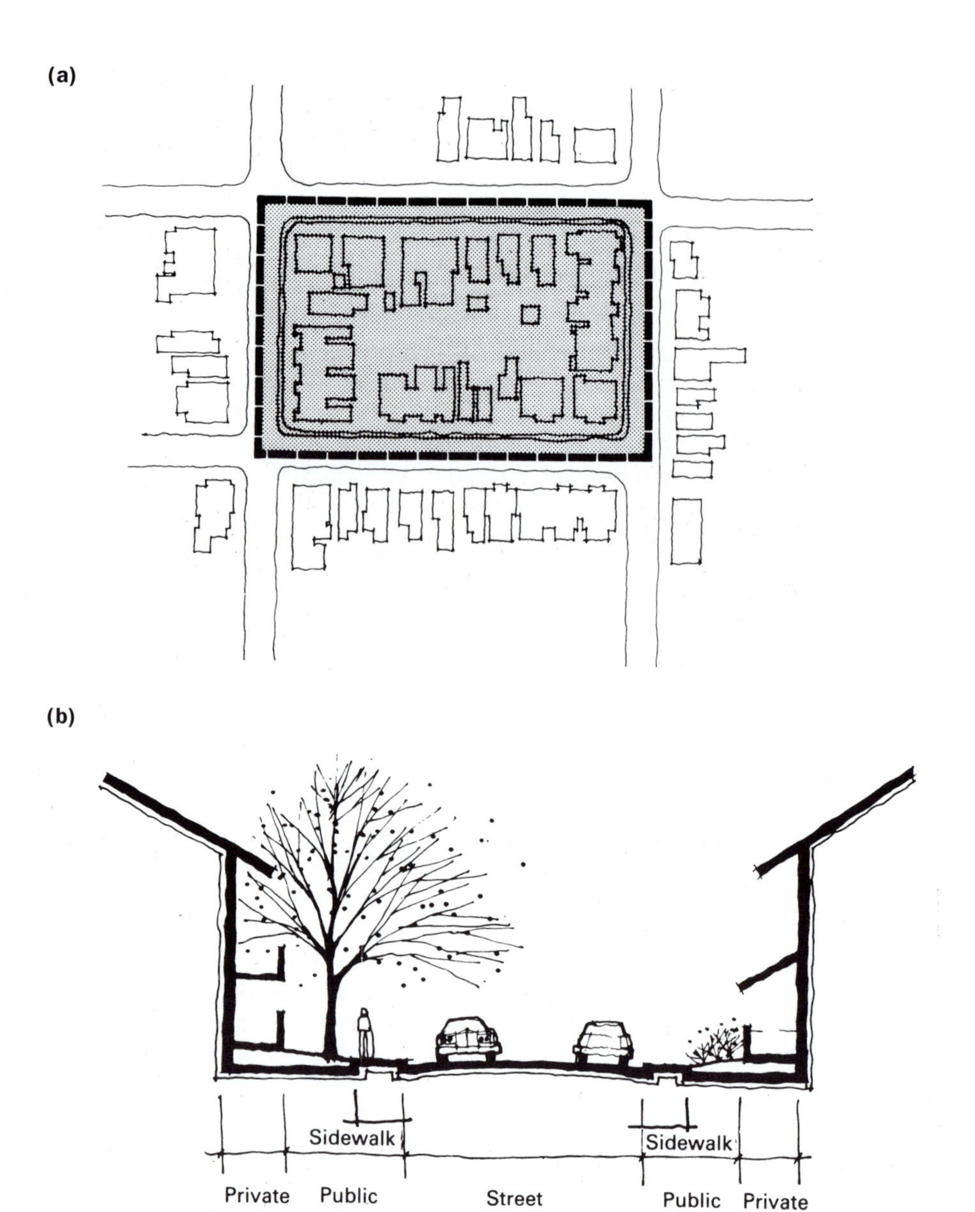

Figure 3.5 Open space of residential areas, LeBreton Flats, Ottawa. The analysis of space in a typical medium-density residential area shows how much space is officially delegated to cars. People, however, do not recognize demarcations of space in such inner-city areas where planning gives way to common usage. Space allocation in residential block (a) (Delhousie neighbourhood, Ottawa): housing 30 per cent and backlots 23 per cent (no community recreational activities); front gardens 14 per cent, pavements 6 per cent and streets 19 per cent (highly active areas, total 39 per cent); driveways 8 per cent. Space allocation in residential street (b) (Delhousie neighbourhood, Ottawa): private space 16 per cent, public space (including pavement) 42 per cent, street 42 per cent.

Source: Hough Stansbury and Associates, *LeBreton Flats Landscape Development*, unpublished report for Central Mortgage and Housing Corporation, Ottawa, January 1979.

Plate 3.11 A residential street in Toronto. Here common usage often supersedes the planning doctrine of separated uses.

> the appropriate response to unpredictable behaviour may involve an understanding that 'play' behaviour is fundamentally about learning, about a structure for a variety of patterns of activity, places for quiet and looking on, for being alone, socializing; avoiding labelling predetermined activities and making flexible places that may be used for whatever activity may occur, or be needed, at any point in time. It is about the need for keen observation of what children do, how and where they do it.

The task of designing public spaces has more to do with creating the physical and institutional frameworks within which people can shape their own environments in accordance with their changing needs over time. It is this combination of community and history that gives a long lived-in neighbourhood its distinctive sense of place.

Vacant lands

Vacant waterfront land, or the economic decline of industrial areas, permeate the urban landscape and contribute to neighbourhood deterioration. Vacant patches of land between buildings where colonizing grasses or trees may be found, corners around parking lots and school yards as well as designated park spaces, provide important opportunities for creative learning. In some cities, residents have begun to assume responsibility for transforming unused land into productive open space for recreation, play and community gardens. The Federation of City Farms in England and Europe, the Trust for Public Land, San Francisco's League of Urban Gardeners, are examples of grassroots organizations that are concerned with these emerging social and environmental objectives and with practical ways of achieving them (see Chapter 5, pp. 184–8).

The Camley Street Natural Park

So-called waste spaces play a key role in environmental education. This is particularly important for city children who are typically deprived of the experience of seeing and understanding nature in wild rural places. In Britain, considerable efforts are being made to provide such opportunities in the city itself.

The Camley Street Natural Park near King's Cross and St Pancras railway stations in London is one such place. It was created from a disused coal yard and purchased by the Greater London Council in 1981. The London Wildlife Trust recognized the remarkable diversity of plants that had colonized the site and suggested that it had great potential as an environmental educational centre. The proposal was approved in 1982 and work began the following year.

The site is adjacent to the Regent's Canal and surrounded by industrial development and rail tracks. Its development included a large wetland fed by the canal, woodland, meadow, a pathway through the property and observation decks associated with the wetland. The nature centre provides environmental education for school children of all ages, who come from local schools. The centre is staffed by a full-time teacher who discusses life processes, urban development issues and field studies. The centre is visited by some 3,500 people a year and is reported to be fully booked by mid-March each year. There are also adult training courses for such subjects as risk assessment and conservation. British Waterways monitors water quality in the canal and wetland that is reported to be of a relatively high standard.

One reaches this little park, of less than 1 hectare in size, after a long walk from the nearest public transit station through an uncompromising, pedestrian-hostile, industrial environment. However, once through the Camley Street Natural Park gates, a completely different place opens up, a sensory world of enclosing woods and meadows, water and aquatic plants, a low, modest, pitched-roof shelter that fits here, but is clearly not an

Plate 3.12 As an urban nature centre servicing London schools, the focus on children has broadened to include adult education in such topics as risk assessment. For children its fundamental value is about *experiencing* in a completely different environment as an essential basis for learning. The Camley Street Natural Park in London has marvellously varied places for so small a learning centre: the wetland and gathering area for aquatic study.

urban building. It is this feeling of contrast, the experience of isolation from the city yet clearly within it, that gives this sensory-rich place a significance as an outdoor natural sciences centre and as a place to visit. It is where children can come to *experience* first and *learn* after. It is this combination of experiencing and learning within the city that forms the basis for formal knowledge of natural systems. There is a delight in the wonder of children as a dragonfly alights on the leaf of a water plant, or a tiny tadpole swims beneath the surface. The cry of a young Londoner as he peered into the water that 'blimy, there's something livin' down there' will always ring in my ears.

Communications links

Transmission lines, railway rights of way, canals and highways consume enormous quantities of land and provide considerable potential for urban space. These links, in particular abandoned railway lines, major utility rights of way and canals, have several characteristics that will determine the opportunities for alternative functions:

- they provide physical and potential biological links through the city to the surrounding countryside;
- for the most part they have little active use, being regarded in many cities as 'waste-lands';
- in many cases public access is restricted for reasons of security or ownership.

Right-of-way lands are often colonized by naturalized plant associations that have succeeded on their own where there has been little or no disturbance for some time. Many harbour plant species that are not found elsewhere on lands subjected to horticultural management. These characteristics indicate that communications links have environmental and social value as corridors for the migration of plants, wildlife and people. Their planning is therefore associated with recreational access, education and as reserves, and management should reflect these values. Much highway land is taken up with interchanges that are often maintained as mown turf. Reforestation and naturalization of these areas, within the limits of traffic safety, reduces maintenance and provides improved wildlife corridors, as practice in North America, Britain and Europe clearly shows. The potential contribution to recreation uses and pedestrian linkages of urban highways is often overlooked. Many could make use of underused rights of way to provide pedestrian and cycle connections through the city, a common practice in some European cities that integrates pedestrians and vehicles as basic components of transportation planning. Leftover spaces that result from expressway interchanges could also become useful parts of the park system if connections were made to them. Links by means of underpasses or overhead connections could give such areas new recreational dimensions, as occurs in the urban expressways of Stockholm.

Other linkages include abandoned railway lines (discussed in Chapter 7), the canal systems of nineteenth-century industrialization and electrical transmission rights of way that are increasingly recognized for their potential as connectors, both through and between cities. Electricity rights of way may pose height limitations for reforestation and are frequently interrupted by roads and other obstructions that limit their value as wildlife corridors. At the same time they are potentially a major provider of education and recreation space, often supporting a diverse group of pioneer plants and valuable as wildlife habitats. In 1967 the Central Electricity Generating Board in the Midlands region of England opened its first nature trail, field study centre and nature reserve on a substation site, and since then others have followed, developed in association with county councils, education authorities and naturalist societies.[60] These corridors also provide much-needed park space, market and allotment gardens, particularly where they are associated with residential areas. At the same time there is considerable debate and controversy over the effects of magnetic fields on human health. Some research has suggested that power lines pose risks of cancer, while others have found no such evidence. While there may be no clear-cut conclusions on this issue, communications links of all kinds have wide value for forestry, wildlife habitat and recreation when they can be integrated into the corridor planning network of the city.

Rooftops

The flat rooftops of many commercial and industrial buildings are an example of often highly visible, but publicly inaccessible, places. They could, with some adjustment to roof design, provide the ideal sites for upper-level urban wetlands and other habitat types. A few centimetres of water that can pond in some areas and allow marsh vegetation to establish itself could, when designed with the same care as decorative rooftops, provide stop-over places for migratory birds and nesting sites for resident species. The rooftop wetland could, to some extent, replace some of the natural habitat lost to urbanization (see Chapter 4, pp. 153–7, and Chapter 6, pp. 201, 215–17).

Industrial lands

Industrial lands account for a major proportion of metropolitan areas, with large parts serving little productive use. A number of characteristics that are of importance to alternative uses are pertinent here.

- Many industrial complexes are high-security operations, and are fenced and inaccessible to the general public. Industrial buildings are often surrounded by large areas of closely manicured turf (particularly those that represent 'flagship' monuments to corporate aesthetic taste). Their potential for ecologically appropriate alternatives reflecting another set of environmentally based values needs to become common practice. In some situations this could be associated with various types of livestock such as sheep or geese, providing returns in maintenance, visual amenity and public awareness of rural occupations. Oil storage installations and municipal sewage and water filtration plants that have security fences are examples of these high-security operations.
- The heat generated from many industrial operations has potential for greenhouse market gardening, linked directly with industrial plants, or indirectly on adjacent land. In some places the integration of industry with agriculture has potential economies as an alternative approach to maintaining agricultural soils within the expanding industrial edge of the city.
- Much land is unused or abandoned, often near waterfronts or in inner-city areas. Fortuitous colonization often combined with poor drainage has, in many cases, created areas of special botanical, wildlife and heritage interest. These are also often naturally protected from intrusion by security fences. Redevelopment usually ignores the rich natural heritage that it replaces, and reclamation often replaces natural diversity with 'green desert' recreational developments and aesthetic amenities. Many of these areas have the greatest value left as they are. Alternatively, some of their natural assets could be incorporated into new development. Design policies should recognize the inherent opportunities they represent to enrich the city's landscape and provide alternative places for the study of natural processes and history.

Industrial heritage and contaminated land

There are more than 400,000 brownfield sites in North America, and about 750,000 sites across Europe suspected of being heavily contaminated from former industrial and military activities.[61] The legacy of the industrial era, therefore, is a pressing issue faced by all Western cities. The great industrial environments of work – the ships, railways, grain silos and industrial steel mills – represent both an opportunity to celebrate past heritage, and a liability in terms of the contamination of soil and water they have left behind. More often than not, efforts to remediate polluted soils leads to the destruction of industrial heritage.

The complex problems of soil and groundwater contamination have to be addressed at many levels – biological and technical, jurisdictional, financial and legal. The legacy of hazardous chemicals, heavy metals and floating hydrocarbons at groundwater level involves the resolution of many issues. Among them are appropriate clean-up technologies, acceptable levels of site mediation, liability, appropriate policies and regulations, and risk assessment that weighs such issues as environmental conditions, new land uses, regulatory requirements and financing.[62] There are also issues of human health and the impacts of contamination on natural systems. For instance, the transfer or pathways of contamination into the food chain through plant uptake which is then passed on to wildlife is not well understood. Approaches to remediation vary in response to site, types of industrial pollution and environmental, legal, jurisdictional and political agendas; the need to develop an integrated ecosystem-based, rather than a piecemeal, approach; the need for public involvement. Methods for dealing with these problems include leaving the material where it is (such as contaminated river sediments), excavation and disposal (shifting the problem from one place to another at considerable expense), treatment on site, where the soil being restored is left in place, and phytoremediation involving the use of plants where contaminants can be broken down organically through micro-organisms. In Tacoma, Washington, strong environmental legislation can impose heavy fines for failing to clean up a site, and casts a wide net for liability that includes 'current owners and tenants without regard to fault, transporters of contaminated wastes, and lenders'.[63]

A major concern is the frequent conflict of objectives that occurs in urban renewal between rehabilitating contaminated soils while protecting industrial heritage and the complex natural succession that one finds in many old industrial areas. But the brownfields occurring in all major cities are central to sustainable approaches to future development, a role that will have a major impact in restricting outward growth. An examination of the regional and local issues being implemented in the Emscher area of Germany's Ruhr Valley is discussed in Chapter 7.

Cemeteries

Cemeteries are often among the most valuable open spaces in cities. For some they may be regarded as a forbidden yet challenging playground for small children in search of horse-chestnuts and adventure among the tombstones. For others they are places of quiet and repose, away from the noise of the city. Old established cemeteries, with their narrow roads, varied topography and vegetation, provide a secluded haven for walking, jogging, nature study and meditation. For many people old tombstones provide the best and most interesting records of local history.

The long-term cost of maintenance also poses problems that could be turned to advantage in these often pastoral urban landscapes, with sheep to keep the grass mown. Situations such as this have occurred in the past, as we shall see in Chapter 5 on urban farming, and they still occur today in church cemeteries and graveyards in some English rural towns. The role of the cemetery in the conservation of wildlife habitats is also significant, since they enjoy seclusion from intense human activity and often provide conducive environments for animals and birds.

Native and fortuitous natural habitats

Wetlands

Natural wetlands and open areas of water contain rich associations of wetland species and are susceptible to damage through groundwater depletion and water pollution. Abandoned industrial sites may also support rich communities that have occurred through a combination of water impoundment and natural succession. Being enclosed or off limits, they survive as precious natural reserves. These sites are often durable features in the city landscape, but in private hands are prone to demolition. As created marshes, sewage lagoons are rich in nutrients and support diverse associations of plants, invertebrates and birds (see Chapter 2).

Woodlands

Remnant natural woodlands often contain locally rare trees, shrubs and ground vegetation, and are important habitats for a variety of animals and birds. Naturally occurring regenerating woodlands in such places as vacant lots, lanes and old residential streets are tough and resilient, and provide climatic and social benefits to less favoured parts of the city.

Figure 3.6 Wetlands and open bodies of water contain rich associations of wetland species and are susceptible to damage through groundwater depletion and water pollution.

(a) *(right)* Ravines and valley lands. Those that contain steep banks and remnant native vegetation that are highly susceptible to erosion and destruction often contain highly diverse groupings of plants. Valley floors in many ravines are often disturbed by the removal of the original vegetation for reasons of access, park development and so on. Many of these are resilient and can withstand considerable use.

(b) *(below)* Remnant mature woodlands often contain locally rare trees, shrubs and ground vegetation as well as birds and other animals.

(c) *(below)* Naturally occurring regenerating woodlands, for instance, on vacant lots, lanes and so on, are tough and resilient, and provide climatic and social benefits to less favoured parts of the city.

Figure 3.7
Some general categories of habitat type in urban open space: woodlands.

Ravines and valley lands

Those which contain steep banks and remnant native vegetation that are highly susceptible to erosion and damage or destruction from urban land uses often contain highly diverse groupings of plants. They are significant places for passive recreation, trails and education. Valley floors are often disturbed by the removal of the original vegetation due to overuse or inappropriate activities.

Regeneration landscapes

Many landscapes have completely changed ecologically, hydrologically and topographically due to mining or similar types of disturbance. These regenerating landscapes can often sustain high pressures of use from children and adults. As changed but none the less vital, naturally regenerating environments, they are of priceless educational and

(a)

(b)

Figure 3.8 Some general categories of habitat type in urban open space: regenerating landscapes. (a) Many landscapes have been completely changed ecologically, hydrologically and topographically due to mining or similar operations. These regenerating landscapes can often sustain high pressures of use from children and adults. As changed but none the less vital natural environments they are of priceless educational and ecological value. (b) Power line rights of way are the connecting links through the city, support a diverse group of pioneer plants and are valuable wildlife habitats.

ecological value. Former wastelands recolonized by early succession plants are ecologically diverse, resilient, and of considerable educational and social benefit in residential areas. Their preservation and inclusion into new developments would do much to enrich these places.

In this chapter we have examined the natural systems of the city from the perspective of its native, fortuitous and cultivated plant associations. It becomes clear that the conventions which have created the cultivated landscape have been based on technological imperatives that are in confrontation with the realities of urban nature. They relentlessly overpower even the simplest observation of natural process at work from which much can be learned. We have the frequent anomaly of a design process dedicated to urban quality, but instead creating impoverished environments. Naturally diverse areas are replaced by a landscape of turf and cultivated plants that minimizes ecological diversity and social options. While cultivated landscapes are clearly appropriate in the city, the necessary alternatives available to us are rooted in an ecological view of plants as communities; a holistic management framework drawn from environmentally sustainable forestry and agricultural traditions; and an increasingly aware public that is initiating countless naturalization projects in cities worldwide. The concepts of urban forests, planting design founded on succession, grassland management and the larger structure of city spaces that bring together natural process and human behaviour, provide benefits in a more diverse environment, greater economic and environmental productivity and greater social and educational values. These are highly relevant in various ways, not only to industrialized countries but also to developing ones. In addition, the economies over current landscape practice are undeniable, and are increasingly becoming part of an emerging set of environmental values.

It is also apparent from our discussion up to this point that water and plants are indivisible parts of the natural processes of cities that must be seen as a whole. In this way they provide us with a useful and enriching urban environment, and a means of closing the perceptual gap between the city and the larger non-urban landscape. A framework for an appropriate design aesthetic becomes available; one that is responsive to ecological, economic and social criteria.

It is abundantly clear that plants are a fundamental part of the urban scene. Their significance for wildlife, for people and for ecological health varies in proportion to the complexity and variety of their associations. While much has been said of ecologically viable appropriate alternatives to current practice, the urban forest, in its horticultural context, also performs key environmental roles in mediating air pollution and storm drainage. The economies in dollar terms that may be realized from the role of vegetation and trees in cities can be very considerable and provide persuasive arguments that nature plays, in today's economic value-based society, a key role in a sustainable future. Thus the discussion on plants and water in these previous chapters is ultimately about habitat, and habitat is directly related to the existence and diversity of non-human species and so to the quality of the human environment that is the city. Since its health must now be gauged by the sum of its life processes, we must turn to an examination of wildlife.

4 Wildlife

Introduction

There was a time when museums and zoos provided people with the only opportunity of seeing wild animals they had read about in books. They were collectors' items, curiosities to be marvelled at for their size, ferocity, brightly coloured posteriors or strange shapes. Today, they are depressing reminders of a growing number of species that have vanished from the scene. Zoos have become more sophisticated in the way animals are exhibited, and are more enjoyable as entertainment; and some have breeding programmes that are aimed at preserving endangered species. Yet they represent a view of nature that is remote; external and disconnected from human affairs. On this John Livingston comments:

> the role of Nature in the necessary subsidization of the human interest comes into sharpest focus in the use of animals for entertainment. Animals of all sorts, both wild and domesticated, are pressed into service for this purpose. . . . In my view, zoos convey and reinforce not only the 'us' and (undifferentiated) 'them' bifurcation of the living world, but also contrive to feed and nourish the fundamentalist myth of absolute human power and control.[1]

Television and the internet have raised the educational level of the public with a host of informative programmes and websites on the nature of the living world. But one may question whether the urban experience of nature is still not largely focused on the exotic bird in a space frame cage and the elephant and tiger secure without bars behind the well-designed moat. Most of our knowledge comes to us secondhand through the media. Direct contact with nature and the animal world, apart from resident urban house sparrows, starlings and pigeons, is confined to non-urban, or for that matter, non-local experience. It is to be had from the weekend at the cottage, or on occasional school visits to the rural interpretive centre. Most of us know little about wildlife in the places where we mostly live. Can the city itself provide us with a direct experience of wild nature? To what extent does the city provide habitat for wildlife? How are these questions important to the quality of the urban places in which we live? These are some of the concerns that will be explored in this chapter. To do so, we must review some aspects of natural process with respect to food and habitats, how these are altered by the city, the specific requirements for sustaining wildlife species in various urban environments, and how cities have shaped attitudes and perceptions towards the non-human life that shares habitat with us.

Natural processes

Woodland and forest, grassland and meadow, marshes and water are the habitat for wildlife. The diversity, structure and continuing evolution of plant communities, their interaction with land form, soils and climate, dictate the diversity and stability of wildlife populations.

The layering or structure of forest vegetation provides distinct environments that support different groups of species. Some feed and breed on the forest floor, some inhabit the under-storey and others live in the canopy. Warblers have been found to populate well-defined nesting places according to species, thus enabling a large variety of species to occupy the same forest.[2] Different plant associations provide places for different groups of species. Plant succession produces, over time, a range of habitats from open field to mature forests. Each successive stage is home for different associations of insects, birds and animals. Studies in Maine showed how different species of birds are attracted to these different habitats.[3] Open land was populated by savannah sparrows, song sparrows and bobolinks. In low brush these were replaced by field sparrows, Nashville and chestnut-sided warblers. Pioneer forest attracted ovenbirds, the succeeding evergreen forest warblers. Woodpeckers and kinglets appeared with the climax forest. Similarly, in aquatic environments, the increasing productivity of a lake as it proceeds from an oligotrophic (nutrient-poor), to a eutrophic (nutrient-rich) condition attracts an increasing number and variety of wildlife species. Thus wetlands, being highly productive ecosystems, provide habitat for vast numbers of birds and other wildlife. Places that have many different plant associations tend to be richer in species than those that have only a few.

The composition and number of wildlife species are also affected by other factors. The edges between one habitat and another are often more diverse than the interior of the habitats themselves. At the same time large-scale interior habitats are essential for some species that would otherwise be vulnerable to edge predators. Continuity of habitat provides essential migratory routes and helps maintain wildlife populations.

The disturbance of the natural environment through human activity sets up imbalances in plant and animal communities. Equilibrium is maintained by an elaborate system of checks and balances. The loss of habitat on which a species depends for food and shelter and to breed may mean that it has to adjust to the new conditions or abandon its chances of survival. The more adaptable species may survive and flourish. Those less adaptable may disappear. Some modified landscapes where woodland, field, scrub, wetland and open water are associated may enhance species diversity. The old complex agricultural patterns of woodland, hedgerows and fields of Britain and Europe have traditionally been rich in wildlife. The value of hedgerows as a wildlife resource (and for their historical associations with ancient parish boundaries) has become increasingly recognized in Britain, and is reflected in grant-aid schemes for hedgerow planting and management begun in 1992.[4] Yet overall the rural landscapes of Britain have been severely impoverished by industrial farming. As Chris Baines has observed, the disappearance of ponds, meadows and broad-leaved woodland from industrial agricultural practice has meant the loss of butterflies, wildflowers, lapwings and skylarks.[5]

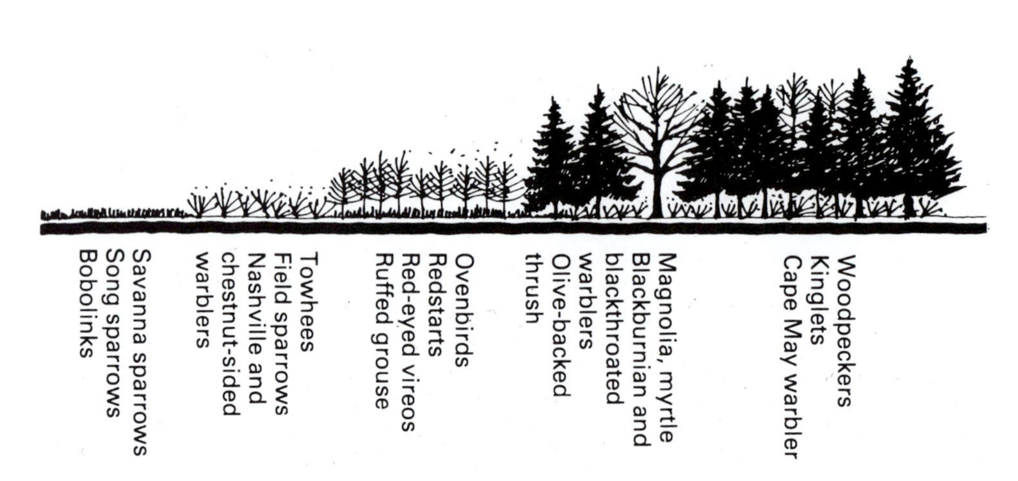

Figure 4.1 Habitat: the layering or structure of forest vegetation provides distinct environments for different groups of species. (a) Different birds live at different levels of the forest depending on their nesting and feeding habitats. Generally the greater the layering of foliage on a vertical profile the greater the diversity of species. (b) Rojer Tory Peterson's study of bird fauna in Maine showed how different species inhabit different successional stages of forest from open land, low brush, pioneer forest, mature spruce forest.

Source: Peter Farb, *The Forest*, New York, NY: Time-Life Books, 1963.

Urban processes

Urbanization and wildlife

Urbanization has radically altered both natural habitat and wildlife communities. Studies in the USA of the effects of the urbanization of agricultural land on wildlife have documented the changes that take place.[6] Farmland, field and woodland species declined drastically as urbanization advanced and as suitable habitat was reduced. A few species, however, increased dramatically. House sparrows and starlings, virtually absent prior to urbanization, became the most abundant species. The primary effect of land-use changes is to fragment forested and other types of habitat, and to convert extensive areas into isolated islands within predominant urban environments. This disrupts the functions of natural areas and inhibits wildlife interaction and gene flow among habitats. The concept of island bio-geography has amply demonstrated that large islands support more species of plants and wildlife than do small islands. It is now realized that the same principle can apply to habitat patches in terrestrial environments. Forest fragmentation is now considered by many researchers to be one of the most important environmental issues (see pp. 228–32).[7] Taken as a whole, the urban environment is a patchwork of many habitats. The relative numbers, distribution and diversity of animals and birds in various parts of the city are directly related to the diversity, area and structure of vegetation, which determine habitat quality. The plant groups discussed in the previous chapter therefore provide a useful guide for an understanding of this fact. At the same time, nature in the city is thriving, particularly in the old industrial landscapes of Britain and Europe.

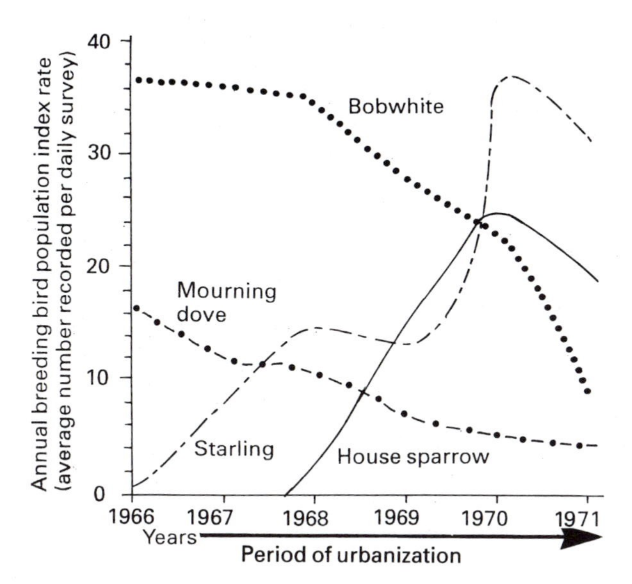

Figure 4.2 The impact of urbanization (new city of Columbia, MA) on bird populations. The typical farmland species such as bobwhite and mourning dove showed rapid decline. Starlings and house sparrows, virtually absent from the area before development, showed the most striking population increases.

Source: Information for the graph from Aelred K. Geis, 'Effects of Urbanization and Types of Urban Development on Bird Populations'. In *Wildlife in an Urbanizing Environment Symposium*, Co-operative Extension Service, Amherst, MA: University of Massachusetts, June 1974.

The cultivated landscape

The habitat of heavily built-up areas is an environment of buildings, paved surfaces and cultivated spaces. The ornamental and biologically sterile tree and lawn landscapes of downtown parks offer little in the way of food, shelter or breeding places for wildlife. The herb and shrub layer of natural woodland becomes the paved or grassed floor of the city, trampled by people and continually hunted by cats and dogs. The intermediate layers have been eliminated. The canopy also disappears where trees have been planted as separated individuals. Ornamental plants often lack insect life, fruit or seeds on which birds can feed. The disappearance of native animals and birds favours a few aggressive species that are readily adaptable to this habitat. Thus pigeons, sparrows and starlings all thrive in large numbers, feeding off the city's wastes and living and breeding in buildings. In North America, the house sparrow was introduced from Europe in the mid-nineteenth century. The starling was imported and liberated into New York's Central Park in the 1890s. Old world rats and mice arrived in ships accidentally. It is estimated that there may be as many rats in a large city as there are people.[8] The ancestral form of the domestic pigeon is the rock dove. Urban structures, being a human-made substitute for its original windswept Scottish cliffs, provide it with ideal places to roost and nest.[9]

As the density of building reduces beyond the city core, however, more vegetation survives, or is planted along the streets and gardens of residential areas. Robin, redwing, blackbird and grackle are common birds in gardens and local parks in north-eastern American cities. Even though these places are heavily populated by people and domestic pets, the more adaptable native species have gained a foothold. The North American black squirrel, raccoon and skunk have flourished. Raccoons dine on the contents of rubbish bins and take up residence in attic and chimney. Ducks and geese in many waterfront parks have become resident and proliferated, feeding in summer on grass and handouts from park visitors, and wintering on patches of open water created by an industrial plant or generating station. Other species, not normally tolerant of urban conditions, have taken to the city, where the warmer climate and altered habitat provide the right environment for survival. Rooks have formed colonies and bred in urban areas in England; the main shopping area in downtown Cheltenham is an example, with people inhabiting the ground plane, the birds the tree canopy above. Hawks prey in most cities on the small animals that inhabit waste places. The English black redstart, one of Britain's rarest birds, has adapted to urban industrial areas where it now chiefly lives.[10]

Native habitat areas

As we saw in the last chapter, many cities still retain elements of native vegetation diversity which may be found in ravines, cemeteries, railway embankments, campuses and waste places. Woodlands of mature and dead trees, open grassland, scrub and young growth, water and wetland are often present, supporting a variety of native birds and animals. Whereas North American prairies are dominated by native plant species, old fields contain a large proportion of non-native plants, with introduced species such as clovers, ox-eye daisy, common St John's wort and cinquefoils. In city centres about half the vascular plants are aliens, arriving with either human or non-human assistance.[11] Many of these places, however, are remnant environments, disconnected from the rural landscape and from each other by the city. And so, where they remain isolated and small, the chances of maintaining high wildlife diversity, particularly animals, are relatively small.

Plate 4.1 Natural habitat structure: complex associations of forest, meadow, wetland and open water.
Source: Steve Frost.

Plate 4.2 Changes of habitat structure. A hostile environment for native wildlife is created when turf is substituted for natural ground layers, intermediate layers are eliminated, when the canopy disappears and when native plant species are replaced by non-food-producing ones. Diverse native populations are replaced by a few exotic species.

Problems and conflicts

The imbalances and stresses created by urban activities have many repercussions. Diseases transmitted from wildlife to humans and from humans to wildlife occur more easily when people and animals share limited space in the absence of natural controls. Some diseases are carried by wildlife and transmitted to people. For instance, rabies is widely distributed in rural areas in North America, skunks and foxes being the principal maintenance hosts. The free movement of animals into cities makes the disease a continuing threat to urban and rural people and is expensive to control. Salmonella occurs in English sparrows during cold weather and can be transmitted to pets and people.[12] Some diseases may be carried by pets as well. Toxoplasmosis is particularly infectious to children and is picked up from the faeces of domestic cats in sand-boxes and play areas.[13] Urban pets create problems of health and security in crowded city conditions. Feral or stray dogs have been known to attack people; defecation is injurious to plants and is a major source of stormwater pollution. Most major cities regard this as a health hazard, since many older drainage systems discharge stormwater untreated into receiving rivers and lakes in heavy rainstorms. Consequently, overpopulation and uncontrolled breeding, particularly among strays, have become major management problems. The conflicts between people and wildlife range from questions of safety to aesthetics. Birds are hazardous around airports, squirrels damage telephone lines, pigeons make a mess of building cornices and statues in squares, and in Venice they are known to contribute to severe deterioration of stone ornamentation and sculpture on historic buildings. Large flocks of urban gulls can temporarily close down lake beaches because of the pollution they create by defecating. This raises questions about what is valuable or a nuisance, and thus of our perceptions of wildlife in cities.

Perceptions and values

Our discussion in the last chapter focused on the inherent conflicts of values and priorities that are reflected in the design and management of the urban landscape. These concerns are equally relevant to our discussion of wildlife since we are dealing with inseparable issues. What makes a pet poodle loved and a house mouse persecuted has the same relationship as the garden rose to the dandelion in the lawn. While in the new century the environmental movement continues to change perceptions, the acceptance of nature is still a function of how it conforms to a predetermined set of values and to what extent it is under control. It is tolerated on our own terms, within the limits of convenience, deep-seated acculturation and aesthetic conventions. The conflicts are heightened, however, with wildlife. Animals, birds and insects are more difficult to control than plants. Their presence is more obvious. They are potentially more damaging to human health and welfare. Getting rid of aggressive weeds is less frustrating or demanding than trying to stop raccoons spreading the contents of a rubbish bin over the street on pick-up day. Those same raccoons make charming pets as babies, but fully grown they are a menace in the attic or with the pet goldfish. Pigeons being fed in the square have become an inevitable part of the urban scene and a favourite subject for photographers, but in and around buildings they are messy and a nuisance. Life in the city has not been conducive to promoting an understanding of natural cycles. Indeed, it is ironic that wild animals, like most species in nature, have value when they are rare or endangered, but not when they are abundant or successful, a fact that is reflected in government wildlife and habitat protection policies.

What gives rise to these conflicts and human attitudes towards nature is the subject of much philosophical debate in the literature. Thomas More suggests that the literary encounters with wildlife by urban children greatly influence anthropomorphic values and popular misconceptions about non-human beings.[14] Donald Worster has shown that responses to nature over the past two centuries have shifted from the Arcadian view of peaceful coexistence to one that expresses the utilitarian view of nature as a resource.[15] John Livingston has taken this further, suggesting that our understanding of nature is informed by an ideological insistence that domination is somehow 'natural'. The social theory of the Industrial Revolution, for example, lay behind Charles Darwin's explanation of evolution as a competitive 'struggle for existence', a still prevailing view among biologists that non-human animals compete among themselves for territory and social rank.[16] An informal enquiry I once conducted on what wilderness meant to different people showed how differently this can be interpreted. Many people said that it's 'a state of mind rather than a reality'. Evan Eisenberg has compared the city to a jungle – of 'being lost in something huge'.[17] On being asked the question, an Australian naturalist suggested, after some thought, that 'wilderness was the places where you could still be eaten', and Oscar Wilde once defined nature as 'a place where birds fly around uncooked'.[18]

Of immediate importance to the thesis of this book is the search for ways in which the city and nature may be brought more closely together to promote alternative values. First, we can act on the notion that environmental literacy in cities involves an understanding of wildlife as an integral part of natural processes and the relationship of life systems to people, and what it can teach us about coexistence. Second, there is the question of diversity and choice. If choice is one of the key factors necessary to a healthy social environment, then a rich and varied wildlife environment is one way of achieving it, since the diversity of human cultures and interests that make up cities should be reflected in a wide variety of urban places. Third, there is the more utilitarian view of wildlife and plants as indicators of urban health. Monitoring and reproduction of shore birds may give us a clue to the condition of the city's aquatic environment. The presence or absence of lichens is an indicator of habitat conditions and may reflect its suitability for people. How these concerns become part of urban design must be examined in the light of the strategies outlined for plants and the inherent opportunities the city provides.

Alternative values: some opportunities

The last chapter suggested that the integration of objectives for land-based rural occupations, when conceived as environmentally sustainable practices and adapted to the city environment, provides an alternative framework for urban design. The same principles are also applicable to wildlife and to plants. While the preservation of rare or endangered species has become a major preoccupation of government policy, these policies tend to ignore the fact that the maintenance and enhancement of representative associations, those that are common or ordinary, are also vital to the biological integrity of our cities. They may also be achieved more realistically in culturally shaped landscapes that have become a patchwork of remnant habitats. This view is expressed in the science of landscape ecology that explores the integration of ecology and human activity.[19] It describes how a heterogeneous combination of ecosystems, such as woods, meadows, marshes, corridors and human settlement, is structured, functions and changes. It encompasses all environments, from wilderness to urban landscapes, and includes the distribution patterns of landscape elements, i.e. energy flows in soils, the movement patterns of plants, animals and nutrients, and ecological changes over time. Landscape structure includes three major spatial and visible elements of the landscape: patches (woodlots surrounded by agriculture, or open space surrounded by urban development), matrix (homogeneous areas containing distinct patches within it), and corridors (wooded streams, power line rights of way or transportation routes). The interaction of living organisms and habitat within the city environment provides us with the basis for understanding the significance of nature as a whole. And, while the protection and preservation of important natural places is where we begin the process of reshaping the city environment, we continue with opportunities for their restoration and re-creation. These considerations form the basis for the following sections of this chapter.

Natural areas and fortuitous habitat

The habitats with the most obvious wildlife potential are the rivers, streams, woodlands, canals and pre-urban landscapes that still exist within the city's boundaries (Chapter 3). Those that retain sufficient species and structural diversity of vegetation and are linked to larger habitats are likely to have the greatest faunal diversity. They are therefore among the city's richest and most precious places. But the very process of urban growth involves interference with the natural processes of succession and the creation of a complex patchwork of wildlife habitats. It is these that provide the unexploited opportunities for urban design. They are to be found in the less obvious places that lie behind the city's façades and public travel routes, and that often go unnoticed.

Some are around us in the patches of regenerating land and vacant lots that have been colonized by the plants of the city. Associations of plants and animals are everywhere in evidence. Weedy places abound with voles and mice together with the hawks that feed on them. Dock leaves are food for the painted lady butterfly; stinging nettles are food for the larvae of the red admiral butterfly;[20] monarch butterflies associate with milkweed. Old rooftops may combine standing water and soil and patches of vegetation that become fortuitous wetland habitats for visiting and resident birds. The window ledges of high-rise towers may become a nesting place for peregrine falcons. The natural or contrived impoundments that result from stormwater run-off in many urban areas create new wetlands and wet meadows. Explore the old industrial sites in many a downtown area and you will find unexpected and surprising landscapes behind the chain-link fences, where a combination of a poorly drained site and the natural colonization of aquatic plants has created ideal marshy breeding places for musk rats, ducks and geese.

We must look to those places where energy is concentrated to find other rich wildlife habitats; to environments that have been created by the processes and functions of the city. The waste heat that is dumped into lakes and rivers from storm drains, sewage treatment plants and electricity generating stations, while usually polluted, maintains open water in cold climates and an aquatic habitat for shore birds. Rubbish disposal sites attract rodents and other small mammals and gulls. The availability of food, shelter or heat may provide a breeding place for a species of bird, or benefit migratory flocks and winter visitors. Lagoon systems for sewage treatment adopted by smaller municipalities concentrate nutrient energy and attract large numbers of wading and shore birds, creating habitats that are often richer in species and bird diversity than undisturbed shore lines.[21] As urban habitats, lagoons provide a place for native species of wildlife that naturally inhabit marshes and ponds, and may be seen as one of the city's greatest interpretive and educational resources for naturalists, school groups and the community at large.

Wildlife and regenerating waterfronts

The Outer Harbour Headland on the Toronto waterfront is a spit of land nearly 5 kilometres long enclosing the city's outer harbour on Lake Ontario. Begun in 1959, it was intended to accommodate the increased shipping expected from the St Lawrence Seaway. The shipping boom, however, failed to materialize, and for many years the alternative future of the Headland was a source of continuing debate. There were plans to turn it into an aquatic park, complete with hotels, a marineland, sailing clubs and parking; visions of constructing a maritime pioneer village, with wooden quays, cobblestone streets and shops.

From the beginning, however, the Headland started to evolve into a complex natural environment and is a fascinating example of the regenerative processes of nature. Its soils consist of an unconsolidated mixture of fill and rubble brought from city construction works and from sand dredged from the lake bottom. This sand was dumped along the sheltered side of the lake, forming four new low-lying embayments and protected lagoons. Over the years on the unprotected lakeside, water, wind and wave action have been busy grinding concrete, bricks and stone along the shore into rounded pebbles. Up to 1977, when I recorded its evolution over eighteen years in *City Form and Natural Process*, a total of 152 vascular plants had naturally colonized the barren soils, brought by wind, birds and people. Of these, eighty-eight represented introductions into southern Ontario with the remainder being native.[22] Some were rare within the metropolitan Toronto area, and two species were identified and recorded by botanists for the first time in the region – golden dock (*Rumex maritumus*) and sticky groundsel (*Senecio viscosus*).[23] By 1992, some 400 species had been identified in seven major plant

Plate 4.3 The Outer Harbour Headland from the air, City of Toronto. An extraordinary emerging productive habitat for fish, animals and birds, protected as an urban wilderness.

Source: Toronto and Region Conservation Authority.

Plate 4.4 A place to contemplate and experience natural forces in an urban setting.

communities[24] including two nationally rare, one provincially rare and four regionally rare plants. A forest of cottonwood appeared early in the successional process, and by 1992 had reached a height of 10 to 12 metres.

This evolving landscape soon began to attract migrating, nesting and wintering birds, and became a significant staging area and migration corridor. By 1976 some 185 species of birds had been sighted at the Headland. Of the mammals, raccoon, skunk, muskrat, rabbit, Norway rat and groundhog had been recorded.[25] With spectacular rapidity the Outer Harbour Headland became host to great numbers of nesting gulls and terns. In 1973, ten pairs of ring-billed gulls attempted to breed. By 1977 some 20,000 pairs were nesting on the peninsula, growing to 80,000 pairs by 1982, along with other breeding birds that included the herring gull, common tern and the rare Caspian tern. By 1992 the total number of bird species had grown to 290, of which forty were breeding at the site. Five species of water birds nest in significant numbers: the herring gull, common tern, black crowned night heron, double-crested cormorant and ring-billed gull – one of the largest colonies in the world. The Caspian tern, for reasons unknown, has disappeared, but in April 1993 coyotes were sighted on the Headland, travelling down the Don River valley corridor to take up residence. Conservation Authority staff found evidence of den sites, which suggested that a pack was established and breeding.[26]

This remarkable place has become one of the most significant wildlife habitats in the Great Lakes region in an environment where industrial growth has destroyed many of the habitats that birds require, and has rendered others toxic. Before large-scale harbour developments began in 1912, Toronto's waterfront was one of the richest and largest marshes on Lake Ontario. These were filled in to create the harbour, industrial lands, railways and expressways that are the legacy of the city today. But the creation of the Headland has, by accident, re-created some of the very habitats that were destroyed earlier in the century. Its existence has permitted much wildlife to recolonize. Without the construction of this peninsula biologists believe that many species such as the common tern would no longer be nesting in the Toronto area.[27] And it is almost certain that the coyote, a prime carnivore, would not be residing in downtown Toronto.

The opportunities that occur fortuitously as a result of urban processes have great relevance in our search for an alternative urban design philosophy. I have already suggested that the principle of diversity offers variety of place and social opportunity. It is apparent, if one begins to observe the city with an ecological perspective, that this is occurring unwittingly in many places and in many ways. It needs only to be recognized and used to advantage. I also suggested in Chapter 1 that the principle of least effort is a relevant and important objective. It is perhaps best summed up in an early leaflet distributed by the Friends of the Spit, the group of citizens who, since 1976, have led the campaign to have the Headland protected as a nature reserve:[28]

- What do we want . . . and how much will it cost?
- Keep the place as it is without any major development.
- Let it develop naturally into an increasingly secure wilderness.
- Jettison the (Metropolitan Toronto and Region Conservation Authority) master plan approach.
- The nice thing is that to do these things will cost very little. And millions of taxpayer dollars can be spent instead on more pressing social and economic needs

Further developments since the 1980s

In 1992, in response to a decade of public pressure to protect this urban wilderness, the Metropolitan Toronto and Region Conservation Authority (the agency responsible for managing the Headland at the time) approved a master plan for the area. In addition to protecting its key habitats and wildlife, it also proposed the establishment of a sailing club and a major interpretive centre together with trails and parking. Pressure to revise this plan was exerted by the Friends of the Spit, and two years later a revised plan was brought forward that largely met public insistence on a car-free, no-development, wilderness environment that should be left to evolve on its own.

Plate 4.5 An aquatic habitat creation project by the Toronto and Region Conservation Authority. Crucial habitats are being established for fish and wildlife species, and include shoals, brush bundles, log cribs and a diverse shore line in this embayment.

Source: Toronto and Region Conservation Authority.

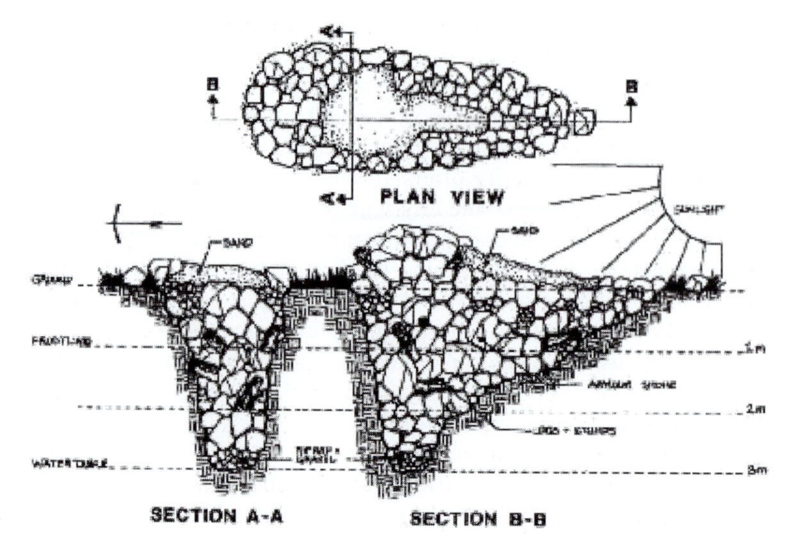

Figure 4.3 Trial rock and rubble habitat structure for snakes.
Source: Toronto Region Conservation Authority.

At the same time, planning for the future redevelopment of the Toronto waterfront in the late 1990s raised concerns that a large influx of new residents could have a major impact on the Spit's emerging vegetation and wildlife. It became clear that some recreational facilities and proactive management would be needed to ensure its continued ecological integrity. In collaboration with the Friends of the Spit and interested government agencies, the master plan was modified to include trails to control visitor access to sensitive areas while encouraging walking, cycling and ornithology. An interpretive centre was proposed to inform people about the Spit's natural processes and issues of stewardship. A major wetland restoration initiative was also begun within embayments on the lakeside of the Spit replacing, in part, the lacustrine marshes that had existed before the building of the 1912 Port. An effort to diversify breeding birds has resulted in open barges being floated out into the lake every spring for breeding terns, allowing people to observe from the land, and also preventing the terns from being crowded out by gulls.

Knowledgeable action on the part of citizen groups is becoming increasingly effective in sponsoring and protecting natural areas in cities. As a consequence of these public actions, the Spit has continued to thrive and provides an unparalleled wilderness experience for the increasing number of people who visit the site every year. Its experiential and scientific value to the city lies in the opportunity to see a complex natural community develop from the ground up. Thus, the Spit has evolved, not only ecologically, but also in its planning and management, the role of the community that had the foresight to campaign for its preservation, and the government agencies who have an interest in its future. The concept of process therefore concerns nature and cities in its broadest sense, and will continue to do so.

Amphibians and reptile habitats

Another aspect of the wildlife scene in urban areas, and one that is often ignored, is the habitat that the city does and could provide for frogs, snakes, lizards, turtles and other amphibian and reptile species. These animals, like the more obvious ones that we have been examining, not only rely on remnant areas of undisturbed habitat, but also on altered sites. Riverine marshes, for instance, may support many species of turtle; the stone-filled wire baskets forming urban stream banks may be habitat for garter snakes; old gravel pits and fortuitous water impoundments are often places where frogs, toads, turtles and snakes may be found. While amphibians and reptiles have been among the least recognized of the urban fauna, it will be self-evident that their continued existence in viable populations is crucial to the maintenance of species diversity and stability. In a study of chorus frogs in the USA it was found that their life expectancy was about four months, with 95 per cent of the population failing to survive the winter.[29] Bob Johnson observes that 'this relatively high rate of mortality may be typical of most of our amphibians which are nurtured in the very productive waters of marshes and swamps. As the frogs are eaten, the energy they have carried from the swamps is passed on to other organisms that otherwise could not exploit the richness of these areas.'[30] Thus to save wetlands means more than saving the amphibians and reptiles that one finds there. It also means that one can save all the trapped energy for use by a whole chain of plants, invertebrates, birds and mammals elsewhere.

Planning and management issues

A strategy for urban wildlife habitat

The naturally occurring wildlife places that we have been examining provide the inspiration for a purposeful design strategy. This takes three essential forms: identifying and planning for what is there now (existing habitats within built-up urban areas), restoring already degraded habitats, and identifying what could be there in the future (a regional planning approach for future urban growth which identifies important habitats that should be protected and integrated into open space systems).

Identifying and protecting what is there now

The city has innate opportunities for complex wildlife habitats. As agriculture has industrialized, there has been a tendency for rural areas to become ecologically less complex, and the city more so. This provides us with the basis of a design strategy. Its fundamental objective should be the enrichment of the city's existing natural wealth; to capitalize on what is inherently there. This is, in fact, one of the significant attributes of the cities of the western Netherlands. The great diversity and richness of its park system, waterways and streets, the mixture of old and new, the contrast between naturalized areas, productive and manicured ones, is in marked contrast to its agricultural hinterland. This is a modern, food-producing landscape of big machines and unlimited views that have little ecological or visual richness, or recreational interest. The Dutch have embarked on a programme of enrichment of the urban landscape, which counterbalances this contrast between urban and rural environments. While limited space is the crux of the problem for the Dutch, elements of such a strategy are highly relevant to other cities.

At a broad level of planning, the urban environments that have special value for wildlife must be integrated into the spatial networks of the city. This must include not only the recreational parks, but essential natural and human-made connections and the exterior environment as a whole. This is particularly significant since many of the most ecologically important urban places are not often included in park planning.

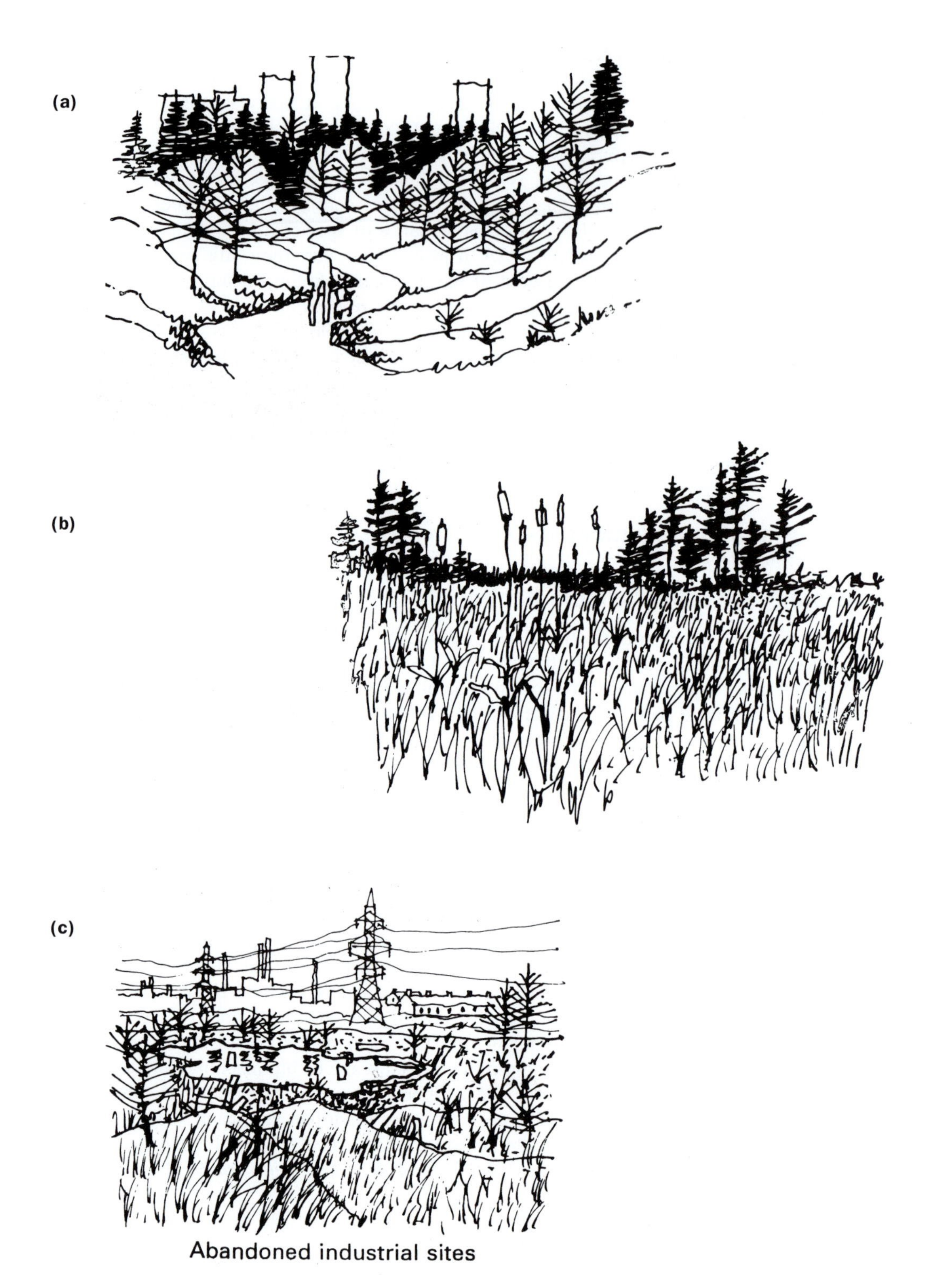

Figure 4.4 Potential wildlife habitat types. (a) Remnant woodlands and corridors within the city's boundaries. (b) Remnant marshes. (c) Human-made resources.

(a) Major city parks that have wildlife potential

Watercourses

(b) Wildlife habitat integrated with stormwater impoundments and watercourses

Golf courses

(c) Places where wildlife habitat and human activity may be integrated

(d) City parks and institutional lands

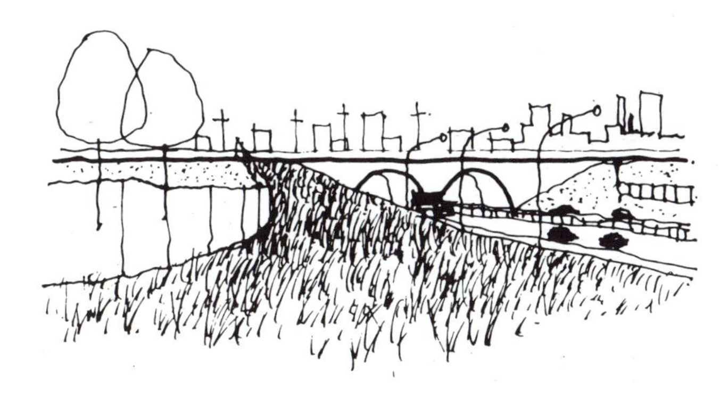

(e) Road edges

Figure 4.5 Potential wildlife habitat types.

The identification of habitat types provides a start for locating highly valued places requiring restricted or carefully controlled access. However, the protection of environmentally sensitive areas has often been based on drawing an artificial zoning line around them, while ignoring adjacent natural areas that lack 'special' features. Consequently, many of the protected 'special' areas are no longer able to support the species and features that gave rise to the designation in the first place.[31] Other habitats may be identified in descending order of significance or sensitivity to disturbance (see p. 95). Special areas include:

- Remnant rural landscapes enclosed within the city's boundaries that maintain an existing community of wildlife species or harbour locally rare or unusual species. Examples include woodland associations, old field and meadow, watercourses, marshes, natural corridors.
- Human-influenced places that have high potential. Examples are the city's sewage treatment lagoons, abandoned naturally regenerated lands, industrial lands, aquatic wintering sites.
- Areas of potentially high wildlife significance and sensitivity where the management of stormwater and wildlife habitat can be integrated. These may occur in association with residential open spaces, major city parks, wetland reserves, floodplains and watercourses.
- Other areas less sensitive to a human presence. These may include open spaces with a wildlife potential that could, with management, accommodate a combination of human activities and wildlife. They may include public parks, cemeteries, golf-courses, institutional lands, industry, public works property and related areas.
- Linear connections linking habitat areas. These may include river and stream corridors, escarpments, continuous woodland, transmission corridors and pipeline rights of way.

The ability of wildlife to survive urban pressures depends on the complexity, productivity, size and shape of habitat. It also depends on the intensity and type of public use and the varying degrees of restriction imposed on the site. The best habitats are often those with the greatest impediments to human use. For instance:

- Places that are open to the public but have little or no direct access. These include travel corridors, such as main roads and urban expressways, abandoned railway lines and canals.
- Places that are restricted to the public for social or security reasons. These include properties and gardens, public works property such as sewage treatment plants, water reservoirs and some high-security industries such as refineries and airports, electricity generating plants and many flat rooftops. Incorporating these into a city-wide wildlife network is crucial to the overall enrichment of the city.

Habitat types

Private property

The long-term stability of private property is a question that has considerable bearing on wildlife and one that makes urban situations different from rural ones. Old cemeteries, where benign neglect has permitted them to evolve into a complex pattern of vegetation, are potential habitat for a variety of wildlife. Golf-courses, often located in valley lands, scenic landscapes or wetlands, are considerably more problematic. Wooded slopes, copses, water, golf greens and brush do create a complex habitat and they incorporate large areas with a low intensity of use. However, there are environmental problems. Fertilizer and pesticide run-off from turf grass and seepage into streams and groundwater, while reportedly negligible during the summer growing season, occurs in autumn and winter.[32] A US Environmental Protection Agency study in the mid-1980s found eight of sixteen pesticides tested in the groundwater beneath four Cape Cod golf-courses.[33] Golf-courses can often disrupt streams and vegetation, and fences inhibit both human and animal movement along valley bottoms. They are, however, among the fastest growing recreational activities in urbanizing areas worldwide. A report in 1992 by the Royal and Ancient Golf Club of St Andrews (the game's controlling body in Britain) concluded that some 690 extra courses would be needed in the UK by the year 2000 to satisfy demand beyond the 1,800 then in existence.[34] Given a universal demand, positive steps towards creating an environmentally benign facility can address the physical and long-term ecological concerns. These might include such actions as creating more naturalized chemical-free landscapes, encouraging diverse plant associations, protecting streams and fish and establishing linkages.

Private residential property that incorporates or creates natural habitat has great potential for attracting wildlife. As Chris Baines has eloquently shown, even small gardens can be made to have a variety of habitats:[35] with a patch of woodland and understorey to attract warblers, squirrels and birds and mammals that prefer habitats close to the ground, long grass and meadow for butterflies, moths and other insects, a pond that supports bulrushes, sedge and water lilies that attract frogs, toads and dragonflies. Food sources may be provided by planting fruit and berry-bearing shrubs and trees, and patches of wild or cultivated flowers that are a source of food for seed-eaters in the autumn and winter. He points out that when the individual garden is seen in context with the surrounding landscape of other neighbourhood gardens, with remnant woods and scrub, naturalized railway rights of way and canals, a connecting framework of natural areas becomes possible, linking up to form a wildlife network of travel corridors. The grass-roots Urban Wildlife Group in England has carried out detailed surveys of much open space in the countryside, and has helped local authorities to prepare policy documents to identify places that need protecting. As Baines has commented, 'we have helped make nature conservation "respectable"', a particularly important factor in a country where 97 per cent of meadows have been destroyed over a period of thirty years, and where 10 per cent of its Sites of Special Scientific Interest are being degraded every year.[36]

City park wildlife

Setting aside wildlife reserves can enhance many city parks and reduce maintenance costs as well. One of the most famous parks in London, Regent's Park, incorporates a protected wildlife area within its boundaries. A densely wooded island in the lake provides undisturbed sanctuary for many species of ducks and other birds. The island is also home for a permanent colony of herons that nest in its mature trees. These normally shy birds have become accustomed to living in a city park environment and the everyday coming and going of park users. But the sanctuary, isolated by water, is close enough to allow people to observe the activities of its inhabitants without undue disturbance. To see these great birds flying low over the water, or feeding their young at treetop level in April, with the sounds of the city in the background, is an extraordinary and quite unforgettable experience.

Plate 4.6 The heron colony at Regent's Park, London. Since the early 1980s when I first recorded the herons they have become immune to the human presence, exhibiting a certain nonchalance to the activities around them.

Plate 4.7 Lakeshore refinery (Petro Canada) in relation to the surrounding communities and the Rattray Marsh (bottom left). The marsh, creek corridor and refinery link has encouraged animals such as deer into these newly created refinery lands.

Source: Steve Frost.

Active industrial sites

The principle of visual access but physical separation for wildlife areas is very relevant to high-security industries. An example of this is the Petro-Canada Lakeshore oil refinery in Mississauga, Ontario. This 225 hectare industrial complex is abutted by urban development on three sides and Lake Ontario on the fourth. In 1975, the refinery underwent an upgrading of its plant. Faced with the realities of having to live in harmony with the communities that surround it, it initiated a landscape plan around its boundaries as part of a plant modernization programme. The plan, conducted in cooperation with neighbouring communities, included the creation of extensive screening and wildlife habitat with earth berms and planting within the refinery's fenced boundaries. This came about at the request of the community, many members of which were keenly interested in wildlife. People had over the years witnessed the continuing deterioration of the environment where they lived.

Analysis of the larger area revealed some interesting features. The site was close to a lakeshore marsh (the Rattray Marsh Conservation Area) which was fed by a wooded creek that extended to its easterly boundary. Thus a reforestation of this boundary strip would create a natural circular link connecting the marsh with the refinery. The site itself had been disturbed by landfill and other refinery operations. It was therefore in poor physical shape. But a natural drainage channel running through the site had created a small area where aquatic vegetation could flourish. There also remained the vestiges of old field communities of wild grasses and flowers from the previously existing agricultural landscape.

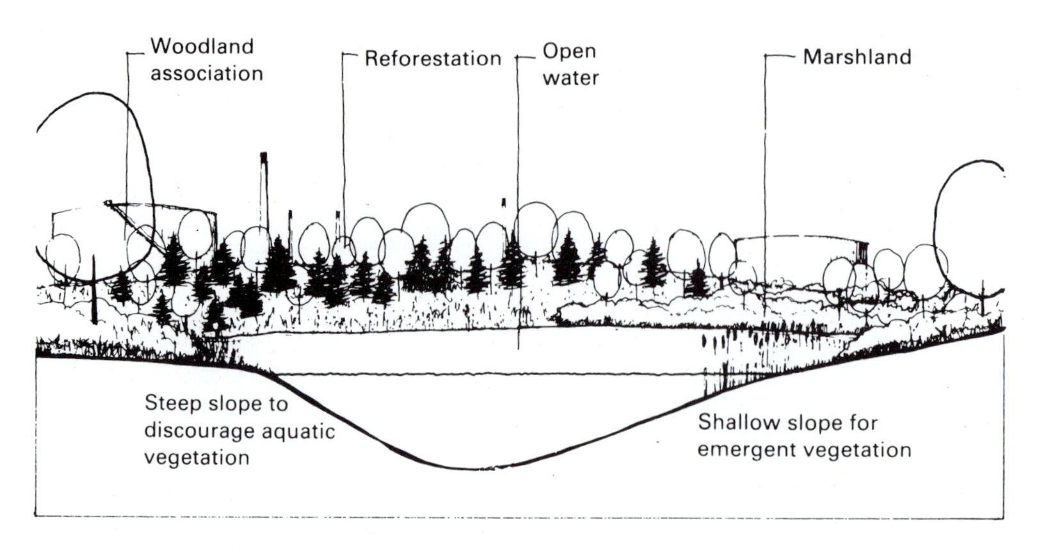

Figure 4.6 Section through the pond showing construction principles.

Plate 4.8 The wildlife pond and wetlands and littoral edges beyond, creating habitat diversity and functioning as a source of water in case of fire.

Source: Steve Frost.

The plan that evolved was intended to maximize diversity and create a number of distinct wildlife environments. A pond was dug adjoining the existing wetland, combining open water with extended aquatic vegetation. The old field community was retained and extended. A variety of small deciduous and coniferous woodland habitats were planted, including the re-establishment of a number of locally native species that had all but disappeared from the area. A linear wooded strip of about 50 metres was planted along the entire eastern refinery boundary to link the wildlife area to the adjacent lakeshore wetland in a continuous corridor.

A crucial element was management. One of the local community's goals was to restore habitat and allow it to evolve undisturbed. To achieve this required three long-term strategies: a minimum of interference from horticultural maintenance; little or no human intervention; and continuing regular public meetings to maintain goodwill and ensure resolution of issues as they arose.

Minimal interference

An initial period of grass cutting during the first several summers after installation was undertaken to reduce competition to shrubs and trees. After that all mowing, with the exception of some boundary areas adjacent to residences, was abandoned.

Human intervention

The management plan was based on permitting visual rather than direct access to the natural areas within the refinery boundaries. Boundary fencing therefore remained as security for the refinery operations with no public access to the reserve, except under special supervised conditions. Outside the refinery's security fence and within the existing park separating the community and refinery, a large hill was built to permit wildlife viewing at a distance while excluding the public from entering the area itself. Other views of the pond were designed to permit closer observation by local residents.

Continuing public meetings

Since the completion of the project these have been held on a yearly or more frequent basis between refinery management and the various neighbourhoods to continue dialogue and resolve problems as they arise.

Developments since the 1980s

Over the period of its implementation, the various plantings have become established and begun to develop distinct woodland, meadow, open water and wetland habitats. A diversity of wildlife has returned to colonize the area. Muskrat have appeared along with raccoons, foxes, ducks, geese and countless songbirds. In 1985 deer set up residence within the refinery boundaries, following the natural corridor from the Rattray Marsh and leaping over the 2.5 metre fence to take advantage of the wooded habitat and food sources, and coyotes followed in the mid-1900s. The hill in the existing park provides views to the plant associations of the reserve, and to the lake – views not previously experienced. Children use its slopes for tobogganing in winter. The Parks Department established a naturalization programme between the refinery corridor and the park, thus extending the natural corridor into the park environment. Maintenance costs for the first four years of establishment amounted to 298 person hours per hectare per year. Following this period maintenance costs were reduced to 38, a reduction of almost 800 per cent.[37] An initial wildlife habitat management plan was implemented in the early

1990s, involving thinning of some woodland trees to reduce competition and promote a canopy and ground flora. Some additional plantings and meadow seeding were added together with a redefinition of mowing along edges adjacent to residential areas.

The fact that this refinery landscape continues to exist may be seen as a continuing commitment that began in the 1970s by the refinery to the larger community, and that benefits both the industry and its neighbours. An evening I spent at the refinery in February 2003, touring the complex and the naturalized area with the Environmental Officer, provided me with some further insights. She told me about the change in production that has occurred, from crude oil, gasoline and asphalt to environmentally friendly products for the food and pharmaceutical industries. A collaborative relationship now exists between the community and the refinery that includes the establishment of several committees that meet ten times a year to deal with issues as they arise. Discussion at one of the meetings revealed that the naturalized area can contribute to the reduction of CO_2 and noise produced by refinery operations, in addition to its other ecological and aesthetic functions.

This is a place where a practical attempt has been made to re-create a natural habitat and effective wildlilfe travel corridor in urban surroundings. The association of wildlife and industry is a symbiotic one. Industry has benefited from a landscape buffer between it and the community that needs little or no upkeep. The community has benefited from the educational and enjoyable environment that is available on its doorstep. Both the industry and the surrounding residential neighbourhoods have benefited from the continuing dialogue that began in the 1970s. This has continued as a regular event ever since, virtually eliminating the sense of confrontation and complaints that once existed.[38] In addition, wildlife has flourished, protected from direct intrusion on its territory by security fences and links to the Rattray Marsh.

Revealing the life in brownfield industrial sites

Deptford Creek, London, UK

Increasingly significant for post-industrial cities in North America and Europe is the reclamation of brownfield sites. As we saw in Chapter 1, much of the ecological diversity that one finds in cities occurs in abandoned industrial lands. Chapter 7 deals with these from various perspectives – industrial and land history, restoration and reuse, and regional landscape planning. But here we need to explore how derelict and often contaminated land must be seen as an asset, rather than a liability, which is how it has long been perceived. The case of the black redstart, protected under Schedule 1 of the Wildlife and Countryside Act 1981, provides that opportunity.

The black redstart's natural habitat is open and rocky mountain areas, but after the Second World War pairs were found breeding in London's bombed areas. As post-war rebuilding eliminated many of these habitats, the birds colonized industrial settings, such as power-stations, gasworks, railway yards and old wharves along the Thames. A highly industrialized tributary of the Thames, the Deptford Creek, is one of these.[39] The black redstart is found along the Deptford Creek and other derelict areas which, in the early 1990s, was plagued with 24 per cent unemployment and declining local industry. Like most declining industrial lands, Deptford Creek is an 'eyesore': with decaying wooden flood walls, piles of rubbish, shopping trolleys buried in the creek's silt that appear at low tide, and the perception of a polluted, muddy and isolated stream. Without revitalized uses the creek, like Toronto's Lower Don River, had no apparent value.

A Creekside strategy was drawn up in 1993 by Greenwich and Lewisham Councils to attract local investment, create employment and promote redevelopment of derelict sites. A major theme of this strategy was the regeneration of the Creekside environment that would:

- Create a visual attraction out of an eyesore.
- Protect existing wildlife and enhance habitats.
- Develop creative solutions for flood protection walls to re-create habitats lost when repairs or replacements were made.
- Remove rubbish.
- Develop a long-term management plan.

The Creekside Environmental Project initiated a rethinking of standard waterfront regeneration practice that had taken place along London's waterfront, such as Canary Wharf. Twenty years ago the banks of the Thames were lined with derelict warehouses and wharfs – empty, decaying and forgotten. With the Thames revival, begun in the 1950s (see Chapter 2, pp. 37–9) to clean up the river, came the return of fish and the beginnings of an urban regeneration that has continued into the twenty-first century. The result, as The Creekside Education Trust notes,[40] has been an ecologically sterile environment whose bland horticultural landscape has precluded the emergence of enriching and diverse habitats. 'The wonderful diverse brownfield sites, a string of ecological pearls, that once lined the Thames went mostly unrecognized and unappreciated until it was too late.'[41] Baseline studies were conducted to determine what wildlife existed on Deptford Creek since very little, if anything, was known about the life systems that might exist here. It soon became clear that development adjacent to the creek would affect the creek itself. Since the river walls were owned by the adjacent landowner, the involvement and collaboration of government agencies and private landowners was both essential and beneficial.

Plate 4.9 Oblique air view of Deptford Creek tributary and the Thames. Derelict industrial areas like this need to be understood as environmental assets rather than as liabilities.

Source: Creekside Education Trust.

Studies showed how ecologically diverse and alive this derelict urbanized tidal river and tributary actually is. One hundred and eighteen species of freshwater and marine fish were identified. The Deptford Creek provides a natural nursery for the flounder and eel; there are over twenty-five aquatic species, and nearly 300 terrestrial species of insects and other invertebrates. Over 140 trees, shrubs and wildflowers grow in the Creek, a high diversity of native and alien plants growing on walls, mud-flats and waste-land places. These plants have adapted to a harsh environment of wind, drought, lack of soil and inundation from the daily tides. The river walls built for flood defences were of particular interest: for the diversity of invertebrates that live in the decaying wood, and their biological significance for birds. Two breeding pairs of redstarts were reported to nest in derelict structures, foraging for insects on the flood defence walls. Other breeding birds include grey and pied wagtail. Linnets, wrens and robins feed on buddleia and other wasteland plants, and water birds such as grey herons, cormorants, coots and moorhens feed in the creek.

Like other long-lived community restoration initiatives (see Chapter 2), continuity is the crucial element if implemented projects and educational goals are to be realized over time. It was for this reason that the Creekside Education Trust was founded in 1999, to provide the organizational structure necessary for continuing the community's work and its association with local councils into the future. The Trust is managed by a board of local people which also includes trustees with such skills as environmental education, community forums, business management and education. A visitor centre, built in 2003, incorporates exhibitions, displays and interpretive information about the creek. It also acts as a focus for visiting school groups and the local communities that are made up of many cultural backgrounds including Vietnamese, Chinese and Caribbean people. The visitor centre will, in the future, be connected to a future floating classroom to be moored alongside the Thames waterfront.

As a community-driven initiative, Deptford Creek has links with other restoration work where the stimulus for action has been proactive rather than reactive. The latter has traditionally been defensive – a fight to prevent a project from happening. The former involves the active initiation of a project by a citizen group who then make it happen. There are important lessons from this Deptford Creek example that challenge traditional views on the rehabilitation of brownfield sites for urban renewal and standards for aesthetics.

1 The summary report *Life on the Edge* makes the point that the term 'brownfields' 'is applied, without any distinction, to landscape features such as derelict buildings, people's gardens and to some of the most diverse wildlife sites (in London). It was coined to set it in opposition to Greenfields, most of which have been made biologically sterile by industrial agriculture.'[42] This comment may also be applied to North American cities, the only difference being that in North America the term 'brownfield' more specifically defines land that has been contaminated by past industrial activity such as soil and water pollution. The need to value the conventional 'waste places' of the city for their ecological as well as their development value becomes increasingly urgent as cities attempt to intensify downtown city districts as an alternative to urban sprawl at the edges (see Chapter 7).
2 Abandoned industrial places such as Deptford Creek challenge perceptions of aesthetic qualities. To the uninformed visitor, the visual image of decaying river walls, rubble and chunks of concrete, weedy plants, rubbish and half-submerged trolleys typifies a place that is run down, a visual mess. Everything about it gives clear signals that this is a derelict and, some would say, just plain ugly part of the

(a)

(b)

Plate 4.10 (a) The biological richness of industrial places like Creekside has been revealed in this centre. Its diversity is in marked contrast to much sterile redevelopment along waterfront lands. (b) The centre. Its rubble roof provides the basis for plants and wildlife habitat to colonize naturally. The building footprint, therefore, involves no loss of habitat.

Source: Creekside Education Trust.

waterfront.[43] Why do so many people still think of environments such as this as places where nature has been abandoned? One answer to the question lies in challenging perceptions about what is significant, beautiful or 'just plain ugly'. There is a need to look beneath the face of abandoned urban areas to gain insights into their past, to see beyond the decaying façade to understand what life lies hidden there. Getting to know the place, which is what happened at Deptford, begins the process of revealing what is there. What was abandoned is now occupied and has purpose; 'there is a distinction between mud and pollution, and although it is brown and opaque the water of the Thames and Deptford Creek is cleaner now than it has been for almost two centuries. Fish have returned to London and about 120 species have been recorded.'[44] Thus, aesthetic appreciation is linked to knowing the place.

3 Ecological restoration usually involves active human intervention to recolonize native plants on a prepared site. The natural history of brownfield sites such as Deptford Creek, however, shows that conventional restoration would be both costly and counter-productive. The question, 'What is the nature of this place?' is the starting point for regeneration. The publication *Deptford Creek: Surviving Regeneration* points out that London is probably the most florally diverse area of its size in Britain, flora that are highly dynamic and capable of colonizing new areas rapidly, even those which seem most inhospitable.[45] The approach to restoration at Deptford Creek is about creating the conditions that are conducive to the colonization of the local flora which then attract insects, butterflies, birds and mammals. Two examples illustrate the approach.

First, the walls lining Deptford Creek are of particular interest to the restoration process. They were built initially to protect much of south London from flooding. Made of steel, concrete, brick and wood, and in various stages of decay, large sections needed to be replaced. The process of decay, however, encouraged colonization of plants and invertebrates, particularly on those sections constructed of wood. The diverse fresh- and salt-water life associated with the tidal stream presented a challenge: the need to protect this rich environment while recognizing a key requirement of the project – the repair or replacement of the wooden walls. The approach taken was to add timber structures to the new walls that would provide ledges where silt can accumulate at appropriate heights, to allow life forms to colonize without compromising the integrity of the supporting walls.

Second, this is the habitat for the nationally rare black redstart which forages on the plants that colonize the stony wasteland. Thus conventional redevelopment would certainly displace this stony habitat. The Deptford Creek Centre's rubble roof demonstrates an elegant solution to the problem. The crushed brick and concrete habitat covered by the building's footprint is now relocated on the roof which, like the rest of the site, will recolonize naturally over time. Thus, the building not only provides habitat for the black redstart and other wildlife, but has now become an inherent part of its site. This solution, incorporated into the new building, has resulted in a policy where planning consent for other developments in Greenwich will be based on their having similar stone roofs. When these are completed Deptford will have the greatest concentration of eco-roofs in Britain.[46]

As an approach to the restoration of brownfield sites, the Deptford Creek project provides the conditions for plants to colonize on their own. It suggests that ecological restoration is a process that requires an understanding of the peculiar conditions of each place. It follows the principle of *Economy of Means*, where the most significant benefits often result from the least amount of energy and effort expended. It also reflects the principle of *Environmental Education Begins at Home*, which is where environmental literacy begins, in the places where most people live, but which have long been the least understood.

Management and introduced species

The imbalance of ecological conditions caused by urbanization creates problems as well as opportunities. An example is the vast increase in exotic species of birds and rodents. The overwhelming reliance on a few, mostly exotic, species of trees for urban parks and gardens is based largely, as we saw in the last chapter, on horticultural and decorative criteria. Their relationship to such factors as insect and animal diversity is rarely considered. So, while they will not clog the drains or cause a problem with telephone wires, the habitat created is generally sterile. A study of the common trees in Britain, together with their history and the insects associated with them, is revealing. The number of insects supported by trees is important since they provide a food source for many urban birds. Research has shown that in general native trees support more insects than non-native trees. For instance, the native English oak supports 423 species, willows 450, birches 334, hornbeam and field maple 51. In contrast, non-natives such as horse-chestnut have nine species and sweet chestnut eleven. Sycamore has only about 10 per cent of the number of insect species found on some native trees.[47]

The impact of wildlife on the city

In central urban areas, buildings have become bird habitat, but the communal roosting and nesting habits of pigeons and starlings create problems. The large quantities of droppings they produce deface buildings and floor surfaces and are expensive to clean up. For many historic cities such as Venice, this is a serious matter. The sculptural surfaces of historic buildings and statues make fouling more obvious. At the same time, one of the great attractions of Venice is its famous pigeons in the Piazza San Marco, and local commercial photographers, with trained cats at the ready to send the pigeons flying, take full advantage of them. It is ironic, however, that studies have shown that guano protects limestone from sulphur dioxide, the industrial pollutant to which Venice and other ancient cities are highly prone. In modern industrial cities the solution is to design buildings to inhibit nesting and perching. Alternatively, crevices, cornices, ventilation shafts and similar holes and projections could be simply protected and bird numbers controlled by reducing available habitat.

Predators may be used to control or reduce the number of common birds that have either become a nuisance or are a hazard in certain situations. For instance, snowy owls are regularly used at some airports where the danger from birds to aircraft taking off and landing is potentially an extreme safety hazard. Peregrine falcons have been introduced into some cities to prey on pigeons, starlings and gulls. Mammals also do well in the habitat and plentiful food supply that the city provides. For instance, male raccoons in the wild claim a territory of up to 5,000 hectares where females are allowed to roam with their young. In the city, however, their numbers more than double. A study by the Ontario Ministry of Natural Resources showed that Metro Toronto has some 10,000 raccoons, roughly sixteen to eighteen per hectare.[48] Toronto has the distinction of being known as the 'Raccoon Capital of North America', but they are found in almost every region of North, Central and South America. In addition, as a consequence of the introduction of a few dozen animals to Europe from North America early in the twentieth century, Germany, France, the Netherlands and parts of Russia now have a growing raccoon population estimated at 100,000.[49] Rats are carriers of diseases such as Leptopirosis, which may be fatal to humans. The house mouse, while carrying few diseases, spoils stored foods.

The impact of urban contaminants on wildlife

Less is known, however, about the impact of the city environment on the health of wildlife. The uptake of toxic chemicals and other contaminants through the food chain, from city sewers, the soils of abandoned industrial areas, groundwater, open water, sediments and air, has been detected in the Port Industrial Area of Toronto's waterfront. Studies were conducted on benthic invertebrates across the Toronto waterfront in 1991, and included caged clams, snapping turtles, sport fish and fish-eating birds.[50] Overall they showed that all trophic levels in the aquatic food chain are contaminated with toxic chemicals, an indication of relative ecosystem health that has potential repercussions on human health. Air pollution is also a well-known health hazard to human beings and domestic animals, but its impact on wildlife has been less well researched. Where population measurements have been taken, significant reductions in vertebrate wildlife have been correlated with industrial air pollution.[51] Population censuses of the house martin in the former Czechoslovakia, for instance, have shown the species to be rare or absent in areas with heavy fluoride, sulphur dioxide, fly ash, cement and nitrogen oxide pollution. In London since the beginning of the twentieth century, there has been a decline and subsequent return of bird populations to the inner city. This has been attributed to the reduction of the once high levels of smoke and other pollutants in central London.[52] The importance of industrial pollution as a factor contributing to the decline of wildlife should therefore not be underestimated and it illustrates the essential links between animal and human health in cities.

Regional connections

Cities are not closed environments, but are connected to rural areas through natural and human-made corridors. Natural corridors include stream courses and rivers bordered by steep banks and wooded valleys. Railway connections, canals, highways and transmission lines have all greatly influenced the migration and perpetuation of wildlife in cities and maintained the links between the city and its natural region. Toronto's valley systems still support white tailed deer, coyotes, red squirrels, chipmunks and a wide variety of warblers, finches and raptors.

At a macro scale there are also the annual migrations of birds that connect the cities to other bio-regions and continents. The connections between cities and nations worldwide and the maintenance of biodiversity can be explained by the behaviour of the eastern kingbird. The kingbird is a large flycatcher and a summer visitor to North America where it breeds and feeds on a diet of insects. In the autumn it migrates to South America where it lives on a diet of tropical fruit. Without the fields, forests and marshes that provide the insects which the kingbird needs in North America, or the tropical forests that supply it with fruit in South America, it would, like thousands of other migrating species, become extinct. Protecting this bird therefore depends on protecting two very different habitats in different parts of the world. Cities are linked to these macro-regions and to each other by the life cycle requirements of the wildlife species which live in them.

In summary, wildlife in the city provides lessons we can learn about the balance of nature upon which all living beings depend. Study of human and natural urban systems is perhaps the best way of coming to grips with the wider issues of vanishing species and pollution. There remains a perceptual dissociation between urban and rural environments that has radically influenced the way urban people view wildlife and nature in general. Cottaging, camping, hunting and most other leisure-oriented activities that are pursued in the countryside are basically exploitive occupations when there may be no long-term investment in nature or the land. Yet our examination of wildlife and its dependence on the other elements of nature's processes shows that many of the city's most biologically productive environments are to be found where energy and nutrients are concentrated. Healthy and productive places are two essential components of sustainability.

This brings us to the question of how this may become a way of regaining an investment in the land. Its productivity, growing food, and our ultimate dependence on the soil for survival, strike at the heart of this issue. What are the implications of growing food in cities in the light of ecological and conserver values – the impacts of its production? What are the connections between soil, plants, productivity and the alternative framework for urban design that we have been exploring? These are questions that now deserve our attention and will be addressed in the next chapter.

5 City farming

Introduction

In the twenty-first century the food we eat, where it is grown and produced, is a commentary on the conflicts that exist between urban and rural values and social perceptions towards the land. To most urban people, the countryside is a recreational resource – a place of escape from the city. The food that appears in the supermarket has little direct connection with the actual process of growing food and how it gets to the table. It is instead dependent on worldwide marketing and distribution networks operating on fossil fuels and based on international trade agreements, but there are signs that patterns of consumption and environmental priorities are shifting. One of the indicators of change may be found in the pursuit of a more healthy diet. This is reflected in the growth of health food stores, farmers' markets, organic food growing and allotment gardening, suggesting there are signs of a return to home-grown versus 'factory-made' food. In this chapter we explore how the links between people, food growing, 'waste' and city space can be creatively re-established as a central aspect of environmental and social values, and how the fundamental links between food and sustainability can, and must, reshape our thinking about cities.

Agriculture: process and practice

Productivity

Agricultural systems are human-made communities of plants and animals interacting with soils and climate. Unlike self-perpetuating natural systems, they are inherently unstable. Cultivation, harvesting and the biological simplicity of a few species inhibit the recycling of nutrients and make them susceptible to attack from pests. The degree to which agricultural systems can be stabilized depends on factors such as soil fertility, the extent to which animal nutrients are recycled and the diversity of plant and animal species under cultivation. Traditional mixed farming practice, while varying widely in the type of cultivation and rapidly becoming extinct in most industrialized nations, has maintained a certain degree of ecological balance. The enclosed field systems of European agriculture relied on nutrient input from farm animals, crop rotation and variety and the natural communities of hedges and woodlands to offset nutrient loss due to the harvest and the depredations of pests. Energy inputs to the system before the age of fossil fuels were limited to horses and men to pull the ploughs, sow the seed and thresh the grain. Fossil fuels provided the fundamental breakthrough. The rubber-tyred tractor replaced human and animal labour. Chemical fertilizers and pest control agents replaced earlier and necessary biological methods of maintaining stability.

Fossil fuels enabled agriculture to increase its efficiency, size and productivity, while decreasing its labour inputs.

The technologies that make this growth possible depend on large land areas and few farms for an efficient operation, a situation that has characterized farming in the industrialized countries. Relatively few, but highly productive, plant hybrids have replaced the more diverse but less productive plants of earlier farming. Worldwide, about 80 per cent of human food supplies are dependent on eleven plant species.[1]

Larger farms specialize increasingly in either crops or livestock and both may be genetically engineered for maximum yield and performance. Animal feedlots and battery poultry have intensified farm output, converting feed into meat and eggs. New breeds of mechanical harvesters have been developed for harvesting genetically modified species of grains, fruits and vegetables with minimum labour. An entire agricultural industry has evolved of which farming is only a part. What is commonly known as 'agribusiness' is based on three major components:

- an input processing industry that produces seed, machines, fertilizers, fuel and related products required for large-scale farming;
- the farm itself;
- the food-processing industry which transports farm products, processes food, and markets and distributes products to wholesale and retail outlets.

Thus modern farming is dependent on fuel energy not only for growing food, but for processing and distribution. Over the past hundred years agriculture has, in effect, evolved from a labour-intensive, low-energy, small-scale and mixed farming operation to a vast industry that is capital- and energy-intensive and requires fewer and fewer people. But the benefits of high production – being able to feed more people cheaply and less grinding labour for the farmer – also bring considerable costs, in environmental and social terms, for the countryside and for the city.

Environmental costs

The first and most basic issue is non-renewable energy. Continued expansion of production, subsidized by an increasingly costly and diminishing resource, is clearly not sustainable. In the 1970s Barry Commoner pointed to the law of diminishing returns where, as cultivation becomes more intensive, greater amounts of energy subsidy must be used to obtain diminishing increments in yield.[2] The cyclical flow of energy through natural systems is simplified in industrial agriculture, which is sustained by non-renewable energy at high environmental cost. High concentrations of fertilizers and chemicals used to maximize homogeneous crops threaten soil life and deplete the humus needed for the maintenance of biological health. Streams, rivers and groundwater receive nutrients and chemicals that lead to water pollution and the destruction of aquatic species and threaten human health. Heavy machinery and tilling contribute to soil compaction and erosion, and consequently a reduction in its fertility.

The complementary benefits to the soil of an animal/crop relationship disappear when crops and livestock become separate industries. Concentrating animals into feedlots involves the use of energy, the disposal of wastes and chemical agents to prevent disease in constricted spaces. This type of industrialized agriculture has been shown to have inbuilt inefficiencies.[3] Feed must be transported to the site. Enormous quantities of manure produced by the animals (estimated at 18 to 32 kilograms per day per animal) are often uneconomical to transport back to the fields and must, therefore, be disposed of. Once fattened, the animals have to be transported long distances to urban markets.

The replacement of manpower by industrial food production decreases the energy value of various food crops. Figures from the New Alchemist Institute for the energy efficiency of different crops grown in California have shown how the amount of energy that goes into growing, shipping, packaging and marketing the food we eat is greater than the energy we get out of it. For instance, the ratio of crop energy over input energy for all raw foods has been calculated as 1.36, whereas the average of all food processed is 0.47. In effect, there does not appear to be a net energy return to society.[4] Research conducted on farmland in upland areas of Britain concluded that as farms get larger they tend to produce less food per hectare, on average, rather than more.[5] This suggests that policies aimed at amalgamating smaller farms into bigger ones in the interests of efficiency and increased production may, in fact, be counterproductive.

Social costs

Biological stability and the sustainability of agriculture are directly related to social issues. The development of modern agriculture has been accompanied by a progressive replacement of human labour with a capital-intensive agricultural industry. The labour-intensive farming of an earlier time maintained viable rural populations, but the relatively low-cost technologies, fertilizers, chemicals and machinery of modern farming have become an effective substitute for labour. An increasing number of people in the developed world have moved from the rural areas to the cities in search of urban jobs. This migration trend is ongoing in developing countries. The reasons for declining populations have been credited to increased competition, more capital required to run economic operations, amalgamation of farms into large units, the centralization of agriculture, increased mechanization needing fewer people to manage more land, and the influences of worldwide trading blocs. These factors, however, work against the maintenance of economically healthy rural communities. Many, once flourishing rural areas have been brought to economic ruin and have been unable to support or house their remaining population.

The steady disappearance of farmland in the face of urban growth has decreased the capacity of the rural areas to supply their local urban regions. City dwellers rely largely on the major food-producing regions of the world for their food. A glance at the origins of the produce displayed in the supermarkets of most Western cities tells us much about the distances (often many thousands of kilometres) that fruit, vegetables and meat must travel, refrigerated, before the produce reaches the buyer. The social costs are also about food that has been engineered for transport over great distances rather than for its taste or nutritional value; the use of chemicals and early harvesting to retard ripening; the promotion of highly processed foods that store (and transport) well. Urban regions are all undergoing development pressures and the consequent loss of farmland. Many cities, in fact, were originally settled in areas of highly productive soils. In the face of such land losses and rising energy and food costs over time, the reliance on imports from distant sources becomes an increasingly short-sighted method of feeding people.

The ecological footprint of contemporary agriculture

It is clear from the previous discussion that the environmental and social costs of current agricultural practices are unsustainable. Another, more dramatic way of understanding their impact on natural systems is to examine them from the perspective of ecological footprints. The discussion of sustainability (see Chapter 1) outlined one way of defining the principle by asking 'How big would a glass dome, or hemisphere, covering a city need to be in order to sustain itself exclusively on the ecosystems contained under the

Figure 5.1 The ecological footprint of a city: How large would a glass dome need to be for the city to sustain itself exclusively on the ecosystems that it contained?

Source: Adapted from Wakernagel *et al.*, *Our Ecological Footprint.*

dome?' Another way of looking at the ecological footprint is to compare growing methods. A British Columbia study compared two ways of growing tomatoes – intensive field agriculture and heated hydroponic greenhouses. In terms of growing area alone, hydroponic greenhouses appear to be seven to nine times more productive than field cropping, and take much less area than the required area for open field production. The study showed that such greenhouses in British Columbia actually require ten to twenty times more ecological footprint per kilogram of tomatoes harvested than high-input open

Figure 5.2 For urbanization, food production and use of fossil fuel, the Dutch use the ecological footprint of a land area over fifteen times larger than the Netherlands.

Source: Adapted from Wakernagel *et al.*, *Our Ecological Footprint.*

field production.[6] The study also illustrates that economic success can be misleading and is not always compatible with ecological integrity.[7] In larger agricultural terms, the ecological footprint may be used to measure the difference in impact on global sustainability which results from consuming produce such as meat and vegetables that are produced locally and those that travel many thousands of kilometres. Contemporary agribusiness, in fact, depends on long-distance transportation, which is subsidized by public taxes and which externalizes costs and environmental impacts such as CO_2 and climate change. Thus environmental degradation results, not only from vehicular fuel, but from the energy used for road maintenance, trucks, airports, and all the facilities necessary to support the long-distance transportation needed to move goods from one place to another. The economic and ecological costs are externalized, from the users and the purchasers of goods to the general public and natural systems. This translates into a question of the *price* of goods at the supermarket versus the *actual cost* of producing it. 'The market has the dollar; the real world has the ecological footprint.'[8]

Global tomatoes and lamb

At a global scale, another ongoing, research project conducted by Professor A. L. Murray and E. Klause examined tomatoes and lamb, both of which are imported in large quantities into Ontario (tomatoes from Mexico and the Southern USA; lamb from New Zealand) but are easily produced in Ontario. Its purpose was to investigate the distance tomatoes and lamb travelled, the ecological footprint they created, and how this would change if they were produced locally or within the region (Table 5.1).

The total quantity of tomatoes purchased in Ontario in 1997 were grown in three ways: 24 per cent from greenhouses, 48 per cent field grown, and 26 per cent imported. Since tomatoes are imported mostly when field production has ceased, the issue is not only their impact on the environmental footprint of transportation, but also involves the energy embodied in field versus greenhouse production. Thus, included in the calculations were the energy embodied in the greenhouse structures, equipment, lighting and heating by natural gas. It was calculated that despite these inclusions, the imported produce had an ecological footprint 2.85 larger for both the market and per individual tomato.

For lamb the study assumed that the environmental footprint of lamb production would be approximately the same in Ontario and New Zealand. Total purchases of lamb from Ontario were approximately 16 per cent versus 83 per cent from New Zealand. The footprint of fresh lamb imported by aircraft (as most of it is) was found to be 411 times larger per lamb than lamb from Ontario.[9] These studies suggest that the overall

Table 5.1 ***The ecological footprint of tomatoes and lamb***

Source	*Mode of transportation*	*Eco footprint/ hectare*	*Ontario value*	*CO_2 metric tonnes*
Tomatoes				
NA imports	Truck	2,867	2.85	221
Ont. Greenhouses	Truck	1,006	1	67
Lamb				
New Zealand	100% air	0.0393	411	0.003

Source: Murray, Alex L. and Kraus, Eric. *The Ecological Footprint of Food Transportation. For Tomatoes and Lamb.* Proceedings from 'Moving the Economy'. An International Conference on Economic Opportunities in Sustainable Transportation. Toronto, ON, Canada, 1998.

benefits to society of imported foods are negative if the *total* costs are accounted for. In addition, it reinforces the view that cities have a key role to play in the growing and production of local and regional food through planning strategies that encourage compact development and limits to urban growth that protect productive local farmland. These issues have major implications for urban growth and its control, and are discussed in Chapter 7.

Urban processes

Problems and perceptions

The problems that confront rural areas are urban based. Rural occupations, such as fruit-growing and dairy farming, which were once common in cities, are largely a thing of the past. As cities grow larger, the vast majority of people have ceased to have any knowledge of rural values and skills. For urban people the countryside is seen as an urban playground, a recreational resource of fresh air and peaceful scenery, not as a working environment for producing food. Many have returned to the countryside to live or retire, others to find a weekend cottage; many more to camp, hike, snowmobile, water-ski or sightsee on holidays. Urban activities in the countryside are frequently exploitive and incompatible in terms of their environmental and social effects. For the farmer, the holiday-maker with his all-terrain vehicle, dog, camper and urban values is a potential menace, interfering with crops, setting fires, and leaving hazardous rubbish that can damage property and livestock. For the cottage owner, the snowmobile, all-terrain vehicle and high-speed water craft disturb wildlife, destroy habitat and are frequently lethal to those who use them. The expanding urban edge of cities brings urban values into confrontation with rural ones, as the new urbanites complain about farm smells and rural activities, the urban edging out the rural. Separated as we are from the sources and processes of our food supply, and the culture that goes with it, we carry around a burden of expectations and ignorance that may often be as destructive as it may be well meaning.

In the cities, pets proliferate. Cats and dogs are cherished, but animals kept for food are regarded with suspicion and banned by local by-laws. Two official recommendations to government pertaining to livestock keeping in cities make an interesting comparison of attitudes at different periods of recent history. The first, during the early years of the Second World War in England, reads as follows:

> [It is recommended] that provision should be made . . . for town dwellers to keep pigs and poultry and in general to continue those rural occupations which have proved to have social, economic, and educational advantage in times of war.[10]

The second, in the city of Toronto in 1981, reads in part:

> a farm animal may be described as any animal commonly or historically used in the production of food. . . . Almost by nature of their purpose these live animals and poultry require extensive space, constant supervision and present obvious problems with regard to defecation and perhaps noise. . . . The typical city home and lot does not provide anything approaching an appropriate setting for such animals. It is proposed, therefore, that ownership of live animals and poultry within the city be banned.[11]

The meat bought at the store was once livestock, but few know of the preparation processes that precede the packaged product. We have little idea of how to handle or manage the animals that are the source of our food and impose sentimental value on 'fluffy baby animals', which inhibits a straightforward acceptance of their use as farm livestock. The slaughter of farm animals for food is an unknown and unwanted experience. It is clear that there is a need for an awareness of where our food comes from for urban children and adults alike.

The city presents biological problems that cannot be ignored without at the same time sterilizing the minds and habits of its inhabitants. Our perceptions of the urban environment and the things we do there have, as I showed in Chapter 1, been moulded by the isolation of modern urban life from what were once rural values, but which still remain one of the fundamental elements of life support in cities.

Redressing the balance

The new millennium has brought a heightened public awareness that diverse and productive cities are a fundamental basis for a sustainable future. Soil conservation, modern versions of traditional small-scale farming, the health problems inherent in chemical food production and concern for a greater control over personal and community destinies are being expressed. A groundswell of organizations is searching for a more humane and integrated relationship with the natural processes that sustain life. Many private and commercial operations have taken to the land to produce 'organically' grown food uncontaminated by chemical additives or preserved in tin cans. So the question must be asked: How are urban issues relevant to the problems of food and farming? An answer may lie in the fact that, as I have previously suggested, rural problems are urban based. There are issues here that affect not only Western cities but, even more crucially, cities in the developing world.

If we look back in history, it becomes obvious that the disconnection between urban people and the land is a contemporary malaise. In pre-industrial cities, necessity dictated a measure of integration between urban and rural occupations. Mumford notes that a good part of the population of medieval cities had private gardens and practised rural occupations within the city. In addition, burghers had orchards and vineyards in the suburbs and kept cows and sheep on the common fields under the care of a municipal herdsman.[12] Pigs and chickens were also kept, and, in the early days, these acted as town scavengers. Until provisions for street cleaning were effected, Mumford notes, the pig was 'an active member of the local Board of Health'.[13]

The New England towns of the USA maintained a similar balance between rural and urban occupations up to the end of the nineteenth century. In Britain prior to the First World War, a thriving dairy industry existed in Liverpool. Many families from the Yorkshire Dales went into the business of supplying milk to a fast-growing urban population, keeping cows in the city and selling the milk to homes on the city streets. The story of these enterprising families is told through one family who stayed in Liverpool for twenty years before retiring to a farm in the Yorkshire Dales that they had bought with the profits of their Liverpool milk business. The business started with two cows and ended with forty-six cows and three horses. The Liverpool house where they lived was at the end of a row, and the front room formed a dairy where people arrived at all hours to buy milk. A short distance away were the buildings housing the cattle. The daughter took a horse-drawn lorry to the local cemetery in the summers to collect loads of grass cuttings that had been put into piles by the cemetery maintenance men. Hay for the animals was obtained from a local farmer who would dump a load on his way to the city market and pick up manure on his way home.[14]

Urban food growing by necessity

While this Yorkshire Dales initiative is an elegant example of self-sufficiency in an industrializing society, the absolute need to be self-sufficient is demonstrated all too clearly in many developing countries. Rapid urbanization and migration of rural populations puts a high demand on cities to provide employment and adequate living conditions. The number of urban poor are increasing throughout Africa, Asia and Latin America, and many of them incorporate agriculture as part of their livelihood strategies.[15] The urban poor can spend 60 to 80 per cent of their income on food, so self-grown food can represent considerable savings on food expenditures. Furthermore, improved nutrition improves people's capacity to work. Urban agriculture can create self-employment and income that can be a significant, but often uncertain contribution to household support. In Nairobi, Kenyan urban agriculture has been found to provide the highest self-employment earnings in small-scale enterprises and the third highest earnings in all of urban Kenya.[16]

At the same time, urban farmers are frequently at the mercy of high land values and the whims of local government. A research study of urban farms in Jakarta conducted in 1994[17] involved plots that were located on open spaces, owned by the city, adjacent to the Ciliwong River. The author made the point that as well as providing a living for farmers, the farms preserved, and gave productive purpose to, the city's green spaces. In addition, growing crops was likely to be relatively secure, since these river lands had been designated as open space. The farmers paid about one-third of their income in rent to the landowner every month. In 1994 the city decided to increase the land tax, which had a considerable adverse effect on what the farmers had to pay in rent. In the intervening years since the study was completed, almost 50 per cent of the study area was converted into either shopping malls or housing as a means of increasing municipal revenues.[18] The argument that economically more valuable land uses will (or even should) replace urban agriculture often underestimates its economic value.[19] The picture becomes more positive if non-market benefits, such as food security, improved nutrition, urban greening and landscape management, are included and urban policies support urban agriculture functions.[20]

The pursuit of rural occupations in many cities clearly demonstrates that it has been based largely on the imperative of necessity. The same may be said of war and emergency. During the German occupation of Denmark in the Second World War, it was the food grown in gardens that saved the citizens from starvation. In Britain, efforts to make the most of limited resources led to the setting up of the Pig Keeping Council in 1939 by the Ministry of Agriculture. Its original objective was to move pigs from the farms, which were threatened by shortages of imported feedstuffs, to the villages to encourage the use of household wastes for feed. These efforts, however, were soon reflected in an *urban* livestock movement. Since the bulk of edible waste came from the cities, pig- and poultry-keeping naturally evolved as a major urban activity. Pig-keeping spread on to bombed sites, in back streets and allotments, and included policemen, firemen and factory workers among the devotees. There was a pig club in London's Hyde Park and another within 180 metres of Oxford Circus. In 1940, in response to the increasing demand for eggs, the Domestic Food Producers Council set up a Poultry Committee that recommended households be encouraged to keep poultry for egg production. The backyard food production movement became officially accepted with an Order in Council which suspended restrictions on the keeping of pigs, chickens and rabbits, subject to certain public health requirements.

By 1943 there were 4,000 pig clubs comprising some 110,000 members keeping 105,000 pigs. While the promoters of the pig-keeping movement originally intended it to be a rural activity, it became urban because the food wastes of modern society were mainly in urban areas. Prior to this, the disposal of waste food levied a heavy charge on public rates. Many of the clubs were organized into cooperatives, livestock being fattened collectively, rather than by individuals, lending greater efficiency and organization to the activity.[21] Other livestock activities included bee-keeping and goats. In England and Wales at that time there were about 30,000 bee-keepers controlling some 429,000 colonies of bees. While bee-keeping, by its very nature, is not an urban pursuit, there were many such operations that were distinctly urban or semi-urban in character. One of the most progressive associations of bee-keepers was in the city of Birmingham, where facilities for an instructional apiary were provided in one of the public parks. It is interesting to note that the planning of parks and gardens was of particular concern to bee-keepers, since the selection of plants in the city's parks would have considerable impact on nectar supply. The maintenance of plant diversity (discussed in Chapter 3) may be seen to be a crucial factor in a functional productive urban landscape. In addition, the value of bees as pollinators for allotment and backyard gardens is, in the opinion of modern bee-keepers, much more important than their value as producers of honey. Goat-keeping was practised intensively due partly to the shortage of fresh milk and partly because the waste on which goats thrive meant that they were not competing with other animals for food. Milk yields averaged over 4.5 litres a day throughout the year from goats that were entirely stall fed and with a yard of some 17 square metres for exercise.

The 'Dig for Victory' campaign during the Second World War in Britain and other countries showed that the production of fruit and vegetables in or near cities could have a significant influence on food production when people are in need. Production from allotment and garden plots in Britain reached a peak when the number of allotments almost doubled from a pre-war figure of 740,000 to 1.4 million. In a parliamentary debate reported in Hansard in 1944 it was estimated that 10 per cent of the food grown in Britain came from this source.[22] During the years when urban food production was at its height, agricultural shows took place in a variety of places, including the basement of John Lewis' shop in Oxford Street, in industrial areas and on football fields. John Green, founder of the 'Back to the Land' Club and in charge of BBC broadcasts on farming and gardening at the time, recorded a visit to Bethnal Green after one of the worst air raids of the war. He found people feeding their ducks on the canals and visited a poultry show at the Working Men's Institute. The inner city had just acquired a new unity with the countryside.[23]

As these examples illustrate, crises and shortages lead to the adoption of alternative strategies for survival and it is in the cities that this has occurred. Since the end of the Second World War, however, an affluent urban society has not had to concern itself with these basic needs as food has become readily available regardless of season or distance. However, with the shift of societal values towards a better relationship with nature, an interest in 'organic' farming, human health and diet has emerged. One of the indicators of these changing values is the rebirth of the farmers' market. Although they had never completely disappeared, pressures from the supermarkets and large-scale farming during the 1950s and 1970s reduced their competitiveness, and suburban expansion replaced the farmland close to the cities that had been the mainstay of their economy. Since the mid-1970s, farmers' markets have revived and by 1993 were reported to number some 2,000 in the USA.[24] Several forces are reported to have led to this revival.

The first has to do with small-scale economics. The uncertain financial returns from selling products to wholesalers have forced farmers to sell direct to the customer. The farmers' market provides an ideal outlet, one that has kept many small farmers in business. The second concerns changing consumer attitudes. An increasingly sophisticated public is looking for variety, taste and pesticide-free produce that the standard supermarket lettuce wrapped in cellophane and 'one-type-fits-all' green peppers cannot match.[25] And in a planning context, urban planners have discovered that farmers' markets make an effective contribution to revitalizing downtowns.[26]

Thus, the opportunities for re-establishing constructive links with the land are tied to the food we eat, and it is in the cities that these links must be re-established. They are the places where alternatives to destructive technologies can be explored and demonstrated through direct experience with soil productivity, the recycling of nutrient and material resources and urban metabolic processes. We can learn about energy and food production, farming practices, market gardening techniques, rural as well as urban affairs, while turning 'waste' into useful products. To do so requires that farming at appropriate levels becomes an integral part of the city's open space and park functions. It also requires that their overall physical and social structure must be reshaped when their essential productive value is recognized. In this way, the ecological footprint of growing food can be shrunk and the move towards a sustainable future can become a reality. The opportunities for doing so should therefore be explored.

Resources and opportunities

Urban productivity in the developing nations

It is apparent that cities have significant potential for small-scale agriculture, so it is here that the task of building a richer and more productive city environment can be achieved. This fact is borne out in cities in the developing nations that must produce food for large populations, often with limited space and energy for transportation, and minimal financial resources to import food. In China, government policy has aimed to create producer rather than consumer cities. At least 85 per cent of the vegetables consumed by urban residents are produced within urban municipalities.[27] Shanghai and Beijing are self-sufficient in vegetables, and many Chinese cities also produce large quantities of poultry and pigs and other essential foods on the edges of towns and cities from where it can be transported at minimum cost to the population centres.[28]

A key factor in the developing nations, where poverty is a major problem, is the need to link land, food production and waste recycling with employment and income. Many Asian countries, such as Indonesia, practise urban aquaculture where fish-ponds provide the opportunity for fish and rice production to be integrated with animal and human waste treatment and absorption. Irrigation with untreated wastewater has been practised for centuries by poor farmers in urban areas. It has also become a widely accepted, though unregulated, practice in many countries due to potential health and environmental risks.[29] At the same time, wastewater also has important economic and environmental benefits. Society may benefit from limiting pollution to localized areas instead of polluting surface waters through untreated disposal. The farmers take advantage of nutrients to increase crop yields and reduce the need for artificial fertilizers.[30] For over 2,000 years Chinese fish-ponds have produced fish fed from grass clippings with animal and human manure feeding the ponds. Fish farmed from these ponds account for 2 million tonnes or just under 20 per cent of total fish production in China.[31] The exploitation of urban organic wastes for food production is to be found in urban farming in the City of Havana. In 2001 more than 80,000 tonnes of compost were produced and distributed to different production centres across the city.[32]

The disposal of unwanted products is a universal problem of expanding cities worldwide, and is often linked to recycling and income by scavenger groups. In 1992, an article in the *Jakarta Post* praised the scavengers of that city for collecting and sorting the rubbish, some 6,000 tonnes per day, and recycling it for a living. Rubbish pickers at the rubbish sites take their finds to nearby junk dealers who buy their old cans, plastic materials and bottles, and sort them out for further recycling.[33] Recyclable plastic, after being cleaned, is in demand by plastic factories where it is reprocessed and used again. Similarly, iron scraps may also be sold for recasting after being melted down. Other materials include used cans that are in high demand by kerosene stove-makers; soy ketchup bottles have a good market value since they can be refilled; furniture shops are grateful for new supplies of crates since it is cheaper to use parts of crates than to buy new ones.[34]

A three-year pilot project to introduce a self-financing Integrated Resource Recovery system in Bandung (the second largest city in Indonesia) reveals the close relationship between environmental enhancement, social organization and job creation and income. The research identified two contradictory waste management systems at work side by side in Indonesia: one that is typical of other developing countries – the 'formal' system operated by local government based on the Western-style concept of collection–transport–dumping of waste – and its 'informal' counterpart – the scavengers who collect materials with a resale value and sell it to industries, which in turn recycle it into reusable products. As the volume of waste increases in the former, so do costs and environmental health concerns, which then require the introduction of increasingly sophisticated technologies.[35] A participatory action research project began in which ways were sought to involve a group of families and scavengers, break down social barriers, develop a sense of common community purpose in the project and establish a centre for organizing common action. Within three months, the initial families who had been trying to survive individually transformed themselves into a dynamic and creative community. By the end of the project in 1986 the community had grown to eighty-eight families earning a living by sorting and recycling recovered paper, glass and metal, composting organic

Table 5.2 *Scavenging Bandung*

Type	*% or number*	*Value (Rp)*	*Recycled into*
Paper	38.9%	70–100/kg	Paper, cardboard, kitchen utensils, various glassware, hoes, caps for bottles, cattlefeed, shoes, soles, etc.
Metals	22.0%	40–300/kg	
Bottles	20.6%	75/piece	
Textiles	12%	200/kg	
Drums (steel)	100/month	150–800/kg	
Drums (plastic)	100/month	5,000/piece	
Sacks	26,000/month	1,500/piece	
Tyres	90/month	125/kg or 500–850/piece	

Source: Hasan Poerbo, Urban Solid Waste Management on Bandung. Towards an Integrated Resource Recovery System. *Environment and Urbanism. Rethinking Local Government – Views from the Third World*, vol. 3, no. 1, April 1991.

Notes: In any one district, or *keeamatan*, there are on average one to fourteen *lapaks* (leaders of scavenger communities). Each *lapak* usually employs around eleven to sixty scavengers, with the total number of scavengers in Bandung estimated at about 2,000 to 3,000. Their level of education varies with 53 per cent having reached primary school level, 25 per cent secondary school level and 22 per cent high school level, these figures include drop outs. The daily income of a *lapak* is between Rp 5,000 and Rp 5,000,000, that for a scavenger between Rp 1,000 and Rp 5,000. The total investment for waste collection in Bandung (recorded) is Rp 408,490,000. The table shows the volume and type of waste collected for recycling and where it is sold.

wastes for sale and intensive urban farming, raising rabbits, improving their housing and other community activities.[36]

This experience became the basis for developing an Integrated Resource Recovery module as the building block of a dispersed system of waste processing. It is one that can be developed incrementally as market opportunities permit, and as a socially and environmentally viable alternative to conventional, large-scale waste management. Of key importance, it serves as an alternative to dumping unwanted materials into the local rivers, an endemic environmental problem in Indonesia. In effect, the two approaches – the formal and the informal – represent two different views of waste. The former considers it as a health and environmental hazard; the latter considers it as an economic resource from which marketable products may be derived. This latter approach thus achieves a number of objectives: it reduces volumes to be dumped; it reduces the need for financing and subsidizing waste management; it creates jobs and income opportunities; it creates social and community cohesion; and it has significant environmental benefits. As Table 5.2 shows, the economic, social and environmental gains of such a model can be considerable when it is developed as an integrated approach to the urban scene.

Productive cities as multicultural places

These examples of how collection, recycling, food production and employment are linked to the city's unwanted products are by no means limited to cities in the developing world. What is a universal condition demanding ecological, social and economic solutions for Jakarta, or Delhi, is also one for New York or Los Angeles. The status of Western cities as multicultural environments has become a fact of life. Multiculturalism is reshaping the physical and cultural character of modern cities. The traditions that ethnic groups bring with them have always enriched the character of inner-city neighbourhoods, with their street markets, residential areas, small industries and restaurants. The productive urban landscapes of mini urban farms and micro vineyards are the hallmarks of Portuguese, Chinese and Italian neighbourhoods. This vernacular is hidden away in alleys, rooftops and backyards in neighbourhoods all over the city. Here one finds an extraordinarily rich variety of flourishing gardens, houses, streets, people, expressing ideas of personal territory, particular ways of using space, what is valued and a rich urban tradition. Behind the front and backyard garden fence one will find productive vegetable gardens, grapevine-covered trellises provide summer shade and grapes for autumn wine-making. These vernacular landscapes reflect still surviving rural skills and cultural connections with the land, and represent very different views of the city in both function and aesthetic priorities.

The most dramatic trends are, however, in the changing social and physical make-up of older suburban areas. Once the domain of Anglo-Saxon, white-collar workers, many of them are experiencing a significant influx of diverse ethnic groups. The urban agriculture of the inner city is moving with them to transform hitherto conventionally landscaped and sterile suburban backyards. Mini farms of small livestock, vegetable and fruit gardens are taking over the previous non-productive lawns. What these trends show is that such changes are driven by the special needs of different groups. The necessities of life usurp what planning by-laws permit. For instance, the richness and mixed land uses of many a downtown ethnic neighbourhood are often prohibited in the outlying suburbs of adjacent municipalities that set rigid restrictions on what may or may not be done to individual houses or properties. Exclusionary land-use controls are ostensibly based on concerns about exceeding current infrastructure capacity, or damaging the environment, but many such regulatory barriers actually stem from communities' efforts

(a)

(b)

Plate 5.1 (a) Every available space is cultivated and used to the maximum in many ethnic areas. Front gardens and backyards and garage roofs supply vegetables for the family – some families being almost self-sufficient. Flowers, religious icons and decoration fill ground plots and rooftops; backyards are sheltered from the sun by grape vines that are grown for wine-making. (b) The market catering to a highly diverse cultural community.

to keep out certain groups of people. Local zoning in the USA, for instance, frequently reflects community elitism or racial ethnic prejudice.[37]

As we will see later in this chapter in our discussion of the Federation of City Farms and Gardens in England, there is a growing recognition that parks systems must become meaningful for poorer neighbourhoods and provide links between urban communities and the land. In the USA, the San Francisco League of Urban Gardeners, a grass-roots non-profit organization, has recognized the realities of North American multicultural society, poverty and unemployment, and is committed to building gardens in culturally diverse neighbourhoods. With 40 per cent of youth under 25 unemployed, the organization functions as a job training programme, employing youths as trainee carpenters, landscape construction workers, tree planters and related skills.[38] The Uhuru Gardens in south-central Los Angeles on a lot still covered with rubble from the Watts riots of 1965 is another example of a project that will demonstrate a restoration of part of the city, by reconnecting people with the land, and by vocational training for jobs in the green industries: 'Uhuru Gardens would marry an environmental ethic to landscape training and internships for community residents.'[39]

Stan Jones, a former landscape architect for the San Francisco League of Urban Gardeners, has noted that '[Multiculturalism] is not a mere acceptance of cultural differences; at best, it actually celebrates those differences – such as when gardeners share produce and learn to eat and even cultivate exotic vegetables'.[40] This comment also points out another reality, that different cultural groups use space differently and have different behaviours and needs. Jones observes that in Chinatown, you will see parks that have a certain number of linear feet of seating and other features following William Whyte's guidelines for urban spaces. But segregated in a quieter corner you will see groups of older Asian women. They don't want to be right up front, they don't want to be near the street, they don't want to be looking at the action – contrary to everything Whyte said.[41]

The tendency for cookie-pattern park design in multicultural neighbourhoods does not recognize that different physical expressions for parks evolve from understanding the cultural, psychological and behavioural needs of the people who use them. In addition, the conventional open space provided by the municipality through public taxes involves high maintenance cost, has little diversity and provides no economic returns. And as Karl Linn (the 1950s pioneer of community design), Randolph Hester and others have shown, this can only be achieved when the people most concerned participate in the design of their own places. The inevitability of the changing urban landscape requires an ecological response that recognizes the fundamentally similar forces of natural and human communities, and helps them develop in their own ways and with their own imperatives and aesthetics. It is necessary to rethink the basic nature and functions of how urban space is used.

Physical and energy resources

It is plain that modern urban tradition has been moulded by a combination of economics, technological and aesthetic influences and values. This is reflected in the fact that vast amounts of nutrient energy and land are conceived of as waste. The idea that the city itself should contribute to the production of such basic essentials as food involves a shift in values and attitudes. From this perspective the urban environment may be seen as a generator of vast *resources* in nutrient energy and land. The leaf litter collected by the parks departments of many cities, for instance, provides potential resources for enriching soils for gardens, parks and rehabilitation sites. The separation of waste and recycling in general has also become common practice. The City of Toronto, for instance, collects and recycles 30 per cent of organic materials generated in the city. Its target is to increase the volume to 60 per cent by 2006 and 80 per cent by 2009, with the ultimate goal of 100 per cent by 2010.[42] Wasted land created by urban expansion, planning regulations and inefficient single-use zoning may, in fact, be regarded as an invaluable opportunity for the future when necessity may well dictate its more productive and intelligent use. The city's spaces, in effect, will be seen as having value beyond the recreational and aesthetic purposes generally ascribed to them.

Land

The availability of urban land is potentially enormous in almost all major Western cities. Railway, public works and public utility properties, vacant lots, cemeteries and industrial lands form a major proportion of unbuilt-on land that has been, and remains, ineffectively used and frequently contaminated (see Chapter 3 on brownfield sites). The amount and type of land naturally varies depending on the peculiar political, economic and urban renewal conditions of each place. But they are reminders of the vast areas that exist as cities expand on to agricultural areas, and as more and more land lies vacant and fragmented within them.

Residential property

At the smallest personal scale, the front garden or backyard provides some of the best opportunities for food growing in terms of energy, efficiency and direct benefit. Commercially grown and processed vegetables and fruit are the most energy-intensive of all crops: it requires more energy to produce them than the energy benefits they return. A comparison between the yearly energy budget of a residential lawn 20 square metres in area and the use of the same space for productive crops has revealed some interesting results. Each unit of energy invested to maintain the lawn, in human work, fuel for the mower, fertilizer and pesticides, returned 6 units of energy as lawn clippings. If the clippings were discarded, the net production efficiency of the lawn could be described as zero. With the same area planted with alfalfa, each unit of energy invested returned 22 units in crop production. In addition, the alfalfa produced enough digestible nutrient per square metre to support the production of 450 grams of rabbit meat.[43]

With respect to diet, the food produced on a residential lot is a small-scale, human energy-intensive operation. For some, less affluent levels of society, growing food has definite economic advantages as the price of food continues to rise. For others it affords opportunities for productive spare-time work and pesticide-free and tasty vegetables. An experiment designed to determine the economics of food growing on the front yard of a small residential lot of 30 square metres was undertaken in the 1980s by the author.[44] Whether the gardener makes money or loses it on the home vegetable plot had seldom been investigated. Total returns for the vegetable patch in the first year were Canadian $136.43 (1980 $). Start-up costs were $159.49, exceeding returns by $23.06. The second year produced total returns of $230.40 with costs of growing seedlings and soil additives being $56.63. Thus a net gain in produce of $173.77 was realized. Calculated in returns per square metre the garden produced $7.68. Four years later this had increased to about $10 a square metre.[45] Analysis of productivity was based on weekly evaluations of the cost of produce at the local supermarket and then applied to produce gathered from the garden. The cost of labour for developing and maintaining the garden was not included. Thus it may be seen that even a tiny plot of land in a city with a relatively short growing season (May to September) can produce a considerable net gain in produce, and increase the margin of self-sufficiency for low-income families. The

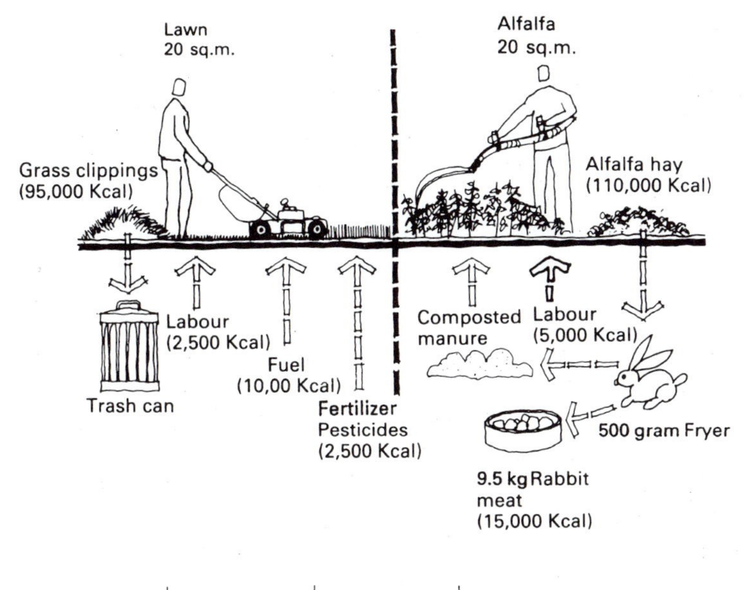

	Energy input Kcal	Energy yield plant	Kcal: Human food
Lawn	15,000	95,000	–
Alfalfa	5,000	110,000	15,000

Figure 5.3 A comparison of two equivalent residential yards of 20 square metres, one producing lawn grass, the other alfalfa. For every unit of energy invested (human labour only) in the production of alfalfa, 22 units of energy were returned in crop production. In the case of the lawn, the net production efficiency (assuming the grass clippings are discarded) is zero. It has been calculated that the rate of energy use for the maintenance of some 6.5 million hectares of lawns in America exceeds the rate of commercial production of corn on an equivalent amount of soil.

Note: All notations represent annual totals.

Source: Adapted from The Farallones Institute, *The Integral Urban House, Self-reliant Living in the City*, San Francisco: Sierra Club Books, 1978.

educational advantages of direct contact with plants and soil are interwoven with a need for greater self-reliance, less dependence on the centralized food systems of agriculture and pesticide-free food. While home food production is a question of personal choice, it is clear from the number of publications and organizations concerned with food in cities that there is a growing interest in the concept of self-reliance and an improved quality of life. Among the organizations concerned with urban problems was the Farallones Institute in Berkeley, California (disbanded in the 1980s), which attempted to demonstrate by practical means how life-support systems can be integrated into an urban house to conserve energy and resources and contribute to the health of the city's environment. Another, the Institute for Local Self-Reliance, Washington DC, provides guidance to local communities in recycling wastes, energy efficiency, urban agriculture and related issues. In Britain, there is the National Federation of City Farms and Gardens (see pp. 184–8) and the Centre for Alternative Technology in Wales. And in Toronto, Canada, various organizations include 'Gro Together', a group of volunteers who organize food drives and food banks for the urban poor. Thus, residential resources for urban farming are considerable and include any space where the individual can grow a few tomatoes or lettuce. Private food growing does not affect public space directly, but the city as a whole becomes richer and more productive because of it.

Wastelands and allotments

The demand for allotment gardens in or near the cities has been growing rapidly for many years, particularly among people who live in apartments in locations where outdoor space is at a premium. Increased leisure time, early retirement and unemployment are credited with the increasing demand for a plot of land to produce organically grown food. Allotments can be fitted to any shape of property. The small-scale and labour-intensive agriculture of the allotment is ideally suited to the small or awkwardly shaped space that has little potential for many other uses. Allotment food production is more concerned with high-value crops such as tomatoes, beans, onions and lettuces that give better financial returns on time invested than do crops such as potatoes and cabbages, which may be cheap and plentiful in the shops. The productivity of such a method is potentially high. It has been estimated that the standard allotment garden of 250 square metres may yield, under experienced management, 20 tonnes per 0.4 hectares or approximately 3.5 tonnes per plot per year. Friends of the Earth calculate that this level of performance, repeated on the 101,200 hectares of land in British cities that were estimated by the Civic Trust to be lying idle in the late 1970s, would yield some 5 million tonnes of food.[46]

Other urban wastelands include the hundreds of hectares of rooftops that for the most part lie desolate and forgotten in every city, yet many present open space opportunities that could be turned to productive use. In Toronto, for instance, the Toronto Food Policy Council is encouraging community gardens on the apartment rooftops that exist in that city.

Heat

A prodigious amount of heat energy is pumped into the city atmosphere from heating and cooling systems and industry, and this has a major impact on urban climate (Chapter 6). This same energy must, however, be regarded as a resource rather than a problem in the context of urban farming. The wasted heat energy from buildings and generating stations may be considered in several ways. One is the question of conservation through better design and insulation, about which considerable discussion and application has taken place in the 1990s and the early years of the twenty-first century. Another is the capture of waste heat that can be made use of for other purposes. Sweden harnesses approximately one-third of the waste heat from its power plants for commercial purposes.[47] The same principle of tapping waste heat has great potential for food growing, particularly in cold countries where the growing season may be limited to four or five months of the year. The need for greater self-reliance in the future is being recognized as food, imported during the non-growing season and based on high-energy inputs, becomes more costly. This potential has applications at various levels, both on the domestic and commercial scale. Self-reliant organizations such as the New Alchemists have for years been developing practical applications of integrated living systems in the USA, where greenhouse agriculture becomes part of a total biological cycle of life systems. In the urban context the heat pumped out from industrial processes may be connected directly to food-growing industries. The use of flat rooftops of many industrial buildings, using lightweight hydroponic (soilless agriculture) techniques, could begin to make more efficient use of both space and energy resources that are normally discarded. The industrial parks of many North American cities could become agricultural

Plate 5.2 Flat rooftops on many industrial buildings offer great opportunities for productive uses, such as gardens, greenhouse production using waste building heat and mini wetlands for wildlife. Compare this North American industrial area with one from Stuttgart, where green roofs have been incorporated into many of the industrial buildings (see Plate 6.8).

as well as industrial producers, combining functions that have traditionally been separated. For instance, whisky distilleries in Scotland cultivate eels (a highly prized delicacy in Europe) in large tanks using the heated wastewater from the productive process.[48] The same principle could have even greater relevance for the city since the costs of transportation can be minimized, and agricultural produce can be grown in direct association with local markets.

Implications for design

The basis for form

Historically speaking, the form of early towns and villages was dictated by their relationship to the agricultural fields where food was produced. The symbiotic relationship between the fields that produced food for the town, and the town that returned its refuse to enrich the fields, was necessary to ensure survival. The role of open space within the town, as Mumford has observed, was primarily a functional one; it was important to citizens for growing useful produce or for livestock.[49] The agricultural cooperative settlements (kibbutzim) in Israel show a similar pattern; a direct relationship between habitat and food-producing land. It was initially for survival in a harsh environment that the kibbutz system arose,[50] but new towns are few and far between, and have little impact on the problems of existing cities. It is with the reshaping of the existing city landscape – with the productivity of its soils – that we must be concerned. The breakdown between people and direct connection with the land began as a consequence of industrialization, growth and mass migration into the cities for non-rural work – a phenomenon that continues in developing countries. The creation of urban parks in the eighteenth and nineteenth centuries was based on spiritual and leisure needs rather than the functional necessities of food growing. This perception of open space – to provide recreational and aesthetic amenity at public expense, thereby satisfying the soul rather than the stomach – has persisted as the main objective of the parks. While recreation is today an essential facet of urban life, it is, as I have tried to show, only one of the many functions that urban space must serve as we attempt to build an ecologically and socially viable future for our cities. Urban agriculture is increasingly becoming a necessary urban land-use function that should be publicly supported as long as poverty, hunger, multicultural traditions and land-connected recreational trends continue to influence city life and city form.

As we have seen, the European allotment garden, located on the edge of the city on railway land and any other piece of ground that can be leased, is a precious resource. High demand and limited land have made it so. The garden plot is the citizen's summer cottage, often within cycling distance from home. Each plot is laid out with loving care, with a shed that includes storage for tools and often living space. On a larger scale, other examples offer fascinating insights into the potential of urban agriculture to make productive use of derelict land while providing fortuitous alternatives to the treatment of unused land.

Farming on industrial land: a commercial venture

An examination of the city's open spaces has shown that industrial lands take up very large areas, particularly on its fringes. These neglected places contribute to the city's visual blight, particularly when they become encircled by development. One such area lies in the borough of North York in Metropolitan Toronto. Once on the fringes of the city, it has seen tremendous growth over the past twenty-five years or more. The York University campus, first located here, was followed by major industrial, residential and commercial development on previous agricultural land. The industrial area adjacent to the university, owned by four oil companies, totals 91 hectares, about half of which have been developed as tank farms and related plant. The remaining lands have not been built on, the consequence of hard economic times for the oil industries. Visitors to this part of the city are greeted by two landscapes as they drive north. On one side is the university laid out with immense spaces of mown turf and street trees lining the road, the product of conventional design vocabulary. On the other are the oil company lands, part of which have been transformed into fields of corn, tomatoes and peas. These lands, surrounded on all sides by industrial and residential development, were leased in the early 1980s to several Italian farmers who have, in the new millennium, continued to tend a thriving market garden operation. The produce, sold to the surrounding community, has established a direct relationship between producer and consumer in the middle of a highly developed urban area. The original stimulus to turn formerly vacant land to productive use of this kind lay in provincial land tax laws. Under the Ontario Assessment Act, land that is in use for agriculture is exempted from full land taxes levied by the local municipality. This act was introduced by the province to protect farmland from the sky-rocketing taxation that always follows urban development. On this basis the oil companies in the 1980s were paying Canadian $200 per hectare for the land not built on and used for farming rather than the industrial rate of $980 per hectare.[51] The municipality, furious over lost tax assessment, took them to court, but was unsuccessful in its efforts to force them to pay full industrial rates, and the market garden has continued to exist.

The anomalies of this case raise some interesting issues. A tax loophole provided an opportunity for turning unused industrial land into a productive, visually pleasing, self-sustaining landscape. It upgrades the general quality of the urban environment and has had direct benefits to the surrounding community in 'pick-it-yourself' fresh produce.

Connections between agriculture and the city may establish potential new patterns of landscape development in urban areas. At the same time, this inadvertent phenomenon may be seen as an unrealistic mechanism for achieving these ends in the existing political framework, where a municipality is denied the taxes it needs to provide public services. There is a further anomaly in the fact that if the oil companies were required to pay full taxes, the primary stimulus for farming their vacant lands would disappear. In this case, the land might either revert to the status of urban blight or be subjected to 'landscaping' to improve visual amenity.

This example points to the fact that intelligent and creative policies are needed to create the useful and productive landscape that will help provide cheap sources of food and point the way to an alternative approach to urban design and suburban growth. For instance, there is a need to protect productive soils and rural traditions on the urban fringe in new development, thereby establishing a new kind of 'urban' that has an investment in the land, rather than traditional development that first confronts, then edges out, land-based industries. It is, in effect, a conflict of two incompatible cultures where the one inevitably degrades the other. Most of the arguments put forward for developing agricultural land are based on the need of today's farmers to sell land in order to survive or retire. Thus, productive land is relegated to a real estate trading commodity. Alternative strategies will depend less on the protection of large-scale farming operations that properly belong beyond city limits, and more on the marrying of development and small-scale and complex economic initiatives. These might include market gardens, greenhouse permaculture, allotments, small mixed farm livestock operations, plant nurseries, pottery and crafts, recycling operations and reforestation for servicing wood-based industries. There is a need for planning strategies that protect agricultural soils and control rising land prices that force farmers to sell their farms to developers (see Fig. 5.4). The increasing number of US cities that have adopted the concepts of Urban Growth Boundaries and 'Smart Growth' policies are beginning to address these issues and will be discussed in Chapter 7.

This multi-purpose and integrated planning approach may be found in the urban forest parks of Zurich, Switzerland, described in Chapter 3 (pp. 110–12). Here we find small-scale commercial agriculture being practised within the city's parks system. Farmers rent space on the common lands surrounding the city and use the land for crops, pig farming and related agricultural pursuits that can be carried out on a small scale. The recreational trail system of the park provides access through these areas as well as through the managed forests. In both Dutch and Swiss parks the concept of farming on a residential and commercial scale has been integrated as a basic function of the parks system. On both these scales farming is privately undertaken with the city providing and leasing out the land. The city's parks, therefore, function as food producers and for leisure and are administered on this basis. The productive processes of food production, of well-kept gardens and soil management become another visible element of urban processes. They enrich the urban experience and provide the basis for an aesthetic that comes out of a true application of sustainable principles linked to the land.

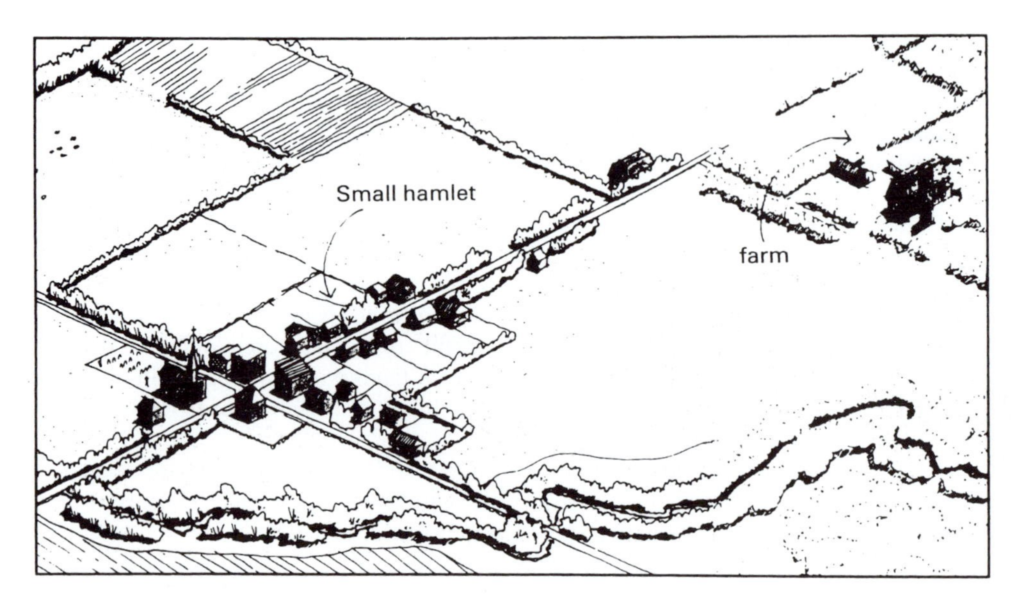

(a) The traditional rural village and surrounding countryside.

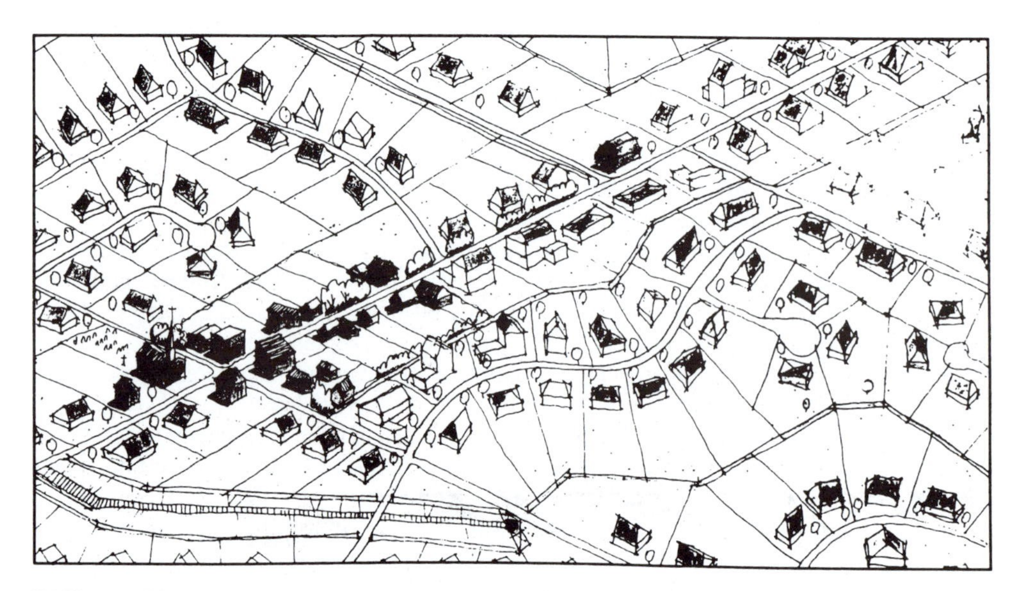

(b) The traditional rural village and countryside typically get swallowed up by conventional subdivisions.

Figure 5.4 The need to protect productive soils and natural features and processes as cities grow involves issues of energy and compact urban form, ecology, soil productivity and rural traditions, and finding new ways of establishing a new kind of urban form that is an integration of city and countryside.

Source: Royal Commission on the Future of the Toronto Waterfront, *Regeneration*, Toronto: Minister of Supply and Services, 1992.

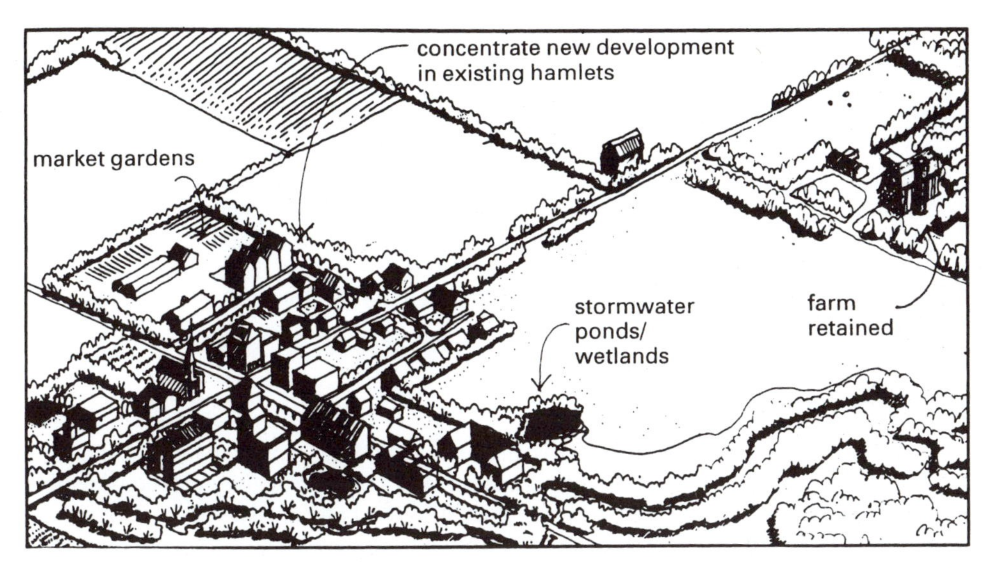

(c) Clustering new development around the village leaves many rural functions, streams and woodlands intact.

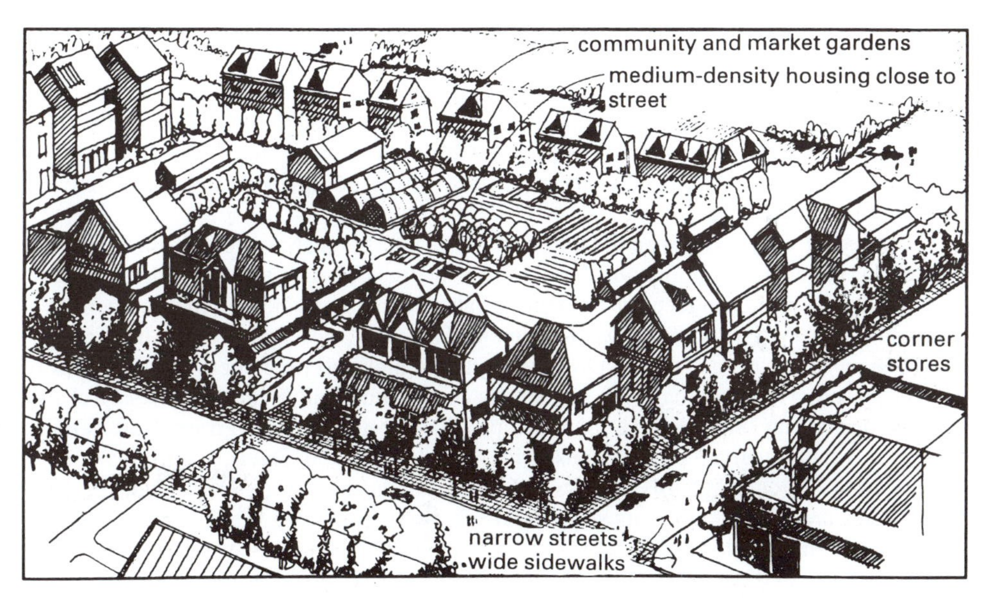

(d) Mixed use development may be associated with small-scale market gardens and other rural based occupations such as plant nurseries, pottery and crafts and recycling operations.

These examples show what long tradition and intelligent planning at the municipal level can achieve – a multi-functional self-sustaining landscape that provides social, environmental and economic benefits. A key concept here is the idea that parks systems can be, at least in part, economically sustaining, contributing in ways other than recreation to the public good. In the context of the realities of poverty and hunger in cities around the world, there are potential social costs of not doing so. The ever-expanding food banks that have come to reflect one of the major issues of contemporary Western society are only a temporary solution and may, in fact, exacerbate the problem by providing no long-term solutions to the major social issues of dependency, lowered morale and self-esteem. Yet there is often public opposition to the notion that parks can, and should, serve such functions. As we shall see on pp. 184–8, these issues have given rise to many community-based initiatives that include the International Federation of City Farms and Gardens.

Food plants and design values

The urban or suburban farmer who is producing food as a part of his livelihood is enriching and maintaining soil fertility, and making land productive with considerable public benefit in produce and amenity value. It is the kind of pleasing environmental quality that we have long accepted in the working rural landscape as a matter of course. But the concept of productive urban land also brings to bear the notion that plants that can be eaten can also have a role in landscape design. The observer of most city gardens cannot fail to notice that few if any have the slightest nutritional value. Conventional landscape planting has no place for cabbages, runner beans, squash or any other edible plant; yet these have aesthetic qualities, too, in texture, form and colour if we stop to look at them with a designer's eye. The consciously designed landscape of the city, as we have seen earlier, is one that has been created by a leisured class which has no need to grow its own food. Food plants, in fact, are usually relegated to the allotment garden. The ornamental flowering crab, cherry and almond replace the plants that used to produce real fruit, ensuring that their purpose in the accepted design idiom remains solely an aesthetic one.

However, the opportunities for using edible plants are just as great as those that are purely ornamental. Examples of the hundreds of trees, shrubs and groundcover plants that produce edible produce for the residential gardener have been well documented in Rosalind Creasy's book, *The Complete Book of Edible Landscaping*.[52] The implications for design, however, go beyond the private garden and into the larger public domain of the city itself. Tree planting along city streets can include fruit-bearing species. Orchards can be grown on unproductive land and provide a return in city-grown fruit. Village Homes, an innovative, energy conserving development in Davis, California, grows apricot and almond orchards and several vineyards in the community's public spaces. Fruit trees are planted along its streets and bikeways, including orange, crab-apple and nut trees, although this practice requires a committed maintenance regime to function appropriately. In the Netherlands old orchards that have been taken over by urban development are integrated into recreation areas. Vines grown on trellises for shade in public places, on buildings, terraces and in gardens may be found in many multicultural communities. They produce crops such as grapes, beans and scarlet runners that have colourful flowers and strong visual appeal.

The aesthetic implications for design also have to do with the very nature of multicultural society discussed earlier in this chapter: It is not about what landscapes should look like, however; it is about process and an understanding that no one aesthetic is better than another. The differences between people of different races and lifestyles are a factor that militates against commonly accepted notions of what is aesthetically appropriate and what is not. Expressions of beauty or symbolic importance to the Chinese community may be regarded as aesthetic kitsch by the designer grounded in the conventions of the European design tradition. The imposition of such aesthetic values on multicultural urban communities raises questions about the role of the designer in the modern city. It requires very different values that have to do, first, with an awareness of the different ways in which people see the world and, second, with a recognition that different cultural groups are the best judges of what is appropriate for them and what is not. What is needed, in effect, is a response to the city's environment that supports its inherent diversity, rather than negating it with outdated aesthetic doctrine; one that provides a framework where different human needs can be expressed in their own ways.

Community action and urban form

The commonly held belief that parks should be provided at public expense by authorities and with little direct public involvement is a legacy of the past that is being re-examined in many cities. This belief dates from the growth of the industrial city in the nineteenth century and the development of the public park by men of social conscience who saw parks as an essential component of urban reform. In England, J. C. Louden's writings advocating public parks contributed to the support by the middle classes of the concept of gardens for the less fortunate. Andrew Jackson Downing and Frederick Law Olmsted in the USA both saw contact with nature as a source of pleasure and benefit to society and a necessary way to improve the cities.[53] Olmsted himself was convinced that:

> the larger share of the immunity from the visits of the plague and other forms of pestilence, and from sweeping fires, and the larger part of the improved general health and increasing length of life which civilized towns have lately enjoyed is due . . . to the gradual adoption of a custom of laying them out with much larger spaces open to the sunlight and fresh air.[54]

Today, we recognize that the nineteenth-century Romantic view of the park as a piece of natural scenery set in the city, but separated from it, for contemplation and spiritual renewal, no longer has the validity it may have once had. The whole physical, technological and social structure of urban life has changed. As I discussed in Chapter 3 (pp. 188–9), changing economic conditions, new demands and attitudes towards recreation and reduced park priorities have made it increasingly difficult to maintain the existing parks systems acquired over more than a century, or to expand them. There is, in addition, a radical shift in the way people view authority and government. Beginning in the late 1980s we have seen a dramatic and worldwide growth of public interest and concern for the environment. A shift in the balance of power in favour of greater public participation in the affairs of the city is occurring – one that is based on positive action, rather than on the negative reaction that was typical of the 1960s and 1970s. The days when expressways, such as San Francisco's Embarkadero, were cutting off waterfronts and creating havoc with its fabric and environment have long gone. The demand for greater control over one's life and surroundings on the part of urban people is reflected in a changing view of parks and the city environment as a whole. New Yorkers feel personally possessive and protective about Central Park. Olmsted himself referred to it as a people's park, his original purpose being to create a public place rather than an aristocratic reserve. This attitude during the 1980s gave rise to a highly vocal public debate with respect to its restoration. As we saw in Chapter 2 (p. 54), in cities around the world, concerned citizens have initiated tree-planting projects in valleys and degraded lands, and the ecological restoration of urban rivers and their watersheds has brought a new era of cooperation between the public and government in achieving long-term environmental and social goals.

The city farms

The determination among people at the grass-roots level to be participants and initiators in decision-making where it affects them and their neighbourhoods, and the increasing incapacity on the part of public authorities to provide the amenities that have traditionally been their responsibility, are creating new conditions in cities. Experience in Britain and North America has shown that the physical decay and social needs prevalent in cities are best tackled by those who have their roots, their families and their futures in the neighbourhood. The development of an alternative type of park or green space, the city farm, was begun in the early 1970s. At that time it aimed to bring derelict land back into use for the benefit of local communities. The city farm used animals as a central part of their activities. The Kentish Town City Farm was one of the first to be launched in 1972 and in 1976 the City Farm Advisory Service was set up with government funding to coordinate and provide assistance to local city farm initiatives.

The successor to the City Farm Advisory Service, the Federation of City Farms and Community Gardens, is a United Kingdom-wide charity. It represents local initiatives that include the city farms, community gardens, a network of school farms, community allotment and community groups involved in park projects. It acts as a support and development organization. Its tasks are to facilitate planning approvals, help obtain funding, and provide expert help and advice to communities. Its primary purpose is to create opportunities for people to enrich and develop their own lives by active participation, provide employment and work experience and make a positive contribution to conservation of the land by encouraging organic farming, and gardening. The Federation receives a £75,000 core-funds grant from the Active Community Unit of the Home Office, and a further £30,000 from the Department for Education and Skills to contribute towards its Youth Development Work. Additional funds come from lottery grants of £100,000, charitable trusts, donations and sponsorship from private companies, as well as 'earned income' from membership fees, room hire, consultancy fees, conferences, and sales of publications.[55] The City Farms are established by independent community organizations, set up and run by local people which allows them to develop programmes that are adapted to the changing needs of their particular communities. They provide knowledge about food growing and caring for animals, and in a range of subjects such as gardening, horticulture and animal husbandry. In addition, a variety of other courses and instruction are offered, depending on the needs of the different communities, such as English as a second language and computer skills.[56] Their income derives from a wide variety of sources, which include local authorities, companies, charitable trusts, and donations and earnings from the farms such as riding school fees.

Contrary to the petting zoos, popular in many cities, which send out the wrong signals about the nature of farm animals and farming generally, there is a basic emphasis on active involvement rather than on observation. There is also a great diversity of activities. Many city farms, for instance, keep a variety of farm animals including sheep, horses, cows, goats, chickens and rabbits. There are garden plots with individuals sharing facilities and communal land management. Many include shops, craft workshops and riding centres, or retain outlets for horticultural produce. Some of the larger and more established farms now have fully fledged community businesses that generate sufficient income to cover extra staffing costs. For instance, the Heeley City Farm in Sheffield has brought investment and created jobs to an area which other businesses have consistently neglected: 83 per cent of the staff were formerly unemployed and 60 per cent live within 1.5 kilometres of the farm.[57] Sanitation and health requirements are monitored by health inspectors who make regular visits to the community.

Developments in the new millennium

At the beginning of the twenty-first century the Federation of City Farms and Community Gardens has grown, in the number of community projects and in the diversity of its programmes. From an approximate total of twenty city farms in 1979, the movement has grown steadily. By 2002 there were sixty-five farms in cities across the country, seventeen of which are located in London.[58] With this phenomenal expansion of community-based initiatives has come changes to, and a broadening of, activities while retaining fundamental community-oriented social goals. Every project is developed specifically for local community needs. For instance, the Culpeper Community Garden in London incorporates forty-eight garden plots that are used by a school, a mental health day service, and an organization for people with learning difficulties.[59] The Kentish Town city farm is the oldest and is, in many ways, a model of these evolutionary changes.

From its early beginnings in 1972, Kentish Town city farm has grown, in area, from approximately 0.8 to 1.8 hectares, and in the scope and complexity of its operations. The buildings on the site were originally industrial, but today the stables have been modernized with new structures that also include a classroom, kitchen and toilets and offices on the upper floor.

The farm animals include sheep, pigs, goats and cows, and there are allotments and family plots. There are also horses and equestrian instruction for hire on Hampstead Heath, all of which have existed here since their early beginnings. But they are only a part of the farm's overall functions. Today, biodiversity and environmental learning have become key factors in the farm experience. Thickets and fruit-bearing shrubs provide diverse habitats on the property for common bird species, butterflies and small mammals; a small bog garden has been planted for examining carniverous plants; herbs abound in various sunny locations. There are large chunks of oak for making seats, and, over time as they begin to rot, for staghorn beetles; there are insect boxes for lacewings; composting; children's mosaics on building walls. The opportunities being made for environmental learning, stuffed into the many corners of this urban site, are endless. The farm has a ninety-nine-year lease on the property with funding from the local authority that provides 80 per cent. The remaining 20 per cent is met through charitable trusts.

The larger neighbourhood of the Kentish Town area has also undergone considerable change, from impoverished vacant land and social need, to recent medium- to high-density housing that has considerable cultural and ethnic diversity. This is reflected in the wide range of educational and recreational opportunities that have emerged in response to community needs. They include classroom activities, courses linked to curriculum targets, in English as a second language, natural sciences and technology, mathematics and skills development. In addition to six full-time staff, an education worker is responsible for these programmes. There are events that celebrate the gardening skills of different cultural groups. Some ninety schools and nursery groups visit the city farm every year, about 80 per cent of whom come from within the larger community.

As an educational resource, the farm itself is diverse and rich in educational experience. There are community gardens where different groups – Bangladeshi, Nigerian, Portuguese – grow food according to their traditions, celebrating different expressions

(a)

(b)

Plate 5.3 Kentish Town City Farm. Activities on the farm have become increasingly diverse, reflecting the needs of the local communities. (a) Making a variety of crafts. (b) Plots for allotment and community gardening.

Source: Mick Magennis, Kentish Town City Farm.

of culture and skill in food growing. There are places for a flock of sheep, and granite rocks for goats and children, a place for plant cuttings where wet and dry compost can be weighed and compared. A number of new sheds have been built since I saw the farm in 1979, for livestock, and for horses that are rented out for riding lessons on Hampstead Heath – a stone's throw away. Children's murals, painted on the shed walls, provide a lively sense of fun, involvement and, often, artistic merit. As a teaching environment, the diversity of the farm has the potential for discussing many topics that include English, mathematics, technology, social development and science.[60]

The city farms concept has, over thirty years, linked community revival and urban renewal in depressed areas of the city, with the growing and production of food that is linked to the restoration of biodiversity, and countless other activities emanating from local communities. The importance of direct urban links to current industrial farming practices, however, is likely to be less relevant than linking government set-aside policies with the restoration of biodiversity and amenity values of the larger countryside. Like many other long-lived initiatives by grass-roots organizations, its success has to do with the vision and commitment of citizens and in the financial and moral support of local government. As we have also seen in Chapter 2, there are parallels with the Don River which suggest that community initiatives are more effective agents of change to sustainable cities than government alone.

As an alternative framework for urban parks, the concept is significant. It returns derelict brownfields land to productive use (discussed in Chapter 7); and through community effort it is self-sustaining, offering facilities otherwise unavailable in inner-city areas. It reduces, therefore, the burden of public expenditure for city parks while at the same time improving health and productivity, which in many housing developments are simply not being provided. In contrast to the normal city park, operating costs are considerably less. Vandalism is reduced to a minimum as a consequence of community action. An interesting comparison was made in the late 1970s between the capital and running costs of a community urban farm and an adjacent London local authority park. While this comparative analysis has not been revised since that time, it still provides an indication of the value that this form of public space provides. The lessons are self-evident. The comparison shows that the community farm is far less costly to build and maintain than the public park under the city authority. It is socially more viable, educationally more pertinent and physically more diverse. This example of a city farm in operation also demonstrates several other things. First, that urban wasteland is a resource out of place and that its creative use does not necessarily depend on high capital investment; second, that the essential connections between community, parks, biodiversity and agricultural productivity are inherently ecologically and socially sustainable when they are brought together. This fundamental concept now has political support in the Mayor of London's bio-diversity strategy[61] and is enshrined in the policy directions that are to be taken for the City. The proposal reads:

> The Mayor will work with partners with expertise in environmental education to improve and secure the long term future of environmental education centres, city farms, and community and cultural gardens throughout London, especially in those parts of London where the need is greatest.[62]

The implications for urban design and management are persuasive. A new approach to our concept of the urban landscape is needed, requiring radical measures to ensure the future environmental and social viability of cities. What we are concerned with is a design philosophy that integrates the ideals of urbanism with nature. This brings us closer to the land and the biological systems from which urban people have been alienated, and gives us the practical tools with which to sustain ourselves in the future. The principles of productivity and diversity, the integration of environment and cultural diversity in the planning, design and management of urban landscapes flow from this philosophy. What logically follows is a view of urban land as functionally necessary to the biological health and quality of life in the city. We need a policy that encourages the creation of both commercial and community gardens, makes productive use of currently wasted energy and land resources, and encourages the perpetuation of self-sustaining urban spaces. Such a policy should also provide real economic benefits to the needy in times of economic depression and high unemployment, and contribute to the maintenance of cohesive and stable neighbourhoods. As working environments, parks must also be seen, to a greater or lesser extent, as economically self-sustaining, providing returns for investment in food and services and contributing to the evolving needs of people.

For many, urban farming is an alternative way of renewing contact with the land and nature through therapeutic and healthy work. For others it is an increasingly necessary method of obtaining food at a reasonable cost. This is particularly significant in developing countries where the imperatives of employment, food, recycling of every scrap of material and economic survival in the burgeoning cities have forced immigrants on to government-owned or wasteland. The concept of productivity has wide implications for urban design. Drawing its inspiration from ecologically and socially based management practice and understanding of human aspirations, it deals with the maintenance of diverse natural and cultural environments. It is from this that its aesthetic inspiration is derived. Urban agriculture also reaches out to the city regions as this affects its potential to shape urban growth, our relationships to the land and to the sustainability of our cities.

The connections between preceding chapters – water, plants, wildlife and urban agriculture – are self-evident. Climate is the great imperative that shapes the land, soils, all forms of life and its adaptations. The next chapter explores this facet of our enquiry at both local and regional levels.

6 Climate

Introduction

The interacting variable forces of wind, precipitation, temperature, humidity and solar radiation are the great climatic forces that have shaped the world's bio-regions, and to which, historically, all life forms including the human race have adapted. Indeed, it may be said that climate, more than any of the natural systems we have examined in this enquiry, transcends all the boundaries of nature and human activities. It pervades and influences water, plants, wildlife and agriculture. It is the fundamental force that shapes local and regional places and is responsible for the essential differences between them. At the same time human settlement has modified micro-climates to suit particular needs and local conditions. Human comfort, and in some cases survival, have depended on the skill with which building and place-making have been able to adapt to the climatic environment. The modern city has had a greater impact on this environment, on living conditions and attitudes, than at any other time. The old arts of creating felicitous outdoor places that take advantage of climatic elements and the material resources of the landscape seem to have been lost. As pressures for energy conservation and the need for civilizing places to live in become more urgent in the last decade of the twentieth century, we must look to environmentally sounder ways of manipulating the climate of cities than the present total reliance on technological systems. My purpose in this chapter is threefold:

- to review the nature of urban climate, and to explore how the outdoors can usefully contribute to urban liveability and conserve the city's energy;
- to bring the various components of natural and human systems into an overall framework for design – seeing the pieces of the puzzle as a whole picture;
- to show how the principle of connectedness, embodied in the influences of local climate, has global implications at every level.

Natural elements and climate

The basic elements of climate – solar radiation, wind precipitation, temperature, humidity – are affected and moderated by the elements of the land, including topography and land form, water and plants. At a macro-scale, land forms create barriers to the movement of air masses. They affect moisture conditions on the windward and leeward sides of hills and mountains. They affect temperatures at different heights of land – temperatures decreasing with altitude. Land forms control the flow and temperature range of air by forming impediments and channels to movement. They create katabatic valley winds that move up during the day and flow down at night, settling in valley bottoms as pools of cold air. South-facing slopes concentrate solar energy and produce different micro-environments from shaded slopes, which affects the growth and patterns of vegetation.

Vegetation controls direct solar radiation to the ground and hence the heat radiated back from ground surfaces. A forest may absorb up to 90 per cent of light falling on it and in general reduces maximum temperature variations throughout the year. It may reduce wind speeds to less than 10 per cent of unobstructed wind and maintain more equitable day and night temperatures than non-forested land. It regulates the amount and intensity of rain reaching the forest floor and affects the deposition of rain or snow and humidity. It reduces glare from reflective surfaces since leaves have a low reflective index.

Water has a profound impact on climate control. Large bodies of water absorb and store a high percentage of solar energy. They heat up and cool down much more slowly than land masses and so act as moderators of temperature on land through the ventilation of onshore breezes. The process of evaporation of water converts energy from the sun into latent heat, reducing air temperatures and acting as a natural air-conditioner.

Urban influences on climate

It is predicted that the energy and resource requirements of cities will, in the foreseeable future, affect not only local, but regional and macro-climates.[1] Global warming, air pollution and related problems are much discussed issues and all originate in cities. A great deal can be done to influence urban climatic conditions locally, however. To do so, we need to examine and understand the influences that affect urban climate.

It is quite apparent that the climate of cities is markedly different from rural areas. Various climatological studies have accounted for five major influences that affect urban climate, based on the fact that energy is the basis for the climatic differences between city and countryside.

- *The difference in materials in urban and non-urban environments.* The impervious surfaces of city streets and paved spaces, and the stone and concrete of building surfaces store and conduct heat much faster than do soil or vegetated surfaces. In addition, urban structures are multi-faceted. Roofs, walls and streets act as multiple reflectors, absorbing heat energy and reflecting it back to other surfaces, so the entire city accepts and stores heat. It becomes, therefore, a highly efficient system for heating large quantities of air throughout its volume. In the countryside, on the other hand, heat is stored mostly in upper layers. In a wooded area the canopy receives and retains most of the heat, while lower levels remain relatively cool. City temperatures are generally warmer than the areas outside. Chandler has found that over thirty years the average temperature in London was several degrees higher than in outlying areas.[2] An illustration of the considerable contrasts in temperature regime of various materials is given by Miess. In relation to a given quantity of energy received, open water is the most constant. Between early morning and midday its increase in temperature may be no more than 3 to 4°C. By contrast, during the same period asphalt may have a temperature increase of 30°C. The temperature of grass may increase by 20°C, but at the same time its temperature drops to much lower levels at night.[3]
- *The much greater aerodynamic roughness of built-up areas than in the countryside.* The arrangement of tower blocks placed individually on their own sites presents a much rougher surface than the open country. This has the effect of

slowing down prevailing winds and increasing localized gusts at street corners and around tall buildings, and diminishing the cooling power of wind in summer.

- *The prodigious amount of heat energy pumped into the city atmosphere from heating and cooling systems, factories and vehicles*. The combination of carbon dioxide (CO_2), chlorofluorocarbons (CFCs), methane and other greenhouse gases used by individual US households per year amounts to the equivalent of 146,800 pounds of CO_2. In terms of greenhouse emissions, this is equivalent to burning nearly 7,500 US gallons (about 345,000 litres).[4] When this is multiplied by approximately 100 million households one gets a sense of the enormous impact that this has on the atmosphere of the planet. In addition, the CO_2 emitted per person is five times the global average.[5] In terms of heating and cooling, in winter large amounts of heat are lost to the exterior, and in summer air-conditioners cooling interior space pump hot air to the exterior, making the problem of high temperatures worse.
- *Problems resulting from precipitation*. Rain is quickly carried away by storm sewers and, in Northern climates, snow is usually cleared from city streets and pedestrian areas. Evaporation converts radiated energy into latent heat, which acts as a cooling process. In the countryside, moisture either remains on the surface or immediately below it. It is thus available for evaporation and cooling. However, in the city, the absence of moisture inhibits evaporation, and so the energy that would have gone into the process of cooling the environment is available for heating, a decisive factor in energy exchange.[6] The capacity of building materials to store heat is greater than that of air by about 1,000 times, and the process of transfer by means of air particles into the atmosphere is much less efficient than evaporation. Only over open water and areas of vegetation does the process of evaporation become fully effective.[7]
- *Air quality*. It is estimated that the major air quality issues of the new millennium are likely to be ozone, particulates and atmospheric carbon dioxide.[8] Increased atmospheric carbon dioxide will likely lead to increased air temperatures and exacerbate ozone problems, with the major emission sources of all three being automobile and industrial processes. Ozone is formed by a photo-chemical reaction of nitrogen oxides and volatile organic compounds in ultraviolet sunlight and moisture, and affects the respiratory tissues and functions in humans. A heavy load of solid particles, gases and liquid contaminants is carried in the urban atmosphere. There are ten times more particulates in city air than in the countryside, which reflect back incoming sunlight and heat, but also retard the outflow of heat.[9] A high volume of particles in the atmosphere reduces the penetration of short-wave radiation in the ultraviolet range. This is biologically important to the production of certain vitamins and the maintenance of health. Motor vehicles are the major source of carbon monoxide in North America and can reduce the oxygen-carrying capacity of blood. The higher the atmospheric concentrations of carbon monoxide, the more serious the health effects.[10] Dilute sulphuric acid emissions are caused primarily by fuel combustion from stationary sources including coal. It is an irritant of the respiratory system which, when breathed in, causes bronchial constriction, resulting in increased respiratory and heart rates. The consequences of air pollution on health have been found, in effect, to be serious, particularly in developing countries where air quality standards may be minimal, or non-existent. A study in Poland, for instance, found new evidence linking air pollution to widespread genetic damage and birth defects. A group of organic compounds in smog caused by burning coal (a major source of energy and domestic heating in the region) called polycyclic aromatic hydrocarbons (PAH) exert a damaging effect on the genetic material in the cells of people breathing air, drinking water and eating food. A comparison

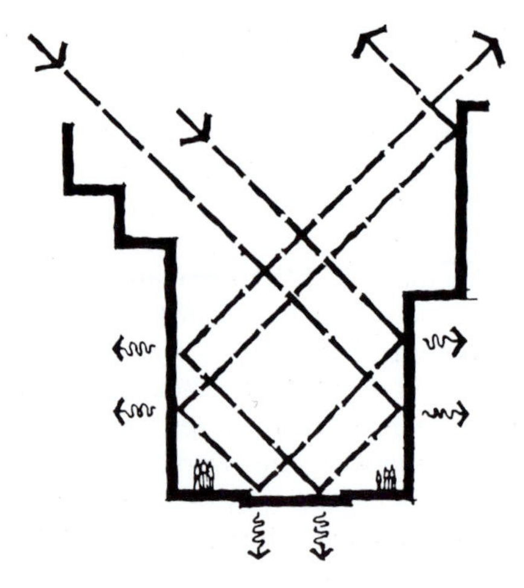

Figure 6.1 In the city, vertical walls reflect solar radiation to the floor and walls of the buildings. Impervious surfaces in walls and floors accept and store heat.

Source: William R. Lowry, 'The Climate of Cities', *Scientific American*, August 1967.

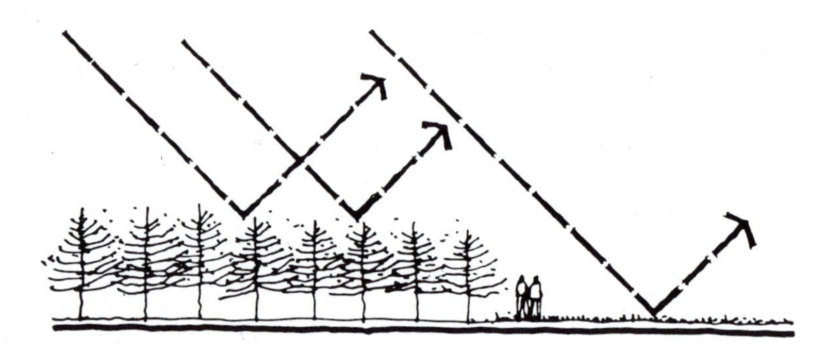

Figure 6.2 In the countryside, solar radiation is reflected back to the sky due to lack of vertical impervious surfaces. Tree canopy retains heat, while lower levels remain cool.

Source: William R. Lowry, 'The Climate of Cities', *Scientific American*, August 1967.

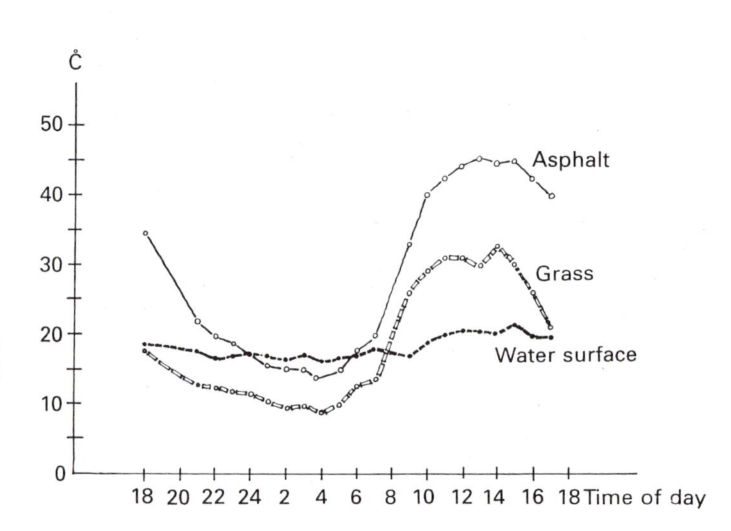

Figure 6.3 Surface temperature of materials. The different physical characteristics of the various surfaces exposed to radiation give rise to a very contrasting temperature regime.

Source: Michael Meiss, 'The Climate of Cities'. In Ian C. Laurie (ed.) *Nature in Cities*, New York, NY: John Wiley, 1979.

between a satellite image of pollution dust in Upper Silesia and a map showing the frequency of cancer in women shows a nearly identical correlation between the two.[11]

The urban heat island

Buildings, paving, vegetation and other physical elements of the city are the active thermal connections between the atmosphere and land surfaces. Their composition and structure within the urban canopy layer, which extends from the ground to above roof level, largely determine the thermal behaviour of different parts of the city.[12] Warmer air temperatures in cities compared to surrounding rural areas are the primary characteristic of the urban heat island. This phenomenon has been studied in some detail by climatologists and is the result of the complex effects of the city's processes on its own climate. Lowry describes it as follows (see Figs 6.5 and 6.6).[13]

Assuming a large city is set in flat countryside with no large bodies of water nearby, the rising morning sun strikes the walls of its buildings, causing them to absorb heat. In the countryside, however, the sun's radiation is largely reflected off the surface with little heat absorbed. As the morning advances the countryside begins to warm up, but the city already has a large lead towards maximum temperatures. The warm air in the city centre begins to rise and gradually a slow air circulation is established with air moving in, rising in the centre, flowing outwards at high altitudes and settling again in the open countryside as it cools. Near midday, temperatures inside and outside the city tend to equalize so that the cycle is weakened. As the afternoon passes, and the sun sinks, much of its radiation is reflected off the countryside but continues to strike building walls directly. Thus the circulation of air is repeated.

During the night the roofs and streets and other hard surfaces of the city begin to radiate heat stored during the day. A cool air layer is likely to be formed at the rooftop level. A stratification of air develops, inhibiting warmer air between buildings from moving upwards. The rural areas, however, cool rapidly at night, due to light winds and unobstructed radiation to the night sky. Although both city and countryside continue to cool during the night, by dawn the city is likely to be 4 to 5°C warmer.

The following day the heat, smoke and gases from the city are contributed to the heat being generated by radiation. The rising air also carries with it suspended particles

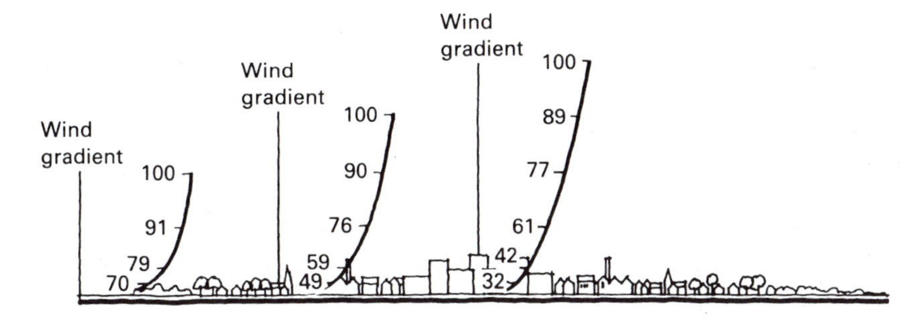

Figure 6.4 Typical wind profiles over built-up area, urban fringe and open sea. Increased aerodynamic roughness of built-up areas causes rapid deceleration of wind compared with open countryside. It has been calculated that wind velocity within a town is half of what it is over open water. At the town edge it is reduced by one-third.

Source: Michael Meiss, 'The Climate of Cities'. In Ian C. Laurie (ed.) *Nature in Cities*, New York, NY: John Wiley, 1979.

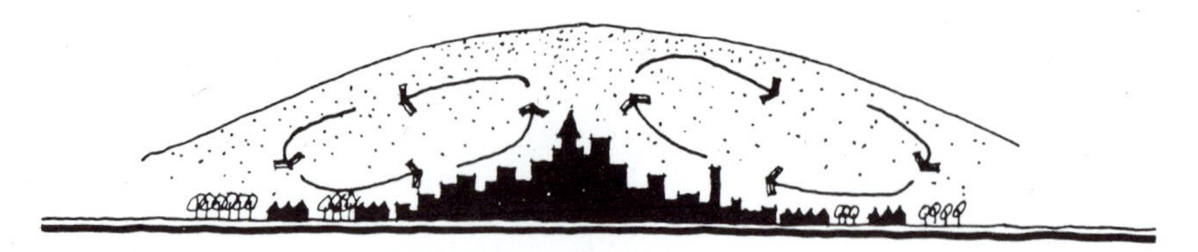

Figure 6.5 The urban heat island. Smog dome over large cities occurs periodically due to urban activities. Air rises over the warmer city centre and settles over cooler environs so that a circulatory system develops. The dome and its effect on city climate may persist until wind or rain disperses it.

Source: William R. Lowry, 'The Climate of Cities', *Scientific American*, August 1967.

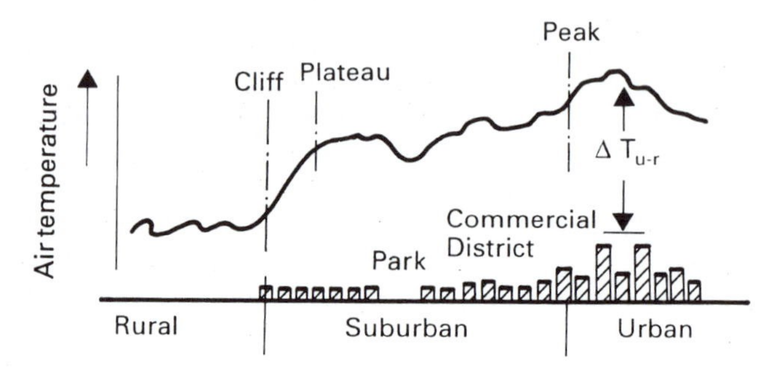

Figure 6.6 Generalized cross-section of a typical urban heat island.

Source: T. R. Oke, *Boundary Layer Climates*, New York, NY: Methuen, 1987, courtesy of Thompson Publishing Services.

of dust and smoke. Over time a dome-shaped layer of haze is formed over the city. At night the particles in the dome become nuclei on which moisture condenses as fog: this fog gets thicker by downward growth and eventually reaches the ground as smog. Smog inhibits cooling of the air and helps to perpetuate the dome by preventing particles from moving out of the system. In the absence of wind or heavy rain, the smog continues to build up. Since less sunshine can penetrate to warm the city in winter, increased fuel consumption adds to the smog buildup.

The process, in effect, is self-perpetuating and is responsible for the severe climatic problems that many cities face. For instance, approximately 3 to 8 per cent of the current demand for electricity for air-conditioning in the USA is used to compensate for the heat island effect because city temperatures have increased by about 1 to 2°C since 1950.[14] The effects of urban heat islands also have wider implications since urban temperatures are increasing worldwide. Comparisons of temperature data from paired urban and rural weather stations suggest that the recent warming trends are due to the heat island effect rather than changes in regional weather.[15]

Problems and perceptions

Mechanical climate control

The search for optimum human climates has been a continual process, particularly in those mid-latitude climates that are less extreme or predictable. Over the past two centuries remarkable changes have been made in the living environment through improvements to mechanical equipment. James Burke describes the chain of events that led to the invention of mechanical air-conditioning, by Dr John Gorrie in 1850.[16] As a physician working in Apalachicola, a small Florida cotton port situated on the Gulf of Mexico, Gorrie had been asked to report on the effects of climate on the population with a view to a possible expansion of the town. Among his recommendations was the need to establish a hospital to treat the fever that sailors and waterside workers endured every summer – an illness that was endemic to the town. Gorrie had noticed that malaria seemed to be connected with hot, humid weather. He began to solve the problem by using ice, circulating the cool air around the hospital wards by means of fans. Since ice was prohibitively expensive, he resorted to a known method for absorbing heat from surrounding gases. He constructed a steam-engine to compress air, which, when cooled, rapidly expanded, and could then be circulated around a room. Subsequently, ice-making and refrigeration machines evolved from this invention and were used to transport food in ships from Australia and also for making German beer. This was followed by the domestic thermos flask and refrigerator for keeping food and drink cold. These inventions became the precursors for the modern air-conditioning unit. One of the first large installations for comfort control of office space was the 300 tonne unit installed in the New York Stock Exchange in 1904.[17] The mechanical climate control of buildings has had a number of fundamental effects on the modern city.

- It has freed buildings from the constraints of weather that were originally imposed upon it. Stylistically, modern architectural form has become an event in its own right, its design responding to the constraints of mechanical engineering rather than to the constraints of site and climate. Modern air-conditioning has permitted the development of the megastructure: great interconnected interior complexes, whose heating, cooling, humidity and daylight are entirely dependent on mechanical systems.
- It has contributed to the radical changes in urban form that have taken place since fossil fuels and other forms of energy have become abundant. The city turns its back on an outdoor environment that has become increasingly unliveable; an environment polluted by dust, smog and exhaust fumes, and alternatively swept by winter winds and cooked by summer heat.
- The preoccupation with internal climate has the effect of denying a climatic role for exterior space. Air-conditioning screens out the products of industrial processes – the chemical pollutants and dust that threaten public health. Unhealthy outdoor climates generate greater reliance on safe, controlled interior ones, and so more and more development provides interior space for urban activities. The subterranean shopping mall and links below the surface of the city is the modern alternative to the open-air market.
- Its effects on lifestyles and perceptions of the environment have been profound. Urban life has become a series of air-conditioned experiences. The home, the office, the school, the bus that takes the children there, the movie theatre, have all been sealed off from the outdoors. It creates a world of its own; separated from the increasing problems of health and comfort in the world outside. It is remarkable how much energy and effort is expended to provide climatic comfort indoors, while at the same time maintaining such unrewarding environments outdoors. At one time it was even proposed that whole cities should be covered by geodesic domes: a suggestion that was seriously discussed for some cities, and which takes the problem a step further to the ultimate technological solution. One of the reasons for the lemming-like flight out of the city on summer weekends is to escape from an oppressive urban climate and the air-conditioning unit for the clean air, breezes and sunshine at the summer cottage.

Plate 6.1 The preoccupation with artificially controlled climate within buildings has ignored the potential for designing with urban micro-climate, making pleasant outdoor places for winter and summer.

As I pointed out in Chapter 1, planning and design doctrine has traditionally been more concerned with conceptual ideologies of built form than with the determinants of natural process. Many early North American towns and institutions, following established planning ideologies, paid little heed to the extremes of climate in the regions in which they were located.

The effort to become independent of the variables of the environment have, by today's standards, been successful. Totally inhospitable places, from the Arctic to the Equator, can now become habitable, with mechanical heating and cooling systems providing uniform interior temperatures. But it has become abundantly clear that the present costs in energy to achieve such a goal are wasteful and, to a great extent, unnecessary. The air-conditioner provides evidence of this fact. The process of keeping cool inside in summer increases the already high temperature outside – a non-productive transfer of heat from one place to another. There are cheaper and more effective ways of achieving similar results. Since the outdoors comprises a large part of the city environment, it can contribute to the modification of climate. At the same time it will be apparent that some problems cannot be solved in the context of natural process alone. Air pollution, for instance, is a macro-problem requiring solutions at the source – the industries and vehicles that create it. This involves many technical and institutional issues that cross regional and national boundaries and are beyond the scope of this enquiry. But there are positive aspects to what may be seen as an environmental problem when urbanism and nature are seen as interdependent. Design, inspired by ecology and laced with a good dose of common sense, provides solutions at less cost and effort.

The vernacular forms of older towns and urban landscapes are examples of adaptation that provide some inspiration and guidance for application today.

Alternative values

Macro-climate, moderated by land form, vegetation and water, has, in various ways, influenced the location and nature of human settlement and uses of the land. Carter has observed that the role of the environment is determined primarily by culture rather than the other way around.[18] There is no doubt, however, that humankind has responded in characteristic ways to climatic influences within its control. To be physically comfortable is a fundamental human need and a comfortable temperature lies in the range of 20 to 24°C. People are affected by climate and react to it automatically even though the response may not be conscious. There is a marked difference in the use of urban places at different seasons. On a winter's day people crowd the sunny side of the street. They seek out spaces protected from the wind. On a hot summer's day they gather where park seats and patches of lawn and trees provide shade or a cool breeze. The manipulation of natural and human-made elements of the environment and solar energy to create felicitous and healthy places to live and work in has preoccupied urban people since the beginning of recorded history.

The business of keeping warm or cool in energy-deficient societies is achieved by the necessity of accepting the limitations of the climatic environment and making the most of its opportunities. This has in the past, and still is in places where traditional technologies are maintained, been done with great sophistication and economy of means. The Maziara cooling jar, for instance, is a traditional water-cooling and purification system, used in rural areas of Upper Egypt and other parts of the world for keeping liquids and perishable foods cool.[19] The action of evaporation absorbs considerable amounts of heat energy (580 calories of energy for every cubic centimetre of water evaporated).[20] Experiments have shown that with air temperatures ranging from 19 to 36°C, water temperature in the Maziara jar remains at a constant 20°C.[21] Cities built in hot, dry climates took advantage of wind for ventilation and cooling. Rudofsky illustrates a

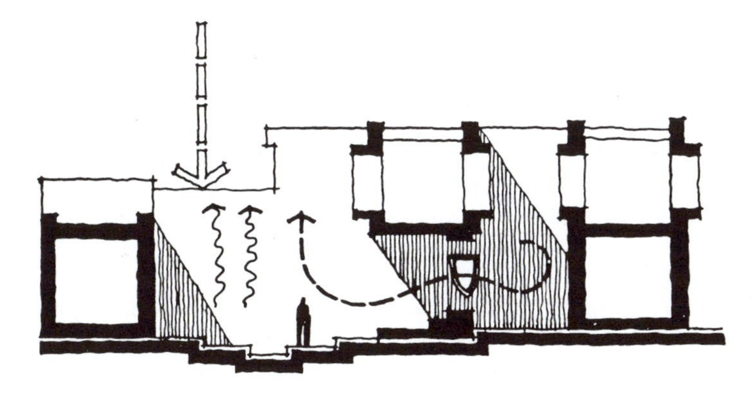

Figure 6.7 The two-courtyard house. In this traditional Middle Eastern design, the deep, shaded courtyard is cool, the large courtyard is warm. The difference in air pressure induces a convection draught from the cool to the warm area. Water-cooling jars placed in the passageway add to the cooling effect of the breeze.

Source: Allan Cain, Afshar Farrouk *et al.*, 'Traditional Cooling Systems in the Third World', *Ecologist*, vol. 6, no. 2, 1976.

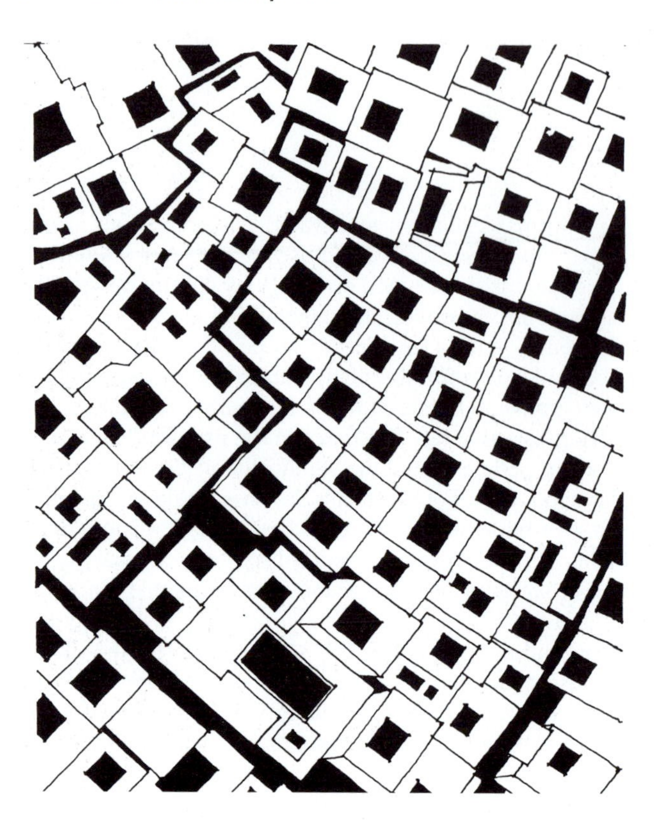

Figure 6.8 Built form of towns in hot Mediterranean climates was often based on courtyard houses closely packed along narrow winding streets to maximize shade.

dramatic example of natural air-conditioning in the Lower Sind district of west Pakistan, where specially constructed wind scoops, installed on roofs, channelled the prevailing wind into every room.[22] Some ingenious passive cooling systems in Iran, described by Bahadori, use wind towers which operate by changing air temperatures and thus its density. The difference in density creates a draught, pulling air either up or down through the tower and through the building.[23] In the county of Kent, in southern England, the oasthouse was at one time built to dry hops. The tapered roof and cowl was designed to swivel in the wind and generate an updraft. These buildings are now being converted into houses, but remain a symbol of a nineteenth-century rural industry characteristic of that county. Courtyard houses and the agglomeration of buildings along narrow streets, typical of Middle Eastern and Mediterranean towns, maintain coolness by trapping cool night-time air and retaining it by day. Many cities in Africa and Spain use awnings or arcades to shield streets from the midday sun. Cities along the Mediterranean coast of North Africa are sited so that their streets, laid out at right-angles to the shore, funnel incoming sea breezes.[24]

Plants and water have long been associated with city courtyards and gardens to provide air-conditioning and places of delight. The Moorish gardens of the Alhambra are a particularly felicitous and well-known adaptation to hot, dry conditions in southern Spain, where the evaporation of water off tiled surfaces and the dappled shade of plants cool its arcaded courtyards. Buildings in cold climates have employed techniques to conserve heat. The traditional Inuit igloo, a perfect expression of adaptation, employs the highly insulating properties of snow and orientation of entrances away from wind to create a habitable micro-climate under the harshest conditions. The hemispherical shape provides maximum volume for minimum surface area, thus minimizing heat loss.

Plate 6.2 A street in Istanbul.

It is said to maintain temperatures of 10°C when temperatures outside are -45°C.[25] The narrow, winding streets, enclosed squares, closely packed buildings and courtyards of the pre-industrial city provide as good a demonstration as any of response to climate and energy conservation. The streets minimized the effects of winds and the open squares trapped sunlight.

In all successful climate control, the siting and organization of built elements and spaces, the use of land form, plants and water has achieved optimum environments for living. Faced with climatic extremes, traditional design methods greatly enhanced urban environments because there was no alternative. Adaptations of these technologies to the modern city are equally necessary if its climatic environment is to be improved. Given the limitations of industrial pollutants, the open spaces of the city perform an important function in the restoration of the energy balance. It is to the resources and the techniques available to achieve this that we now turn.

Some opportunities

Solar radiation and heat gain

The amount of incoming radiation into a city is dependent on its layout and the pattern of its buildings, streets and open-air places. Where sunlight penetrates direct to the floor, for instance, in places that have large plazas and wide streets, radiation is most effectively controlled by vegetation, in particular, trees. The capacity of the forest canopy to absorb large amounts of heat energy is considerable. Short-wave radiation in a closed canopy of maple can be reduced by 80 per cent on a clear midsummer day. The forest can also reduce maximum air temperatures by about 6°C below the temperature in the open.[26] In the city the greater the closure of a tree canopy the greater will be its air-conditioning effect on surface temperatures. Surveys carried out in Germany compared a well-treed square to a comparable area without trees. The daily radiation balance in June of the treed versus the treeless area showed a difference of 256 per cent.[27] Deciduous trees have the great advantage, in climatic regions that suffer from extremes of summer and winter temperatures, of providing shade in hot seasons and permitting the sun to penetrate to the floor in winter.

Wall-climbing vines perform a similar function with respect to south-facing building wall surfaces. While attention may be focused on the considerable area of ground surface in cities requiring shading, it is easy to forget that vertical surfaces vastly increase the area subject to heat gain (see pp. 190, 192). A 1976 German calculation indicated that there was an aggregate of some 50,000 hectares of vertical surfaces in German cities.[28] In energy terms the calculation suggested that vegetation on vertical surfaces can lower the summer temperatures of the street by as much as 5°C. Heat loss from buildings in winter can be reduced by as much as 30 per cent.[29] Biologically, the leaf is an efficient solar collector. During the summer the leaves are raised to take advantage of solar radiation, permitting air to circulate between the plant and the building. It cools, therefore, by means of a 'chimney effect', and through transpiration of the leaves. In winter, the overlapping leaves form an insulating layer of stationary air around the building. Even in climatic regions that are too cold for evergreens to grow successfully, summer cooling may still be an important factor, lending an energy-saving and biological validity to what is generally regarded as a decorative addition to architectural façades.

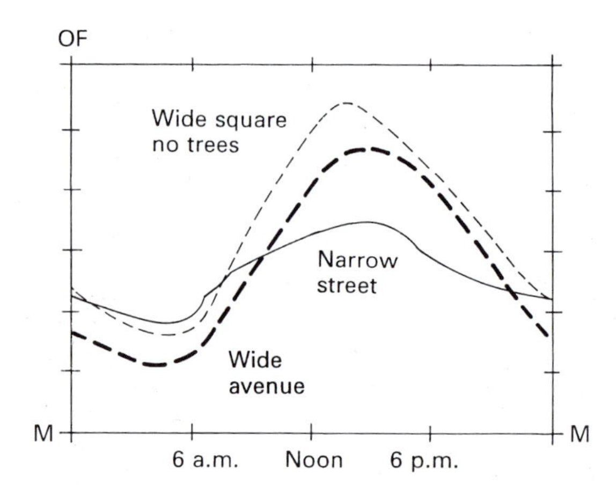

Figure 6.9 Diurnal temperature variation in Vienna, 4–5 August 1931. The graph shows the differences in temperature for a wide square with no trees, a wide avenue with trees and a narrow street.

Source: Gary O. Robinette, *Landscape Planning for Energy Conservation*, Reston, VA: Environmental Design Press for the American Society of Landscape Architects Foundations, 1977.

Plate 6.3 Plants covering building walls do more than look nice. In hot climates they can reduce city temperatures by as much as 5°C: a parking garage in Jakarta, Indonesia.

Rooftops

In dense urban situations rooftops also receive a high level of solar radiation, paralleling the heat of ground surfaces. Rooftops constitute a large proportion of the city's upper-level environment in downtown areas and are, from taller structures, often highly visible parts of the urban landscape. Rooftop vegetation functions in the same way as it does at ground level. Rooftop gardens therefore, can, in addition to reducing stormwater drainage, enhance water quality and create bird habitat (see Chapter 4 on Deptford Creek), and thus perform a multi-functional role in climate control. The limitations to creating this kind of landscape relate to structural support for soil and plants which, while they can be overcome in new projects, may constitute a limitation for the large areas of existing roof spaces in the city if deep soils are a requirement. In cold climates there are also drainage, irrigation, nutrient and frost problems to be overcome. Green roofs, however, present strategies that combat one of the major climatic and visual problems of many urban areas. An examination of many old rooftops will reveal that fortuitous plant communities often gain a foothold. Mosses, grasses and, in places where a small amount of humus and water can collect over time, even adventitious shrubs and small trees colonize these unattended and forgotten places. The issues of fortuitous plant communities have been discussed in detail in Chapter 3 (pp. 112–17), but it is relevant here to pursue their role in climate amelioration. A research programme undertaken by the Parks Department in Berne, Switzerland, on a concrete garage roof, showed that certain plants can be grown on 7 centimetres of soil consisting of pea gravel and silty sand. Various sedum species were used and became well established after a year. Other experiments using grasses and climbers such as Virginia creeper (*Parthenocissus quinquefolia*) have shown that many vigorous plants can adapt to such environments with minimum soil depth or humus content.[30] The city of Dusseldorf, Germany, has regulations requiring large, flat-roofed areas to have roof gardens and Stuttgart actively encourages them (see pp. 215 and 217). These and other examples in Europe suggest that hardy plants adapted to city conditions can survive in hostile environments at very little cost, without adding to the weight of existing or new building structures. The creation of an economical landscape at roof level can contribute to the climatic conditions of the city.

The economic benefits of green roofs have been found to reduce cooling and heating consumption and to provide economic benefits.[31] An energy study for the City of Chicago estimated that peak energy demand would be cut by the equivalent of one small nuclear power plant if all the city's roofs were greened. Reducing energy costs could save $100 million annually. A one-storey green roof structure was found to cut cooling costs by 20 to 30 per cent. In addition, the green roof extends the membrane durability two to three times its life to fifty years or more by reducing winter freeze–thaw cycles. For instance, the Rockefeller Center roof, constructed in the mid-1930s, still retains its waterproofing membranes.[32]

Temperature controls

City temperatures are related to solar radiation and heat gain from urban materials. Temperatures are generally higher than in the countryside. This is accounted for by reduced evaporation, greater conductivity and heat storage capacity of building materials, variations in wind around buildings and the high proportion of airborne pollutants. Temperature can be controlled in several ways.

Water

One of the most effective ways of controlling local climate is through the evaporation of water into the air, particularly in dry climates. This is achieved in several ways: by direct evaporation from open water and by the evapotranspiration of plants. The high run-off coefficients of paved surfaces and the efficient removal of surface water by storm sewers have effectively removed its availability for evaporation and cooling. This function is greatly assisted by reintroducing water into the city. It occurs by design where pools, fountains or artificial lakes are built into city landscapes. But it also occurs fortuitously after rainstorms where water forms ponds in parking lots, low-lying grassed areas and other 'poorly designed' places, where it becomes freely available for evaporation. Surface water introduced into the city's paved and unpaved places by impounding rainwater serves an important climatic function, as well as restoring hydrological balance, and providing recreation, wildlife and aesthetic enhancement. There are also pollution and erosion control advantages in doing so that have been discussed in more detail in Chapter 3 (pp. 107–10).

Plants

Plants evaporate water through the metabolic process of evapotranspiration. The cycle of water is carried from the soil through the plant and is evaporated from the leaves as a part of the process of photosynthesis. It has been estimated that on a single day in summer, 0.4 hectares of turf will lose about 10,800 litres of water by transpiration and evaporation.[33] The transpiration of water by plants helps to control and regulate humidity and temperature. A single large tree can transpire 450 litres of water a day.

This is equivalent to 230,000 kilocalories of energy in evaporation which is rendered unavailable to heat surfaces or raised air temperatures.[34] Federer has compared the effectiveness of this evaporation by a tree to air-conditioning. The mechanical equivalent to the tree transpiring 450 litres a day is five average room air-conditioners, each at 2,500 kilocalories per hour, running for nineteen hours a day.[35] He also points out the important fact that the air-conditioner only shifts heat from indoors to outdoors and also uses electric power. The heat is therefore still available to increase air temperatures. But with the tree, transpiration renders it unavailable.

It will be obvious, therefore, that in energy terms a tree shading a house is more effective. It produces no unwanted waste products from the process of cooling, uses no electric power and continues to work better and better over the life of the tree. A numerical simulation of urban climate has suggested that where at least 20 per cent of an urban area in mid-latitudes is covered by plants, more incoming solar radiation is used to evaporate water than to warm the air.[36] These facts are borne out in Davis, California, where temperatures can reach 38°C. Various studies have shown that neighbourhoods with shaded narrow streets can be as much as 6°C cooler than those with unshaded streets. Moreover, a neighbourhood that is 6°C cooler uses only half the amount of electricity for air-conditioning than an unshaded neighbourhood.[37] Based on these findings, in 1977 Davis City Council unanimously passed an ordinance requiring a minimum of 10 per cent of a paved parking lot to be canopied by trees within fifteen years of the building permit's issuance.[38]

Climatic factors also determine management objectives. The ability of trees to act as sponges of carbon dioxide in the face of increasing concern and debate over global warming has greatly enhanced the attractiveness of urban tree planting. Analysis of the urban forest in Oakland, California, reveals that it currently stores approximately 145,000 tonnes of carbon.[39] Future growth and planting of trees can add to that storage total if the amount of carbon sequestered due to growth and planting remains greater than the amount of carbon lost due to mortality. In addition, it has been calculated that large-scale planting and the use of light-coloured surfaces in cities have the potential to conserve about 2 per cent of the total production of carbon in the USA.[40] The landscape value of trees in raising housing values or in making light industrial areas more saleable is also a significant factor.[41]

Wind

Of all the influences the city has on weather, it is the presence or the absence of wind that has the greatest impact on the comfort of the local climate, as anyone who has walked the streets of a Canadian mid-western prairie city in winter will agree. There is less wind on average in cities than in the open countryside. Meiss states that within a town, wind velocity may be half of what it is over open water.[42] On the other hand, the existence of free-standing towers separated by large open areas and the general layout of streets speed up winds locally, creating the unpleasant, gusty, windswept conditions and drifting snow that are typical of a winter's day. Wind affects temperatures, evaporation, the rate of moisture loss and transpiration from vegetation and drifting snow; all of which are particularly important to local micro-climatic conditions.

The impact of the wind environment around tall building complexes that have become typical of many suburban development projects is well known. The generally uniform low building layout of older towns, arranged along curved streets, provides shelter – the result of lower wind gradients at ground level. When winds meet a building that is considerably taller than its neighbours, the flow patterns change. Air currents divide at about two-thirds or three-quarters of the building height, creating a down draught on

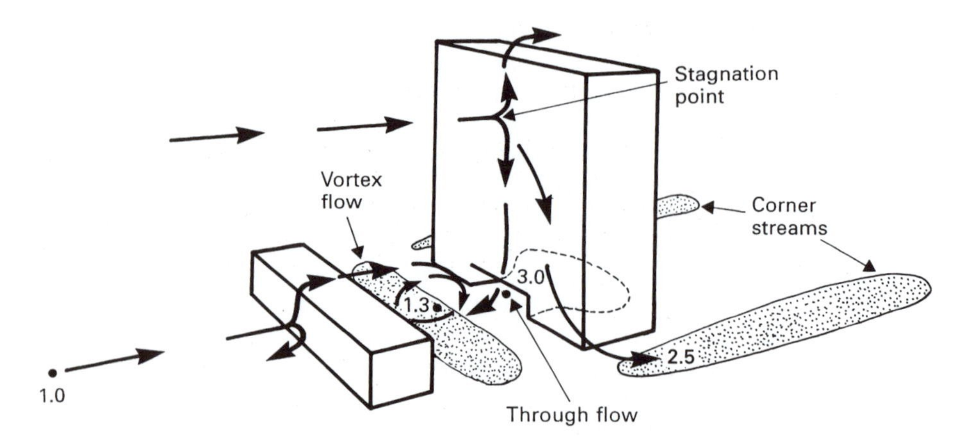

Figure 6.10 Air flow around buildings creates unpleasant and sometimes unsafe conditions in cities.

Source: T. R. Oke, *The Significance of the Atmosphere in Planning Human Settlements*. Figure 4 (page 35) 'Airflow in the vicinity of a tall building with smaller buildings upwind'. From E. B. Wiken and G. Ironside (compilers and editors) *Ecological (Biophysical) Land Classification in Urban Areas, Proceedings of a Workshop*, November 1976, Toronto, ON. Ottawa: Canada Committee on Ecological Land Classification, Ecological Land Classification Series, No 3, Lands Directorate, Environment Canada.

the windward face, and highly turbulent conditions at ground level. Reduced air pressure on the lee side creates suction and high wind speeds around corners and through passageways under the building. Oke has found that since the force exerted by the wind increases as the square of its speed, a threefold increase of speed is associated with a ninefold increase in force; sufficient to knock down passing pedestrians.[43]

Many other studies on the effects of forest cover and shelter belts on winds have been made with respect to the speed of air movement, the protection afforded and the effect of wind barriers on heat loss from buildings. Tree stands are effective in slowing wind; the greater the roughness of the ground surface the more wind velocities are reduced. The smaller an open forest clearing, the less turbulence at ground level there will be. Shelter belts may reduce winds by 50 per cent, the sharpest reduction in wind velocities extending ten to fifteen times the height of the trees on the lee side.[44] According to Olgay, a 32 kilometres per hour wind can double the heat load of a house normally exposed to 8 kilometres per hour wind.[45] The agricultural experiment station at Kansas State University has shown that the heating load on a house can be greatly reduced with the use of windbreaks.[46] Measurements of the effect of windbreak planting around unprotected residential buildings indicate that annual heating costs can be reduced by 10 to 30 per cent.[47] The Ontario Ministry of Housing has also calculated that landscaping around residential buildings in the temperate climate of southern Ontario is capable of producing energy savings in excess of 5 per cent.[48]

Air pollution control

As I mentioned earlier, the only satisfactory controls for the solid particles, gases and other airborne contaminants carried in the city's air are institutional and technical. Much improved air conditions have been achieved in Great Britain due to tight air pollution control laws and regulations. At the same time it is evident that a return to the use of coal would again aggravate the air pollution problem. There is also little doubt that air pollution from vehicular emissions will continue to exist for many years as a reality of urban climate. Where air pollution is dilute, however, an important environmental control device is plants (previously discussed in Chapter 3). It has long been known that plants filter dust in cities. Ongoing research in plant physiology suggests that they do more than act as filters. The surface area of a tree has evolved to maximize light and gas exchange. Trees have ten times the surface area of the soil on which they stand. A hundred-year-old beech, for instance, has been estimated to have some 800,000 leaves, and a leaf surface area of 16,000 square metres per 160 square metres of tree base. Calculations show that the intercellular spaces of leaves (the sum of cell walls) increase the total leaf area to approximately 160,000 square metres.[49] Leaves can take up or absorb pollutants such as ozone and sulphur dioxide to significant levels. Urban vegetation can mitigate ozone pollution by lowering city temperatures and directly absorbing the gas.[50] It has been demonstrated that a Douglas fir with a diameter of 38 centimetres can remove 19.7 kilograms of sulphur dioxide per year without injury from an atmospheric concentration of 0.25 parts per million (ppm). By way of illustration of the effectiveness of trees in removing sulphur dioxide, it shows that to take up the 462,000 tonnes of sulphur dioxide released annually in St Louis, Missouri, it would require 50 million trees. These would occupy about 5 per cent of the city's land area.[51] Measurements taken in 1962 in Hyde Park in London indicated that the concentrations of sulphur dioxide were reduced, partly due to the local circulation of air generated by the vegetation, and partly due to the uptake of gas by the leaves (see also pp. 107–10).[52]

Soil micro-organisms are more effective than vegetation in removing carbon monoxide and assist in the conversion of carbon monoxide to carbon dioxide. It is believed that oxygen released by roadside plants may help in lowering carbon monoxide levels along heavy traffic routes. Nitrogen oxide combined with gases such as oxygen produces nitrogen dioxide, which is then absorbed readily by vegetation.[53]

Vegetation also collects heavy metals. In New Haven, Connecticut, one researcher found that a sugar maple 30 centimetres in diameter removed 60 milligrams (mg) of cadmium and 140 mg of lead from the atmosphere during one growing season.[54] This suggests that vegetated spaces can provide areas where dust can settle out and where air pollutants are diluted. However, it is evident that plant damage occurs when pollutants are excessive. Schmid points out that the severity of plant damage is complicated by many factors, such as the age of the plant, its state of nutrition, moisture when exposed and other factors.[55] Plant species also vary in their tolerance to air pollution and their effectiveness in improving air quality. In effect, plants cannot be regarded as the panacea for ameliorating air pollution problems, but they do assist air purity and serve one other important climatic function – as indicators of air pollution and thus of the health of the people who live in cities. There are therefore highly valid reasons for the reforestation of urban areas in the planning and design of the urban environment.

Energy

While the larger issues of air quality control are beyond the scope of this enquiry, it is now conventional wisdom that a major threat to human and non-human health in cities derives from vehicular emissions. An analysis of the energy efficiencies of various transportation options in terms of the amount of energy used in British thermal units (BTUs) to transport a passenger 1.6 kilometres in Canada demonstrates that the more efficient freight transportation modes are the most frequently used. However, in passenger transportation, the most heavily used mode – the car – is the least energy efficient. This is true for both urban and inter-city transportation.[56] A report on the future of urbanization by the Worldwatch Institute has shown that inner-city residents of New York use only one-third of the gasoline of residents living in the outer regions of the tri-state

Table 6.1 ***Urban density and gasoline consumption in major cities, 1980***

City	*Gasoline consumption per person (megajoules)*	*Overall population density (people per hectare)*	*Inner-city population density (people per hectare)*	*Overall job density (jobs per hectare)*	*Inner-city job density (jobs per hectare)*	*Private automobile travel (per person, kilometres)*
American cities[1]	58,541	14	45	7	30	12,507
San Francisco	55,365	16	59	8	48	13,200
Chicago	48,246	18	54	8	26	11,122
Australian cities[1]	29,849	14	24	6	27	10,680
Melbourne	29,104	16	29	6	40	10,128
Sydney	27,986	18	39	8	39	9,450
Metro Toronto	34,831	40	57	20	38	9,850
European cities[1]	13,280	54	91	31	79	5,595
Frankfurt	16,093	54	63	43	74	6,810
Stockholm	15,574	51	58	34	62	6,570
Paris	14,091	48	106	22	60	4,199
Asian cities[1]	5,493	160	464	71	296	1,799

Source: Robert Paehlke, 'The Environmental Effects of Intensification', prepared for Municipal Planning Policy Branch, Ministry of Municipal Affairs, 1991. Based on Canadian Urban Institute, *Housing Intensification Policies. Constraints and Challenges*, Toronto: Canadian Urban Institute, 1990, p. 13. Compiled from material originally presented in Peter Newman and Jeffrey Kenworthy, *Cities and Automobile Dependence: An International Sourcebook*, Hampshire: Gower Publishing, 1989.

Notes:

1 The figures given for American, Australian, European and Asian cities in the table are average for the cities in those regions studied by Newman and Kenworthy. The data reflect results from ten American, five Australian and twelve European cities. The Asian data are for the three 'Westernized' Asian cities of Tokyo, Singapore and Hong Kong.

2 *Interpretation*: The extent to which people choose to use private passenger cars as a direct function of urban density. Cities which are more compact are significantly less dependent on automobiles. Residents of US cities drive their cars, on average, two of three times as far as do residents of European cities. Toronto is about mid-way between a typical American and a typical European city in this regard.

metropolitan area of New York, New Jersey and Connecticut; Manhattan residents use on average only 400 litres of gasoline per capita each year, a consumption level close to European cities.[57] The relationship between gasoline consumption and urban density in major cities around the world is illustrated in Table 6.1 and clearly demonstrates the implications of current urban form on energy use. For instance, Houston, Texas, with approximately eight persons per hectare, uses four and a half times as much gasoline as Copenhagen with a density of thirty persons per hectare. At the extreme end of the scale, Houston's gasoline consumption is over five times that of Hong Kong with 300 persons per hectare.[58]

Thus, gasoline consumption and air quality are essentially linked to urban form in the low-density suburban and fringe areas of the city, and reinforce the dominant role of the automobile in Western society.[59] It also has radical implications for such issues as the consumption of agricultural land, damage to habitat, biological diversity and depletion of non-renewable energy sources, and the need for producing alternative and more compact urban development patterns (see Chapter 7).

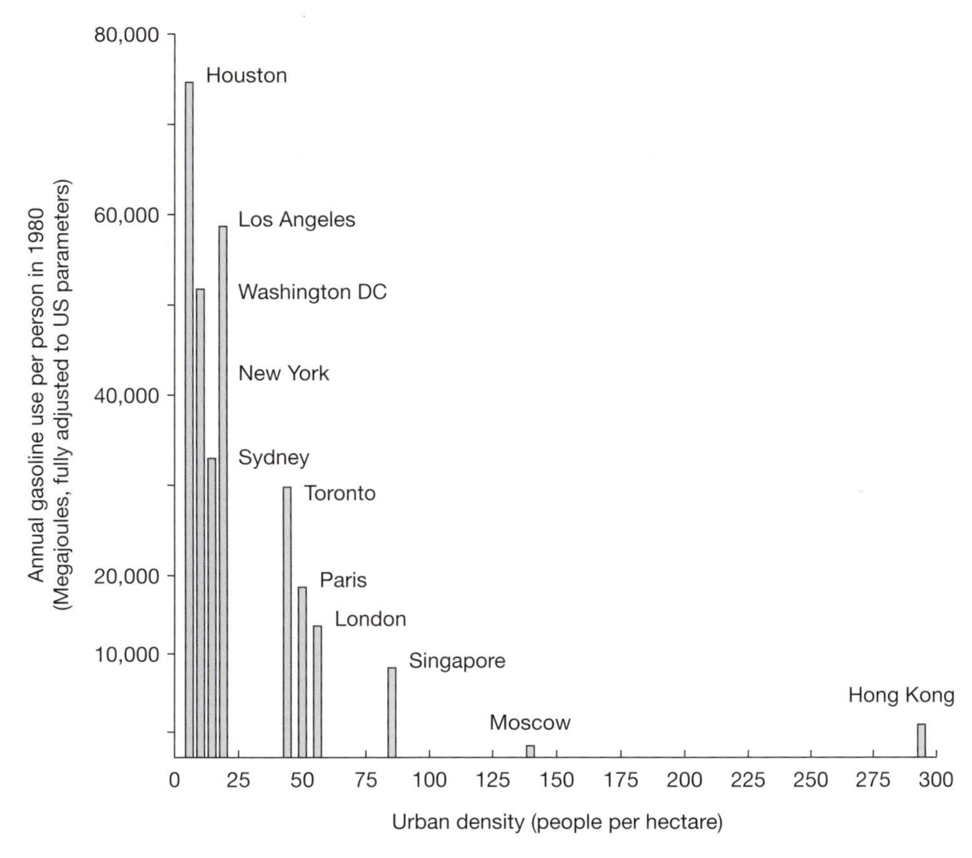

Figure 6.11 Gasoline consumption and urban densities in major world cities. The relationship between increasing densities and reducing energy consumption is clearly evident.

Source: Adapted from Robert Paehlke, 'The Environmental Effects of Intensification', prepared for Municipal Planning Policy Branch, Ministry of Municipal Affairs, 1991, after Peter Newman and Jeffrey Kenworthy, *Cities and Automobile Dependence: An International Sourcebook*, Hampshire: Gower Publishing, 1989.

Implications for design

Green lungs

We have found that water and plants are the important natural elements of climate amelioration in the city. But what is their sphere of influence? How much vegetation or surface area of water is needed to have a marked effect on the climate of the city? Where should they be located? Answers to these questions depend on the climate of the region, the nature and variations of climate between one place and another, the characteristics of the site, its topography and the nature of the built-up area. On a small scale, experience tells us that a sheltered, well-treed environment is a cooler and more pleasant place on a hot day. The sphere of influence of its elements – tree canopy, shrubs and water – would appear to be local, however, particularly if it is an isolated place in the general fabric of streets and buildings. The walled gardens of old European towns create a sphere of influence within them that is considerably cooler, or warmer, than outside. The impact of major spaces, the so-called 'green lungs' of the city, that have long been the ideal of landscape planning, may well be limited in the overall urban climate. Jane Jacobs has observed that the term 'green lung' is applicable only to the park spaces themselves and has little effect on the overall quality of the air in the city.[60] There have been claims that the oxygen produced by vegetation can affect the balance of oxygen in the air. These have evolved from the fact that much more oxygen is produced in photosynthesis than is used up in respiration. There is, however, as much oxygen used up in plant decay and the metabolism of animals that feed on plants as there is released by photosynthesis. The concept of the 'green lung' may be inaccurate in describing the effect of parks on the oxygen content of the city overall, but research has established definite connections between forest vegetation cover, open space distribution and urban climate control.

Meiss observes that the necessary amount of open space in an urban area and its optimum distribution cannot be stated quantitatively.[61] But from a climatic point of view, a fine mesh of small spaces, distributed evenly over the whole city, is more effective that reliance on a few large spaces. These latter areas need to be supplemented by a large number of small parks throughout the built-up area. 'Such a mesh facilitates horizontal exchange of air bodies of varying temperatures, and consequently a balance is reached more quickly and with less resistance.[62] A study of Dallas and Fort Worth found that the heat island reached its peak not in central areas of tall buildings, but along the fringes of the downtown area that contained low buildings and parking lots.'[63] This suggests that efforts to ameliorate climate should be concentrated in these areas. Bernatzky suggests that the effect of the heat island can be partially counteracted by concentric rings of vegetated space to filter and oxygenate the air as it moves inward to the city centre.[64] In summer the heat from increasing densities in the city centre heats air which rises, creating a low air pressure area. This in turn draws cooler air in from the edges of the city. The quality of the air as it moves into the city is increasingly degraded, accumulating gases and particulate matter and being progressively de-oxygenated. Parks and vegetated spaces located in the path of this moving air will alter wind flow, ameliorate air quality and reduce temperature. In Chicago, air flow modelling has concluded that a finger plan with corridors of development and wedges of open space would have the most positive effect on air quality.[65]

Some design principles

There is still much to be learned about the effects of these natural processes on the city-wide climate, and how the scientific data that have been accumulated may be applied in determining optimum patterns of open space. Researchers in the field have varying views on what those patterns might be. We must, however, be careful to avoid the trap that often awaits those seeking answers to such problems – the temptation to create cookie-pattern solutions for every urban situation. This, in fact, is precisely the criticism that has been levelled at many planning theories in the past that have attempted to seek standard solutions for the cities. They ignore the inherent individuality and uniqueness of each place and, therefore, of each city. Thus, while there can be no final or definitive solution to the questions that have been explored here, certain general principles do emerge that are both pertinent and useful to this enquiry.

- The natural patterns of the land, its mountains, hills and valleys, its rivers, streams, open water, forest and grassland, determine local climatic patterns and affect, in some measure, the environment of the city. Although the extent of this influence may be local, the retention and enhancement of natural features for climatic reasons are essential parts of the design process. An example of where climatic patterns, conditioned by topography and vegetation, have had a major impact on city form is the case of Stuttgart, Germany, described later in this chapter.

- Vegetation and water have a major effect on the maintenance of an equable micro-climate within cities. Since the large areas of paved and hard surfaces in the city generate the greatest heat in summer, establishing canopy vegetation wherever possible will reduce the adverse effects of the urban heat island. It will also remove dust and purify toxic gases and other chemicals. Dense canopies that provide maximum shade are more effective than much current practice, which sees trees as individual specimens. In addition to micro-climatic benefits, the continuity of the tree canopy also provides connected environments for some species of wildlife. In South Africa many parking areas are shaded by canvas canopies, to protect vehicles from the summer sun, and also as a protection against frequent and damaging hailstorms. The retention of water and natural ponds in parks and other locations is also crucial to restoring the energy balance by direct evaporation, and they serve many other hydrological and wildlife functions which have been discussed in previous chapters.
- The large roofscape areas of downtown and suburban industrial sites contribute to heat buildup. The development of rooftop planting serves a similarly important role in climate amelioration (discussed previously in this chapter). There is a need for basic practical research into lightweight low-maintenance techniques for establishing plants on existing rooftops. The use of naturalized urban vegetation that can survive with little or no care, and under the most severe conditions, has important implications for climate control.
- In most successful examples of climate manipulation in extreme conditions, the emphasis has been on an urban texture of small spaces and low buildings. In hot summer climates, narrow streets and small living spaces increase shade and reduce the buildup of solar radiation. They are, in addition, easier to control artificially through the use of orientation, shaded canopies, arcades, plants, water and ventilation. In cooler climates they are less subject to cold winds, drifting snow and extremes of temperature. Organization of space can create suntraps by orienting buildings to the south and excluding winter winds. In most cities where built form has evolved with little regard for climatic considerations, the materials of the landscape, as well as new built form, must serve a climatic role. Plants, ground form and building canopies, must be used to create suntraps and sheltered places, to counteract prevailing winds, modify down draught and so on.

The response to urban climate is the first step in the establishment of a vernacular; of a regional character that links built form with the place in which it occurs. Without this response, the designed landscape may be as clearly identified with the now defunct international style as was the architectural style that gave rise to the name, but which none the less continues under different titles. Several contemporary examples of form evolving from climatic determinants are illustrated in the following pages.

Example 1 Climate shaping urban form: a prairie university

In the late 1960s and early 1970s the University of Alberta underwent a major expansion programme to accommodate almost a doubling of its student population. Located within the city of Edmonton, Alberta, the university is very much a part of an urban environment that had one of the fastest growth rates in Canada. One of the most crucial objectives of the development plan was to create an appropriate campus environment for living and working, since the period of greatest activity in the academic year takes place during the severe winter months.

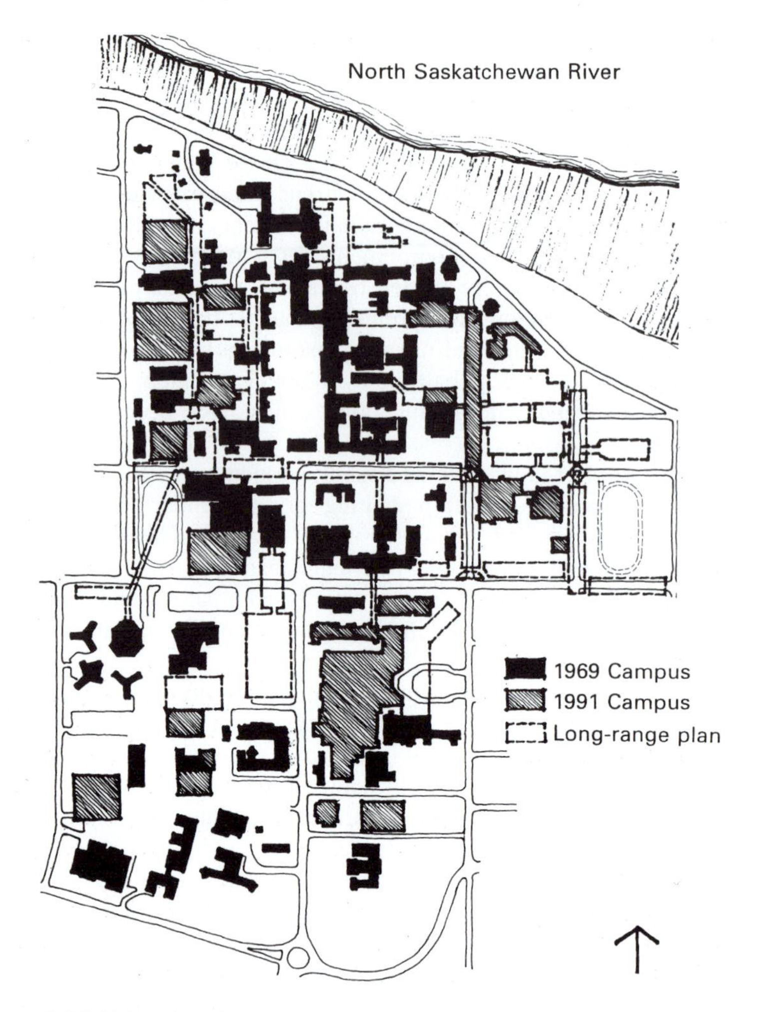

Figure 6.12 University of Alberta campus 1969–91 with possible long-range plan.

Over the years the evolution of the campus had been dictated by conventional, or haphazard, planning. Tall buildings were placed formally in the landscape, isolated from each other by large open areas of unusable space, and reflecting prevailing attitudes and nostalgia for wide open spaces and a rural prairie setting. The inappropriate nature of this model for the climate and environment of the prairies was shown by the way people moved around the campus; cutting through buildings not designed for the purpose in an effort to avoid the unpleasant conditions created by unimpeded winds, gusting around high-rise structures and permanent shade cast by them. An examination of building coverage revealed that less than 15 per cent of the campus was occupied by buildings. The remaining 85 per cent was taken up by parking lots, roads, service areas, manicured turf and residual space that made much of the campus unusable and inhospitable.

An examination of climatic data shows that Edmonton is situated 53.35° north in a region of cold temperate climate. The coldest time of the year, and the period of least sunshine (100 hours or fewer per month between November and January), occurs during the academic year. While prevailing winds are from the south, the highest velocities

Figure 6.13 Design principles. Connecting interior street and vehicular/pedestrian separation with housing over.

Source: A. J. Diamond and Barton Myers, 'University of Alberta, Long Range Development Plan', Consultant report, June 1969.

occur from the north-west. During these periods, the combination of wind and cold temperatures creates, on occasion, a wind-chill factor of more than -45°C, enough to freeze flesh in less than 1 minute. In addition, sun angles at the winter solstice (22 December) are only 13 to 15° at noon. The two most crucial design factors in cold climates, then, are wind and sun. The absence of wind and the presence of sunshine in the shaping of spaces can make the outdoors a very pleasant experience even when temperatures are very low. The development plan for the university, therefore, incorporated the following climatic criteria which became the guiding principles for its growth.

The acquisition of additional land for campus expansion along conventional lines was rejected in favour of a policy of infilling within existing campus boundaries. Land coverage by buildings rose from 15 per cent to 34 per cent and greatly increased the efficiency and economy of land use. Part of the increased coverage took the form of parking structures on many levels to reduce the area taken up by surface parking and to increase usable and contained outdoor spaces. Infill buildings incorporated interior pedestrian streets linking academic departments, housing, food and recreation functions that provided alternative sheltered pedestrian routes during the winter months. The concept of infill and linkage also reinforced the notion of a high mix of uses, accessibility to services and the social integration that isolated university buildings tend to discourage.[66]

- Since the greatest need was to maintain maximum winter sunlight, development was designed to avoid tall structures on the south side of outdoor spaces. Alternatively, indoor/outdoor activity areas could be located on the south side of buildings.
- A limit of four to six storeys was recommended for most new building to reduce down draughts and gusting and increase available sunlight in winter.

These planning principles set a framework for outdoor space that could respond to basic design principles for climatic control at a micro-climatic scale. They included:

Plate 6.4 A courtyard associated with the Student Union provides shelter from wind in winter and gives shade in summer. The courtyard is busy with people during the day and evenings, and has been found to be a safe place.

Plate 6.5 Student Union building. Linked buildings provide interior routes during inclement weather. Outdoor areas function as alternative routes and as gathering places all year round.

Source: Andrew Beddingfield.

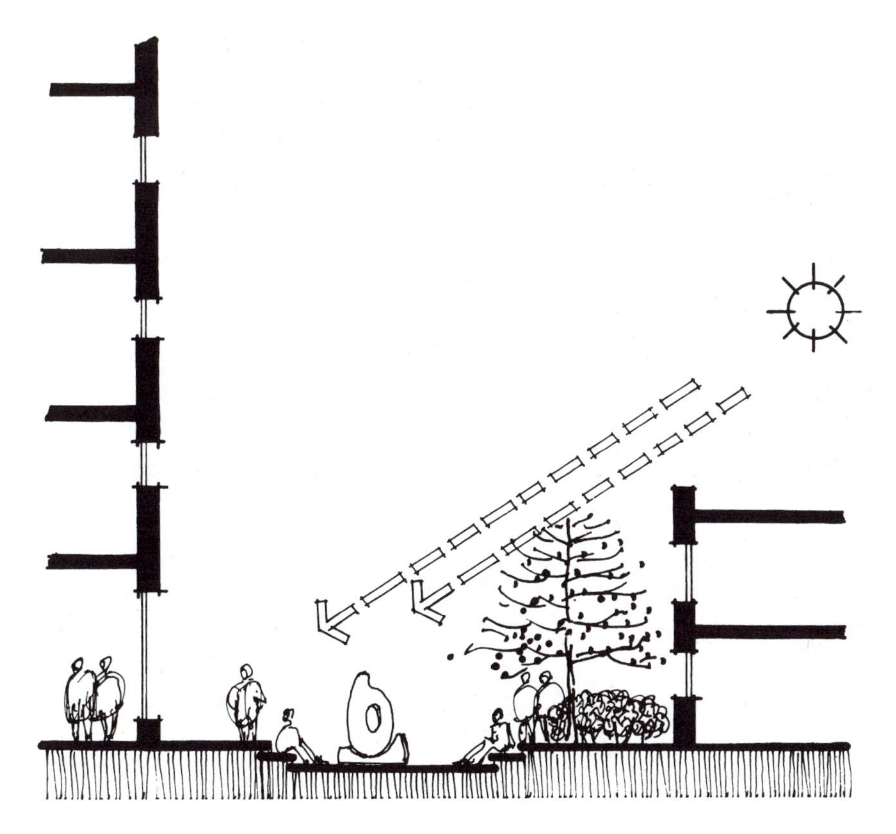

Figure 6.14 Small, well-defined and protected courtyards.

- the creation of small, well-defined, protected courtyards and places for passive use, densely planted to create protected winter environments;
- the use of raised land forms, screens and planting around entrances, sitting and meeting places, to reduce wind chill and create suntraps;
- the use of coniferous windbreaks along pedestrian routes and surrounding large spaces such as athletic areas;
- dense planting or screens at narrow building openings and connected spaces where venturi winds are normally generated;
- reduction of glare from snow and creation of visual interest by the addition of plants and sculptural elements in the winter landscape;
- deciduous tree canopies that provide shade and temperature reduction over paved surfaces in the heat of the summer while allowing full exposure to the winter sun. Dense deciduous canopies also contribute to the filtering of wind from above where a tall building creates down draughts.[67]

Since the 1970s, further phases of the long-range plan have been completed. The densely planted courtyard spaces have matured and the climatic strategies for the newly formed courtyards have begun to fulfil essential functions – creating benign winter and summer environments, active university spaces and the urban design form needed for a true response to the climate of the region. A new subway stop has been located adjacent to the HUB building that has created high activity in the central campus. The university has begun to develop an urban quality for winter and summer conditions that reflects

Figure 6.15 Walkways protected by windbreaks in large open areas.

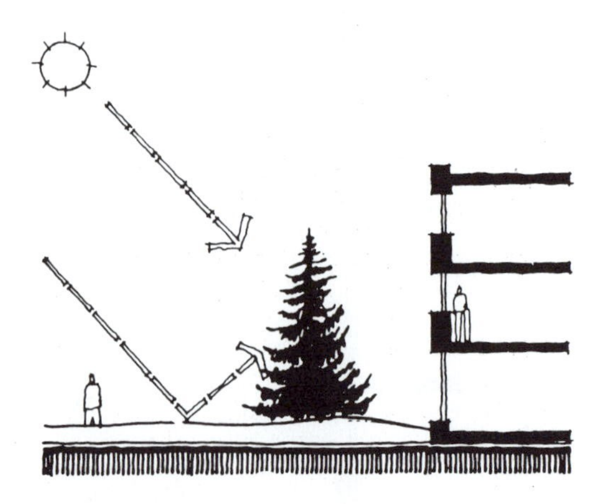

Figure 6.16 Snow glare and intense light from low sun may be reduced by the use of coniferous or dense deciduous plants.

the climatic environment of this prairie region. It should also be noted that design with climate has the greatest significance for sustainability at the scale of urban design, when individual buildings can be integrated within the larger city context.

Example 2 Wind flows shaping urban form: Stuttgart

The links between major land forms, wind, vegetation and urban form are nowhere better seen than in the city of Stuttgart. Located in the centre of an industrial region the city has 590,000 people and was once plagued by air pollution from industrial stacks, cars and burning coal and oil for heating dwellings, a situation greatly accentuated by frequent temperature inversions due to its location. The city occupies two valleys lying at right-angles to each other. The main part of the city is situated in a basin-like valley (the Nesenbach), which is surrounded on three sides by steep vegetated slopes that extend into the centre of the city. The rest of the urban area is located along the more open valley of the Neckar, Stuttgart's major river. Air quality became worse when urban expansion started to occupy the valley sides, replacing vineyards and forests with residential development. This interrupted normal katabatic air flows associated with valley formations, from the surrounding vegetated hills to the city that had originally acted to minimize air inversions. The unhealthy conditions that frequently occur because of the city's location has influenced city planning since the sixteenth century, although climate-based planning did not begin until 1938.[68]

It was apparent that there were links between the distribution of parks and open spaces, climatic phenomena related to valley topography and wind flows, and a healthy city environment. The prohibition of coal and oil for home heating was included in 150 local plans in the Stuttgart region and replaced with natural gas, reducing sulphur dioxide emissions by over 100 tonnes per year. As a planning tool it determines what slopes – in forest cover or vineyards – must be protected for unimpeded air flow and where urban development should be located. The use of land for construction and other non-open space purposes is regulated in accordance with German planning laws at the Federal level.[69]

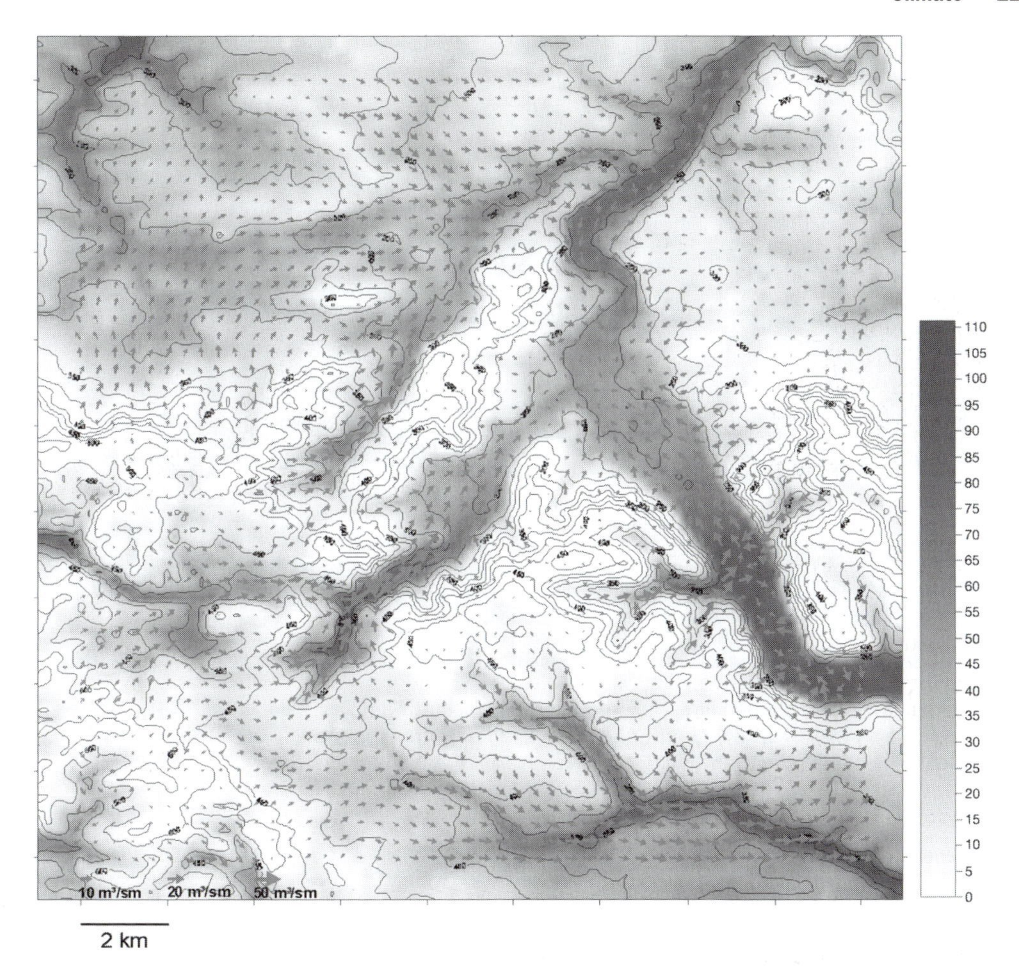

Figure 6.17 Diagram of Stuttgart's cooling system showing directions of katabatic winds down vegetated slopes and valleys into the city.

Source: Professor Dr J. Baummuller, Office of Environmental Protection, Urban Climate and Planning, Stuttgart.

A network of parks was established, linked to each other, to the river valleys and to the forests and vineyards on the valley slopes, to the city centre and the larger regional landscape. Within the 200 square kilometres of Stuttgart, green open spaces – forests, vineyards and city parks – account for 60 per cent the total municipal area.[70]

The structure, function and character of the city's spaces are very diverse. Stuttgart's natural topography has much to do with this. The city's central parks are located at the bottom of the valleys and those parks and vineyards further out occupy the steep slopes which provide breezes down wooded topography and a dramatic settings for the city. The traveller arriving at Stuttgart's railway station will be greeted with a view of the city centre and vineyard cascading down the valley slope. Within the parks themselves (covering some 490 hectares), plants, lakes and water sculptures provide places full of refreshing and varied sights and sounds. Rooftop planting is also an important factor in reducing air pollution. There are many examples of roof gardens in residential areas, office buildings and factories, and Stuttgart has been encouraging owners to incorporate them into buildings, assisting owners in their planning and contributing 50 per cent of the costs.[71] Other initiatives include the construction of low-energy buildings that

Plate 6.6 Heavily shaded linear parks maintain cool environments. Climate studies in Stuttgart have shown that there are twenty-seven days of heat stress in the city but only six days in the forested areas. Pedestrian bridges over roads link the city's parks together.

Plate 6.7 Oblique view of Stuttgart from the air. These forested slopes are a key factor determining downtown ventilation and air quality in the city as a whole.

Source: Office of Environmental Protection, Dept. of Urban Climate and Planning, Stuttgart.

Plate 6.8 An industrial green roofscape. Green roofs are a very important component of Stuttgart's urban climate strategy.

Source: Garten und Friedhofsamt, Rooftop Planting organization for Stuttgart.
Photo: Ute Schmidt Contag.

reduce energy costs by 30 per cent. Alternative energy sources are also being promoted that include solar collectors for hot water and photovoltaic solar panels. The city's first wind power plant is offering its customers special rates for electricity derived from solar and wind energy.

On both a citywide and human scale, the parks and working landscapes within and surrounding Stuttgart are among the most climatically functional, socially useful and aesthetically pleasing of any modern city in the Western world. From a landscape planning perspective their influence on its climate is matched by their key function as 'green infrastructure', giving form and identity to the city. In the twenty-first century, however, the major improvements in air quality will be determined by the reduction of car emissions. The goal is to reduce CO_2 and other greenhouse gases by a minimum of 20 per cent by the end of 2005. The main function of wind-based climate control will be to ensure that air quality remains at the same enhanced quality.[72] Yet 800,000 cars pass through the city every day and expanding the road network is restricted by topography, issues that have induced Stuttgart to join forces with the Federal Government and other partners to develop new approaches to transportation planning. Stuttgart 21, a major transportation and redevelopment project, is intended to transform the existing station into an underground through station for high-speed train service. The reduction of travelling time in long-distance and local traffic through Stuttgart and environmentally acceptable travel by rail will become more attractive.

Concluding reflections on urban climate

One may be tempted to comment that the value of plants, land form and water to the creation of beneficial micro-environments has been 'rediscovered' by the contemporary science of climatology. It has begun to measure something that people knew by trial and error and have made use of for generations in pre-industrial built environments. Applying traditional methods of climate control, in effect, now has the blessing of the scientific method.

But there are climatic phenomena that the older cities rarely, if ever, faced. Atmospheric pollution, urban heat islands and down draughts from tall buildings, drainage systems and hydrological imbalances are a creation of large industrial cities. Creating favourable habitats by natural means combines traditional wisdom, modern science and intelligent planning. The use of plants and water on the walls, floor and roof surfaces of the city can create natural climatic control and in large measure restore the energy balance through evaporation of water into the air and the metabolic processes of plants. The arrangement of built and landscape elements can reduce the impact of wind, take advantage of sun and create favourable micro-climatic environments. Following the principle of economy of means, the natural patterns and materials of the landscape can be made to work for the city environment in new ways and at less cost, compared to the energy costs currently incurred to maintain highly inhospitable conditions. The question 'How can human development contribute to the environments it changes?' provides us with a positive and proactive basis for action that can restore healthy climates, biological diversity and productive urban soils. It also provides us with a foundation for internalizing nature in human affairs. And if the natural and human processes that sustain life are integrated and visibly parts of everyday life, we enrich immeasurably the urban environment.

The interdependence of human communities and nature, enshrined in Barry Commoner's principle that 'everything is connected to everything else', illustrates that the more one learns about one's home place, the more relevant it becomes to larger regional issues, and the greater the need to protect biological diversity beyond local boundaries. Thinking in terms of these larger frameworks as a basis for action and policy is discussed in Chapter 7.

7 The regional landscape: a framework for shaping urban form

The city region

In the late nineteenth century, possibly the most influential of Patrick Geddes' ideas was his conviction that the city and its surrounding landscape were interdependent.[1] In the 1880s urban planning was based on the belief that the city and its rural hinterland were two entirely separate sets of problems. Geddes insisted that to understand the city one had to understand the connections between the two. To demonstrate this idea, Geddes took over a disused observatory equipped with a camera obscura, which was near the castle – Edinburgh's highest point. The camera obscura was a projection device mounted at the top of the observatory tower that overlooked the city. It projected a panoramic image of Edinburgh and the rural landscape surrounding it on to a large flat white table below. Its purpose was educational, to show people the essential connections between the city and its region which Geddes described as 'any geographic area that expresses a certain unity of climate, soil, vegetation, industry, and culture'.[2] His great planning innovation came out of this idea: the concept of regional planning, and the basic three-part components of the city region, which he described as the interrelationships between place; folk; and work (place being the environment, folk being society, and work being the economy). For Geddes, regional study gave understanding to an 'active, experienced environment', and in his book *Cities in Evolution* he drew attention to the fact that the new neotechnic technologies such as electric power and the internal combustion engine were already causing cities to disperse.[3]

A hundred years later, in the 1980s, Philip H. Lewis, an American landscape architect and regional designer of great distinction, began analysing patterns of urbanization across the USA from satellite photos taken at night. Based on the observations he made at his Centre for Environmental Awareness, at the University of Wisconsin, Lewis showed that the growth of urban systems in North America could be guided from rings of urbanization to protect a variety of significant natural features needed to sustain the cities. This view of urban systems has, in his words, 'greatly altered our view of how urban areas should be perceived and managed'. By identifying all counties from satellite imagery that are bright at night, the Centre discovered configurations of twenty-three urban constellations. Estimates from the 1980 census for these counties indicate that 80 per cent of the American population lives within these constellations.

In effect, where settlement in Patrick Geddes' day in Scotland was at the centre surrounded by its regional landscape – a sort of urban hole in a rural doughnut – today, in the urbanizing regions of the USA, it is the natural resources of the regional landscape that must form the centre; a rural hole in an urban doughnut. Like our first views of the earth from space that suddenly brought home the reality of a finite planet, these satellite observations are dramatically altering our view of urban growth.[4] It brings home the urgent need to protect environmental systems within the city region as the basic

(a)

(b)

Figure 7.1 (a) Satellite photo of the USA at night. (b) Population centres of the USA. Cities of over 20,000 people were connected in a manner that made the least impact on identified critical natural resource patterns.

Source: Environmental Awareness Centre, University of Wisconsin. Also in Philip H. Lewis, Jr, *Tomorrow by Design*, New York, NY: John Wiley, 1996.

structuring element of macro-urban form. It is evident, therefore, that urbanization is the most crucial developmental trend affecting the regional landscape, with repercussions for rivers, wetlands, groundwater, soils, vegetation and wildlife. In the following pages I discuss some of the different approaches being taken to various regional themes. They include:

- the concept of greenbelts and their applications to British cities;
- urban growth boundaries and new urbanism in the USA;
- natural landscapes as frameworks for urban growth;
- ecological planning and restoration of brownfield sites: the industrial Ruhr valley;
- linking cities with parks and areas of high ecological integrity in the macro-region.

The greenbelt

Urban growth and identity

The preceding discussion illustrates that the issues of urbanization in a North American landscape are very different, in scale, character and history, from those of Britain. In a small country with a very large population, the problems of urban growth may be said to be greatly exacerbated. In his book *Garden Cities of Tomorrow* (1902), Ebenezer Howard turned to decentralization as a solution to the expanding city. New towns located in the countryside would have a fixed population limit of 32,000 people living on 405 hectares of land (1,000 acres), about one and a half times the size of the historic medieval city of London. It would be surrounded by a permanent greenbelt five times that size and contain farms, and a variety of urban institutions. As the towns grew, others would be started a short distance away.[5] As Peter Hall notes: 'If regional planning provides the framework, the garden city provides the "civic objective", not as a temporary haven of refuge but as a permanent seat of life and culture, urban in its advantages, permanently rural in its situation.' Later, Sir Patrick Abercrombie's 1944 plan for Greater London was based on Howard's garden city – a gigantic greenbelt around built-up London, with satellite towns within it and green wedges running inward from the Greenbelt into the heart of London. The New Towns Act was passed in 1946 and the eight towns were well on the way to completion by the mid-1960s.[6]

The continuing protection of the English countryside, clearly a different environment from the cities, is one of the country's much admired characteristics, particularly for North American visitors who are familiar with low-density suburbs that sprawl endlessly across the landscape. Greenbelts surrounding cities have, since the 1960s, been central to planning policy in the UK covering about 1,556,000 hectares that include fourteen greenbelts and about 12 per cent of the country's land base. They are backed up by a commitment to sustainable urban development that has been a cornerstone for greenbelt planning since the early 1990s.[7] Sustainability is defined, in part, as the conservation of energy and minimizing car emissions, encouraging accessibility without mobility (being able to walk or cycle to local places); in part by public transit that contributes to sustainable patterns of urban development, and access to the countryside. In practice this means that urban growth needs to be denser and more compact, focusing on small neighbourhoods, and combining homes with jobs and services.[8] Greenbelts have a number of purposes: to protect watersheds, groundwater, aquifers and other biophysical components of the landscape; protect the countryside and farmland from urban encroachment; prevent neighbouring towns from merging together; preserving the physical setting and special character of historic towns; and assisting in urban

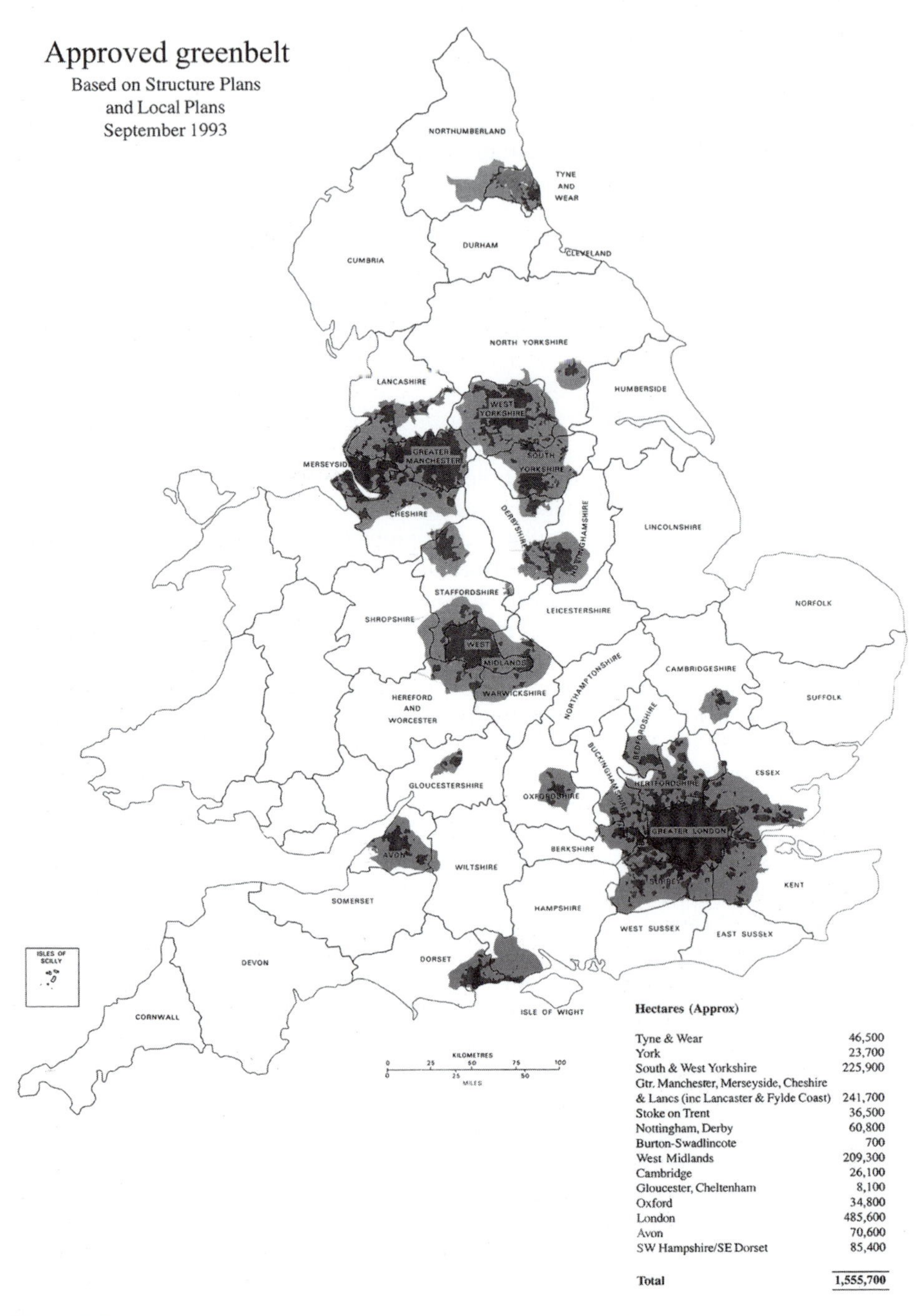

Hectares (Approx)	
Tyne & Wear	46,500
York	23,700
South & West Yorkshire	225,900
Gtr. Manchester, Merseyside, Cheshire & Lancs (inc Lancaster & Fylde Coast)	241,700
Stoke on Trent	36,500
Nottingham, Derby	60,800
Burton-Swadlincote	700
West Midlands	209,300
Cambridge	26,100
Gloucester, Cheltenham	8,100
Oxford	34,800
London	485,600
Avon	70,600
SW Hampshire/SE Dorset	85,400
Total	**1,555,700**

Figure 7.2 The current fourteen greenbelts in the UK totalling 1,555,700 hectares.

regeneration by encouraging the reuse of derelict urban land. Boundaries also need to be clearly defined to ensure the long-term viability of agriculture, recreation and amenity, by using physical features such as roads, streams or woodland edges.

The government policies for the countryside are set out in the White Paper *Rural England; A Nation Committed to a Living Countryside*,[9] requiring that the countryside be managed sustainably. This entails accommodating necessary change to rural areas, encouraging economic diversity and employment, reducing travel by car, maintaining the character of the countryside and the quality of rural towns and villages – in effect, safeguarding distinctive landscapes.

It is at the local county level, however, that policies and principles are implemented, and that local needs and issues and the peculiar environmental and political conditions of local districts come face to face with national greenbelt policies. This was borne out in 1996 hearings in which representations were made to a panel that was conducting a public examination of the County of Dorset Structure Plan (to the year 2011). With regard to the issue of settlement patterns, a question under debate was whether the boundaries of the south-east Dorset greenbelt should be adjusted to allow for additional development. The following are examples of the comments from participants on this question.

- One proponent said that the greenbelt as an absolute constraint was flawed. Some flexibility was needed to achieve community benefits in some settlements, while at the same time not compromising the principle of sustainable development.
- The representative of one parish commented that the emphasis on infill in the policy had resulted in high-density estates, changing the character of the village.
- The long-term retention and protection of the greenbelt was advocated. As greenfield sites became exhausted, the greenbelt would, for the first time, play a significant role in limiting the spread of regional urbanization. The release of land around towns would result in even more dispersed settlement patterns than already existed and increase long-distance car journeys. Flexibility with greenbelt boundaries would be the thin end of the wedge.
- The New Forest District Council supported keeping both of the existing south-east Dorset–Hampshire greenbelts since they had a significant role in controlling human pressures on the New Forest that now had equivalent status as a National Park.
- English Nature and other environmental organizations had concerns about the national and international importance of the sensitive heath land and aquatic habitats in and around the larger builtup areas. They emphasized the importance of not relaxing greenbelt edges since this would thwart extending and linking some important wildlife habitats.
- Approval by the Secretary of State of the south-east Dorset Structure Plan was contingent on containing the outward spread of urbanization and preserving the separate identity of settlements. At the point where further development conflicted with the definitive limits of the greenbelt, they would need to be accommodated in other ways. It was likely, however, that the anticipated restraint on urban growth would come into force during the planning period, which would require meeting sustainability objectives and the deliberate intention to provide less housing than would be needed to meet demand. Such a constraint posed questions about whether maintaining the greenbelt was compatible with a strategy that sees urban growth as the engine that will drive the growth of the county.
- It was stressed by several representatives that development in their settlements would be greatly intensified by infilling and redevelopment. Pressures on local services had created profound changes in the character and amenity of the settlements. The tight

> restraint of the greenbelt around them would require that facilities be provided, such as schools, village halls and sports grounds, or development that might provide some local employment.[10]

As a policy mechanism for limiting growth outwards, the greenbelt is subjected to a variety of economic, environmental and development pressures and conflicts that emerge at a local level. There is much ongoing discussion about the future of the greenbelt as a planning concept. The Town and Country Planning Association, for instance, pressed for a reappraisal of the roles, purposes and extent of greenbelt policy. For instance, greenbelt policies currently restrict opportunities for alternative forms of urban growth that could conform better to sustainable development principles. Future expansion of some urban areas should be well connected to public transportation corridors, and accommodate not only housing but a mix of related land uses and employment. Yet such extensions can conflict directly with greenbelt policies. The restrictive nature of greenbelts inhibits the potential for encouraging new forms of development in rural areas and for diversifying their local economies and communities. To make settlements more sustainable, the land around towns should become a zone for ecological and sustainable uses. Examples might include smallholdings for organic gardening, community woodlands, composting projects, wind farms and biomass power for district heating.[11] A review by the Royal Town Planning Institute and other organizations on the future of the greenbelt concluded that: there was a failure of greenbelts to keep pace with the changing planning policy agenda of recent years; public perception of the role and purpose of greenbelts was out of step with reality; there were conflicting aims and objectives in the application of greenbelt policy. The Country and Land Business Association recognized the importance of preventing sprawl, but said that sustainable development aims should also be met in rural areas. Current guidance on greenbelts did not, however, provide a framework for achieving this.[12]

Linking town and country

Inherent in government policy is the maintenance of regional identity. It emphasizes the essential differences in character and function, between what is urban and what is rural. At the same time, urban and rural environments are interdependent, particularly in relation to greenbelts closest to urban areas – the urban fringe. There are pressures to build 4.4 million new houses between the 1990s and 2016. A national traffic growth of 38 per cent is expected for the same period. Resolution of these conflicts and ensuring the most appropriate sustainable use of the land for farming, forestry, conservation, recreation and sports have become crucial issues.[13] Thus achieving sustainable regional and economic development involves links between urban and rural areas.[14] And given that over 50 per cent of countryside visits take place within 8 km of home, these factors place the urban fringe in a key position to establish links between town and country. Thus the countryside in and around towns should make a major contribution to the quality of life, providing a place for leisure, the arts, sports and recreation. The need to understand the needs of different cultures and active participation is also emphasized. A programme proposed by the Countryside Agency in 2002 was intended to realize the social, economic and environmental potential of these fringe areas. Some of its recommendations include:

- encouraging local authorities and other agencies to establish countryside management services which are active in urban fringe areas;
- establishing 250 country parks close to the urban areas;

- establishing greenways as important new access routes between town and country, and safe links between recreation areas;
- encouraging collaboration between rural and urban local authorities.[15]

Other factors that are relevant to the urban fringe areas include developing rights-of-way plans for walking and cycling and improving links between town and countryside, and investigating how new development can enhance, rather than detract from, the quality of the landscape.

Sustainable land management

While there are many places in Britain where the distinctiveness between town and country is clearly defined, the diversity of its habitats, its plants and animals, that has traditionally given the countryside its distinctive character, have, since the Second World War, continued to diminish. Traditional farming was once responsible for the countryside's cultural and natural diversity. Today, its role, and the countryside as a whole, is undergoing a transformation of its traditional functions. Farming occupies around three-quarters of the land surface of England and remains the major use of rural land.[16] Yet agriculture in the UK is threatened. Over 20,000 jobs were lost in 2001. Foot and mouth disease ravaged parts of the country in 2001 and it was feared that traditional livestock farming would disappear in some worst hit areas. Confidence in farming as a business has been reduced and 40 per cent of farms in the new century were sold to people interested in a country home in an attractive rural setting.[17] The rural landscape is being shaped increasingly by interests outside the industry, from changes to rural communities, recreation and tourism and changing uses of agricultural land that involve non-food crops, forestry and amenity planting. In response to these changing conditions, the Countryside Agency proposed a strategy for sustainable land management. Four principles guided the development of policy and practice.[18]

1. The land should be managed to deliver numerous benefits in addition to the production of food and fibre that include the conservation of soil and water, restoration of biodiversity and wildlife habitats, employment and businesses in the local and larger rural economy and opportunities for tourism and recreation.
2. Land management should reflect principles of sustainable development by maintaining primary environmental resources (soils and water), ensuring strong long-term local economies.
3. Land management must be integrated with rural development. For instance, the supply of products directly and indirectly, from the management of the landscape, on which rural tourism is based.
4. A framework which can reflect regional and local needs and aspirations should be established. People need to have a greater voice in shaping land management to deliver public benefits at both local and regional scale, but which recognizes the fact that much of the land will remain in private ownership.

Land management, in effect, should contribute positively to the economy of rural areas. It places less emphasis on mass production of food for the global market and more on processing and marketing of produce locally. This has several benefits. It strengthens the links between product, place and landscape character, and it reduces food miles (see Chapter 5). Improved effectiveness of local and regional supply chains can also retain more of the value of products locally. Also realized are jointly produced services: for example, paying farmers to use land for water catchment or flood alleviation where that saves investment by public or private companies who provide infrastructure.

The potential benefits for the environment are increased effectiveness of water management, woodland expansion for biodiversity and sequestering carbon, and restoration of degraded landscapes. Social benefits include a better understanding of city people for rural occupations and functions (a critical issue in North America), a greater range of rural employment and training opportunities, and improved quality of life for all, for example, availability of locally grown food and improved opportunities for access to the countryside.[19]

Realization of this agenda over the short to medium term involves the application and upgrading of a number of programmes. The agri-environment schemes, for instance, that exist in the United Kingdom developed for a number of reasons. The absence of any system of publicly owned land, such as the Crown Lands in Canada, meant that little control existed over the actions of land managers. In fact, in the past, British planning laws worked to ensure that while agricultural land was protected from urban development, there was no control over normal agricultural operations, even when they damage the environment. The increasing intensification of production methods since 1945 has resulted in the loss of many important landscapes and wildlife habitats. Public concern at these losses has led to a search for methods which could protect the great variety of landscapes, natural habitats, historic features and recreational opportunities that are so valued in the UK. The Royal Society for the Protection of Birds in its publication *Futurescapes* maps out nine key habitat types across Britain where restoration should be located, each natural region having its distinctive character. The habitats across Britain, for instance, include salt marshes, Caledonia pine forests, heath land, downland, pastoral and arable farmland, and wetland.[20]

Three schemes that have financial incentives encourage farmers to follow certain approaches to management. Environmentally Sensitive Areas (ESA) and Countryside Stewardship (CS) schemes are both operated by the Department for Environment, Food and Rural Affairs (DEFRA). ESAs are designated areas where it is considered that traditional farming practices have created distinctive landscapes. There are currently forty-one such areas in England, Scotland and Northern Ireland. Within the designated areas farmers can apply to enter a management agreement with DEFRA for which they receive grants. The management agreements last for ten years with a break clause after five years. A farmer must enter all land within the ESA into the scheme and this often means that the whole farm is included. Farmers can choose their 'tier' of entry; typically the higher the tier the higher the payment but the more stringent the environmental management required. For example, in the Pennine Dales ESA the payments and management prescriptions include:

- *Tier 1A: Arable and improved grassland, £20/ha/yr*
  Avoid damage to the sward by overgrazing, undergrazing or poaching. Limit use of organic and inorganic fertilizer to no more than current levels.
- *Tier 1B: Meadows and pastures, £135/ha/yr*
  Set limits on fertilizers, lime and herbicides. Set limits on stocking numbers and on cutting dates for hay meadows.

The Countryside Stewardship scheme in contrast is available to farmers and managers of land throughout England. The scheme is voluntary so that individuals decide whether to apply, choose which aspects of the scheme to follow and how much of their land to include in an application. However, the scheme is also competitive in that applications for the scheme are submitted during a set period each year and the schemes compared. Those proposals that appear to offer the most benefits to landscape, wildlife, history or public amenity are accepted into the scheme and are given an agreement that lasts for ten years. There are some compulsory elements – the so-called cross-compliance elements which include maintaining public rights of way and not damaging existing field boundaries. However, the menu approach of the scheme means that agreements are tailored to individual site requirements. Thus the countryside is intended to remain both economically productive and actively managed to maintain the cultural and environmental values that, at one time, were the result of traditional farming practice.[21]

It is interesting to note that as the environment of the countryside has deteriorated biologically, abandoned areas in the cities have become increasingly diverse, due largely to neglect. Chris Baines has remarked:

> ironically, the post-industrial inner city is far greener than it was when I was young. Nature has healed the scars of heavy industry and much of the derelict land is wild and wooded once again. The urban rivers are less polluted than they've been for years. . . . And all those millions of suburban gardens have matured into a rich mosaic of flower borders, ponds and shrubberies: an ideal habitat for songbirds, butterflies and frogs.[22]

The rural areas are also changing, however. An example of how such programmes can help farmers contribute to the ecological enrichment, and to wider environmental benefits, of the English countryside, was the 2003 winner of the 'Farmcare Silver Lapwing' Award, a prestigious UK award for 'conservation on the farm'.[23] Nick Tremlett, a commercial farmer in the chalk landscape of the Wessex Downs in southern England, grows commercially arable crops such as wheat, rape, peas and barley. His efforts to diversify the functions of his property have included the conversion of old dairy units into offices, renting redundant farm cottages and establishing sports facilities, all of which have enabled conservation projects to be undertaken. Parts of the property that have been set aside have been used to increase wildlife diversity. A large block of land serves as a sanctuary for wildlife, and other, smaller blocks have been linked to a network of grass and wildflower field margins. They include wildflowers that attract butterflies, birds and mammals. The River Wylye, a well-known chalk stream river, has been managed to improve brown trout habitat and other wildlife. A wet meadow has been modified to encourage curlews and snipe to return to the area. He comments, 'Farm conservation is a long-term investment. Farming is not easy at present but stewardship payments help fund conservation measures which we should see as part of sound business practice. They add to the value of wildlife and the value of the farm.'[24]

An ecological analysis of a greenbelt: Ottawa

The influence of planning measures to counter the destructive advance of cities, as Geddes predicted in the 1880s, resulted in the British greenbelts and, as we shall see later in this chapter, the North American urban growth boundaries. While effective in controlling urban sprawl, their fundamental planning motivation has been social and economic rather than ecological, although the protection of farmland and natural features is certainly an important factor in greenbelt policies. The concept of a greenbelt surrounding Ottawa – Canada's capital city – proposed by the French planner Jacques Greber in 1949, was greatly influenced by Sir Patrick Abercrombie's London greenbelt. Its extensive linear parkways within the city were also linked to the larger regional landscape through the Gatineau Park, forming an impressive green structure for the city that includes 16,200 hectares of public lands acquired during the 1960s. However, importing such a planning concept direct into a North American landscape, one that is very

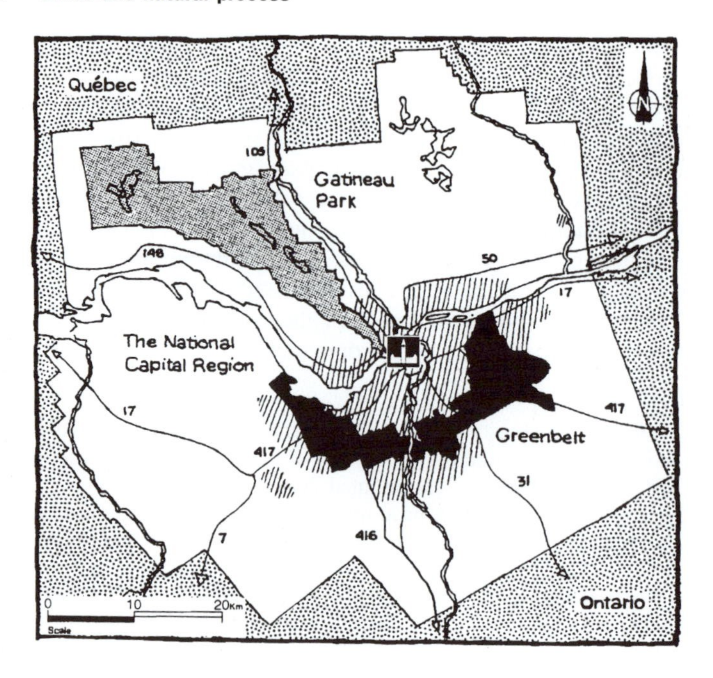

Figure 7.3 The Ottawa greenbelt, a concept imported to the National Capital Region from England via a French planner, Jacques Greber. To what extent did the greenbelt protect the ecological integrity of its natural region?

Source: National Capital Commission.

different from that of Britain, raises an important question. To what extent has the Ottawa greenbelt actually protected the ecological integrity of its natural region?

In 1991, an ecological study of the Ottawa greenbelt system was undertaken for the National Capital Commission.[25] The study's purpose was to examine the biophysical functions of the greenbelt and its environs, and to illustrate how an ecologically based vision would help guide future urban growth and greenbelt planning. Of particular significance was the very different philosophical approach that it took to the greenbelt compared to how it was conceived in the late 1940s. The study was based on principles of landscape ecology as a framework for analysing the greenbelt. It involves the flow of energy – water, nutrients, plants and wildlife – among natural areas.[26] It is concerned with the spatial distribution of natural areas in a human-dominated environment and is one of the principal tools relevant to landscape planning in urban areas. It combines a science-based approach to how natural systems work, with a place-related planning approach to its implementation on the ground. It is concerned with the physical patterns of landscape, the corridors and the patches of habitat generally inhabited by wildlife. Landscape ecology suggests that the primary consequence of human development on the natural environment is fragmentation, which converts extensive or continuous areas of forest, wetlands, meadows and other types of habitat into isolated islands.[27] Typically, human activities disrupt ecological functioning and inhibit interactions between the systems, nutrient and gene flows among habitats. The fragmentation of biological communities, in fact, is recognized as one of the most crucial environmental issues affecting urbanizing regions and ecologically diverse protected areas and parks beyond the city, an issue that will be discussed later in this chapter.

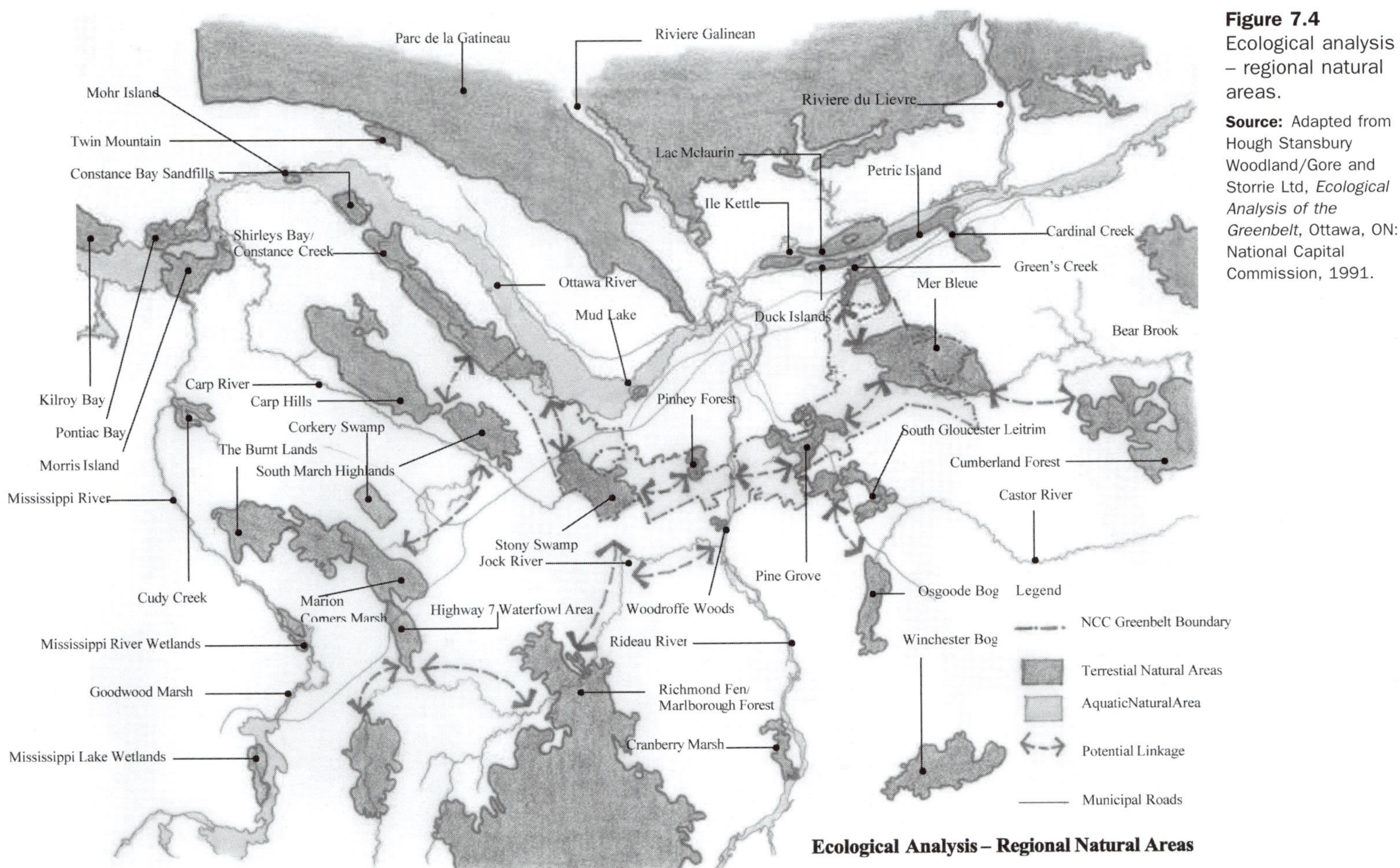

Figure 7.4 Ecological analysis – regional natural areas.

Source: Adapted from Hough Stansbury Woodland/Gore and Storrie Ltd, *Ecological Analysis of the Greenbelt*, Ottawa, ON: National Capital Commission, 1991.

In order to measure the naturalness and significance of the region's habitats, the study used a matrix approach, designed to identify wildlife species that could occur in various habitat types. These included watercourses, old fields, marshes, fens and bogs, and forested habitats. It also examined those that were contained within and outside the greenbelt boundaries, and the degree to which its boundaries protected them. The Ottawa/Carlton urban region as a whole has a highly diverse and extensive body of natural areas. They maintain an extensive genetic pool, and serve as headwaters for numerous rivers and tributaries. Development pressures in recent times have included the establishment of satellite communities associated with municipalities around the outer edge of the greenbelt. Also included are transportation corridors, servicing and infrastructure, recreation and buildings, advancing from outside the greenbelt as well as from within. Many linkage and buffer areas have consequently been adversely affected. Proposed urban development in the region will continue to add pressures to surrounding natural environments by encroaching into their boundaries and zones of influence. Additional infrastructure will consequently be required to service these new developments, which will result in a continuing degradation and fragmentation of natural systems. Of particular importance was the fact that the majority of significant natural areas in the Ottawa region were either unprotected, or at best only partially protected, by existing greenbelt boundaries.

An example is the Mer Bleue, the largest and best example of an undisturbed bog in both the region and eastern Ontario, and a provincially significant wetland. It is the habitat for a population of provincially and nationally significant spotted turtle; Henslow's sparrow, which is threatened across Canada, and the least bittern, which is

Plate 7.1 The Mer Bleue, a large pristine bog of great ecological significance within the Ottawa greenbelt, was found to be only partially protected by greenbelt boundaries.

Source: National Capital Commission.

rare in both Ontario and the country as a whole, and declining, due to loss of suitable nesting habitat. There are two nationally rare, and seven provincially significant, plant species, and seventeen regionally rare birds. The bog is the headwater for the tributaries of two streams and provides a natural linkage between a large forest to the east and parts of the greenbelt to the west. It has significance, therefore, in a regional context. It is considered to be in excellent ecological health, and minimally affected by human activities.

Potential threats to the Mer Bleue from development and fragmentation largely concerned water quality and quantity since bogs are extremely sensitive to changes in water levels and nutrient concentrations. For instance, leachate from fill adjacent to the bog would likely increase nutrient and sediment concentrations. A proposed residential development is a potential threat to changes in the quantity of water to the bog. Private ownership and potential drainage and peat excavation in the eastern portion of the bog could have serious impacts. The greenbelt boundary, however, cut through the Mer Bleue bog, severing the natural drainage connection between the bog and forests to the east (Cumberland Forest), putting water quality at risk and fragmenting links between the two.

In order to better integrate human settlement patterns with functioning natural systems, the report to the National Capital Commission recommended that modifications to the greenbelt were based on three essential components:

- areas of high biological significance within which development is prohibited and recreational use restricted;
- buffers that protect primary areas from incompatible land uses and maintain large contiguous areas of habitat;
- linkages which connect primary natural areas and act as movement corridors for flora, fauna, water and people. These were given levels of priority relative to their biological significance, from a first priority that maintains essential links between significant functioning natural systems, to a third priority that includes major recreational corridors or greenways that also provide habitat linkage.[28]

The land-use policies formulated for the ecological plan for the Ottawa greenbelt were consistent with the intention of the Federal Land Use Plan 'to reflect the special functions of the Capital so as to enhance the Capital's image'. One of the underlying policy principles of the Federal Land Use Plan notes that the capital should 'display its key institutions, facilities, and symbols on prominent visible sites . . . [to] ensure a built environment that is in harmony with the natural setting'. The significance of regional parks and the greenways and greenbelts that link them therefore lies in the application of ecological principles for developing a landscape structure that will guide future urban growth, and where the interdependence of human and non-human systems is a primary goal. About 22 per cent of the greenbelt has been designated a Core Natural Area.[29]

Developments since the mid-1990s

The *Greenbelt Master Plan* was published in 1996. Its purpose was to provide strategic policy guidance for land use, programming, landscape planning and character to ensure the long-term relevance and quality of the greenbelt into the future. It incorporated a series of core natural areas (approximately 22 per cent of the greenbelt) of major significance as key components of a continuous natural environment into the greenbelt, to be used extensively for ecological research and environmental education. They are sheltered by adjacent buffer zones and linked to each other and to natural areas beyond the

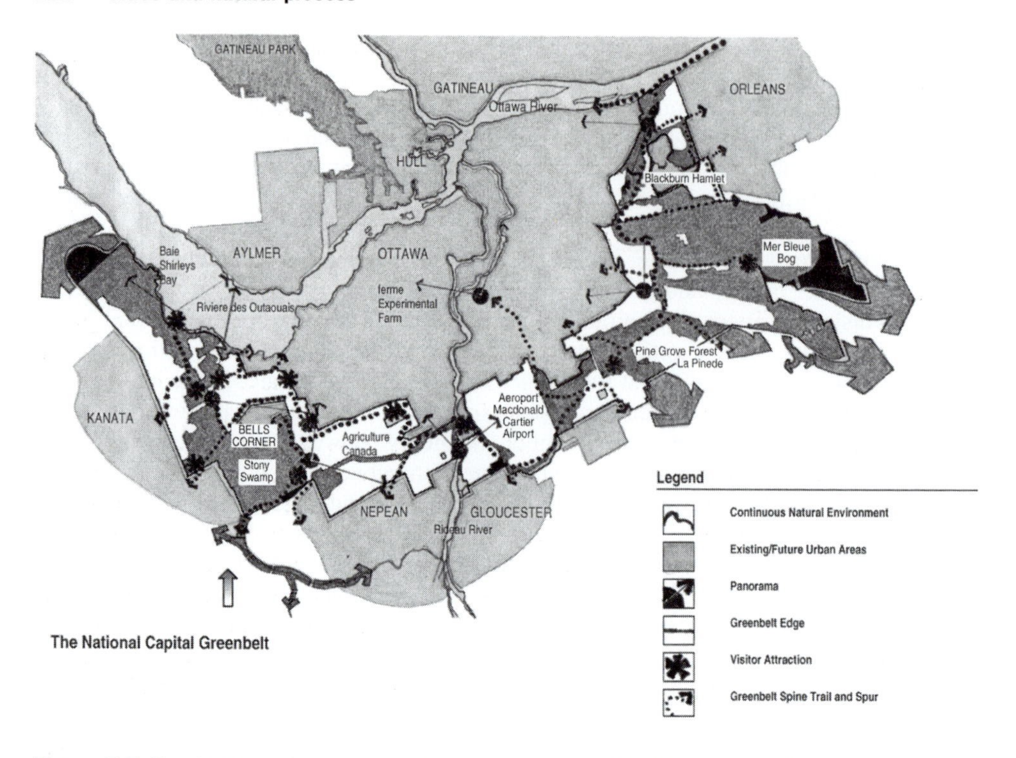

Figure 7.5 The National Capital Greenbelt concept plan. Modifications to the greenbelt. A commitment to ecological principles links natural areas together and extends the greenbelt in unprotected areas. Note the Mer Bleue (see Plate 7.1).

Source: Adapted from National Capital Commission Master Plan, courtesy of National Capital Commission.

greenbelt, allowing movement of plants and animals, and enhancing ecological health and resilience of individual natural areas and of the region as a whole.[30] These and other changes in response to urban growth and transportation networks have consequently added approximately 650 hectares to the original greenbelt.

Introduced in the 1950s from Britain, the circumstances that led to the creation of the greenbelt no longer exist. The Federally owned lands were acquired in the 1960s at a time when the economy and the political will were exceptionally strong and could not be replicated today. This has given rise to the 'conservation of a remarkable diversity of landscapes and human activities in the symbolic heart of the nation that are unique among federal government land across Canada'.[31] Today, the significance of the greenbelt also lies in ensuring that its natural landscapes continue to function ecologically, and to act as an expression of the capital's unique regional character and sense of place. This objective has been reinforced over several decades by National Capital Commission efforts to naturalize and reforest Ottawa's greenway system from limitless turf back to a landscape of regional diversity.

Greenways

The notion that isolated parks should be connected has existed since the early days of the parks movement in North America. Its underlying focus was essentially a social one. It had to do with leisure and recreation, providing people with the opportunity to experience a continuous landscape uninterrupted by roads or boundaries (see Chapter 3, pp. 118–119 and 123). In Britain the right of way across private and public lands has been in existence for centuries. In North America proponents of the connected park in the latter part of the nineteenth century include W. H. S. Cleveland's open space system for Minneapolis St Paul that connects a series of lakes with the Mississippi River into a continuous green system, a circuit of some 65 kilometres around the city. Similarly, Frederick Law Olmsted's plans for the Niagara Parkway Gardens was conceived to protect the American and Canadian lands surrounding the Niagara River from uncontrolled tourism. The present parkway, implemented by the State of New York and the Ontario Provincial Government in the 1890s, has retained much of Olmsted's original plan.[32] And at a macro-scale, Philip H. Lewis' mapping of the natural and cultural features of Illinois and Wisconsin in the 1950s led him to the conclusion that the corridors that emerged needed protection in planning legislation.

Throughout the 1950s and 1960s, the rapid suburban expansion and new town development, the concept of the continuous pedestrian open space system, linking housing with schools, churches and shopping centres with residential areas, became an integral part of new town planning. It introduced the idea, novel for its time, that vehicles and pedestrians should be physically separated. The satellite towns of Radburn, New Jersey, and Wildwood Park in Winnipeg, Manitoba are examples. And as we saw earlier, Greber introduced the concept in his plans for Canada's national capital. The significance of greenways, however, lies in their potential for regional landscape connections and for using large-scale infrastructure links and integrated systems, such as abandoned railway lines and other rights of way.[33]

Since their construction in the nineteenth and early twentieth centuries, railways have become significant natural corridors. The decreasing importance of railways in some metropolitan areas and the increasing number of abandoned lines provide opportunities for cycle and pedestrian access to link residential communities, parks, schools and commercial areas. The growing North American 'rails to trails' movement is an example of where walking and cycling trails are establishing links between cities. The Rails to Trails Conservancy movement in the USA has been helping to bring about track conversions since the 1970s, and 6,436 kilometres of tracks are reported to have been converted to bike use during this period.[34] These greenways provide a network of trails along abandoned railway rights of way throughout the country. One of the first conversions was the 1967 Elroy–Sparta State Park Trail that crosses 51.5 kilometres of dairy country in south-west Wisconsin.[35] The considerable landholdings of railway companies that have lain idle for many years provide significant social and productive value. In North America and Europe thousands of kilometres of railway tracks have gone out of service which gave rise to the Rails to Trails movement and the conversion of the rights of way into trails linking parks and natural areas, historic sites, towns and villages. There are more than 500 greenways in the USA. The Bay and Ridge trails around San Francisco are each about 650 kilometres long. Planning and implementation of the Ridge Trail has involved citizen groups, municipalities and agencies. Similarly in Oregon, the Willamette River Greenway runs through nine counties, and nineteen municipalities.[36] And in Ontario, the Lake Ontario Greenway Strategy involves trails along both the north and south shores of the lake linking Ontario and New York State. Connections north to the Niagara Escarpments, Bruce Trail, the Ganaraska and the Rideau trails also connect with the Oak Ridges Moraine, the source of the river systems that flow south to the lake.[37]

The greenway may be seen to be an old idea, dating back centuries in Europe and to the nineteenth and twentieth centuries in North America. At the same time, their functions have diversified. Today, they represent ecological, economic and community-based

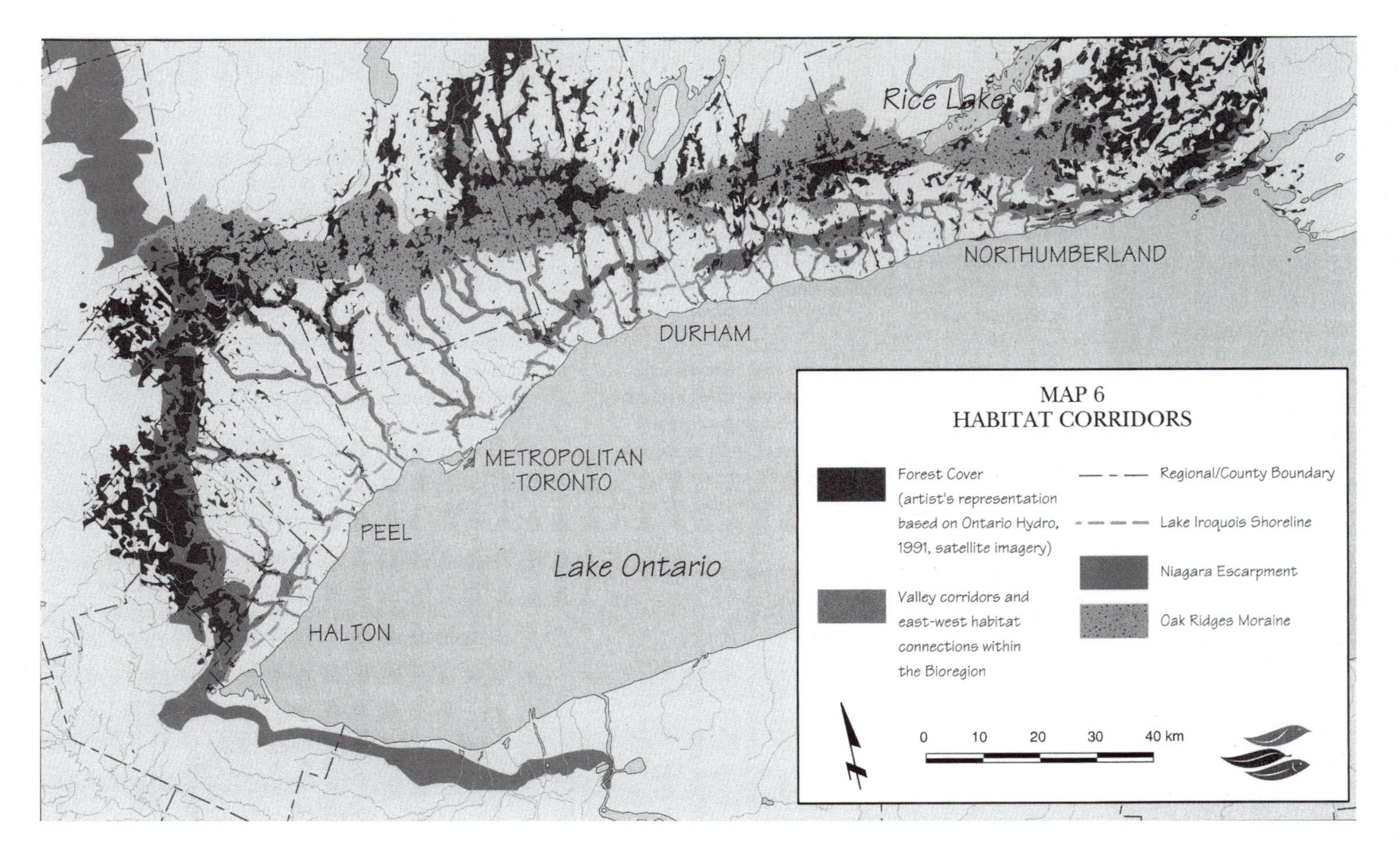

Figure 7.6 The Lake Ontario Greenway Strategy. North–south and east–west connections forming a major natural and cultural system. A co-operative venture by municipalities in the region to buy into the concept and fund the trail was a major achievement in coordination by the Waterfront Regeneration Trust.

Source: Lake Ontario Greenway Strategy, Waterfront Regeneration Trust, 1995.

ideas that play a key role in integrating natural habitats, parks, towns and cities. Their value lies in establishing regional connections, not only with urban areas, but also with the countryside in ways that accentuate their differences as places and ensure their sustainability.

Smart growth

A North American approach to the greenbelt

Unlike the UK, which is a small land mass with a large population, North America's uncontrolled urban growth has been associated with the myth of limitless space. The relationship between the land areas covered by development and increases in population is interesting. Between 1982 and 1997 the amount of urbanized land used for development in the USA increased by 45 per cent. During this same period, however, the population grew by only 17 per cent.[38] The consequences of sprawl for farmland, river valleys, watersheds and woodlands, from atmospheric pollution, costs of infrastructure and the placelessness that results from the separation of land uses by zoning, are well known. A study commissioned by the New Jersey legislature concluded that low-density development consumed 130,000 more acres than a more compact urban form, at an additional cost of $740 million for roads and $440 million for sewers and water infrastructure.[39] A study from the Environmental Justice Resource Centre at Clark Atlanta University examined the impacts of urban sprawl on communities in the metropolitan region. It found that sprawl-fuelled growth is widening the gap between the region's 'haves' and 'have nots' and is pushing people further and further apart geographically, politically, economically and is also heightening racial disparities.[40] Several responses to this predicament have emerged – the concept of 'smart growth' and 'smart solutions to sprawl', which, fundamentally, means controlling outward urbanization, promoting compact development and embracing comprehensive and ecologically sound approaches to urban growth.

The urban growth boundary has parallels with the English greenbelt as a means of containing growth. The boundaries mark the separation between urban and rural land. They bring certainty to knowing which lands are to be developed and serviced, and which will remain rural and protected. In North America a number of cities have incorporated urban growth boundaries including Vancouver, British Columbia, Washington, Boulder, Colorado, Lancaster County, Pennsylvania, Portland, Oregon and Minneapolis St Paul.

In the State of Oregon, urban growth boundaries were created for Portland and each of the state's 241 cities in the early 1970s, as part of the state-wide land-use planning programme.[41] The cities are responsible for planning within their boundaries, the adjoining county is responsible for planning outside of city limits, and special districts provide urban services. The state's Land Conservation and Development Commission ensures that the UGB meets the requirements of 'Goal 14' which lists a number of criteria that must be considered, including the amount of land needed for growth. An important aspect of a boundary's location has to do with an efficient use of land that protects agricultural land at the city's edge, and ensures cost-effective public services within the boundary.[42]

As a consequence of this clear definition between what is town and what is country, land is, in general, more valuable inside the boundaries. Densities and housing potential are consequently greater after establishing the UGBs and the agricultural economy is strong. The state-wide farm tax deferral policy prevents the sky-rocketing taxation

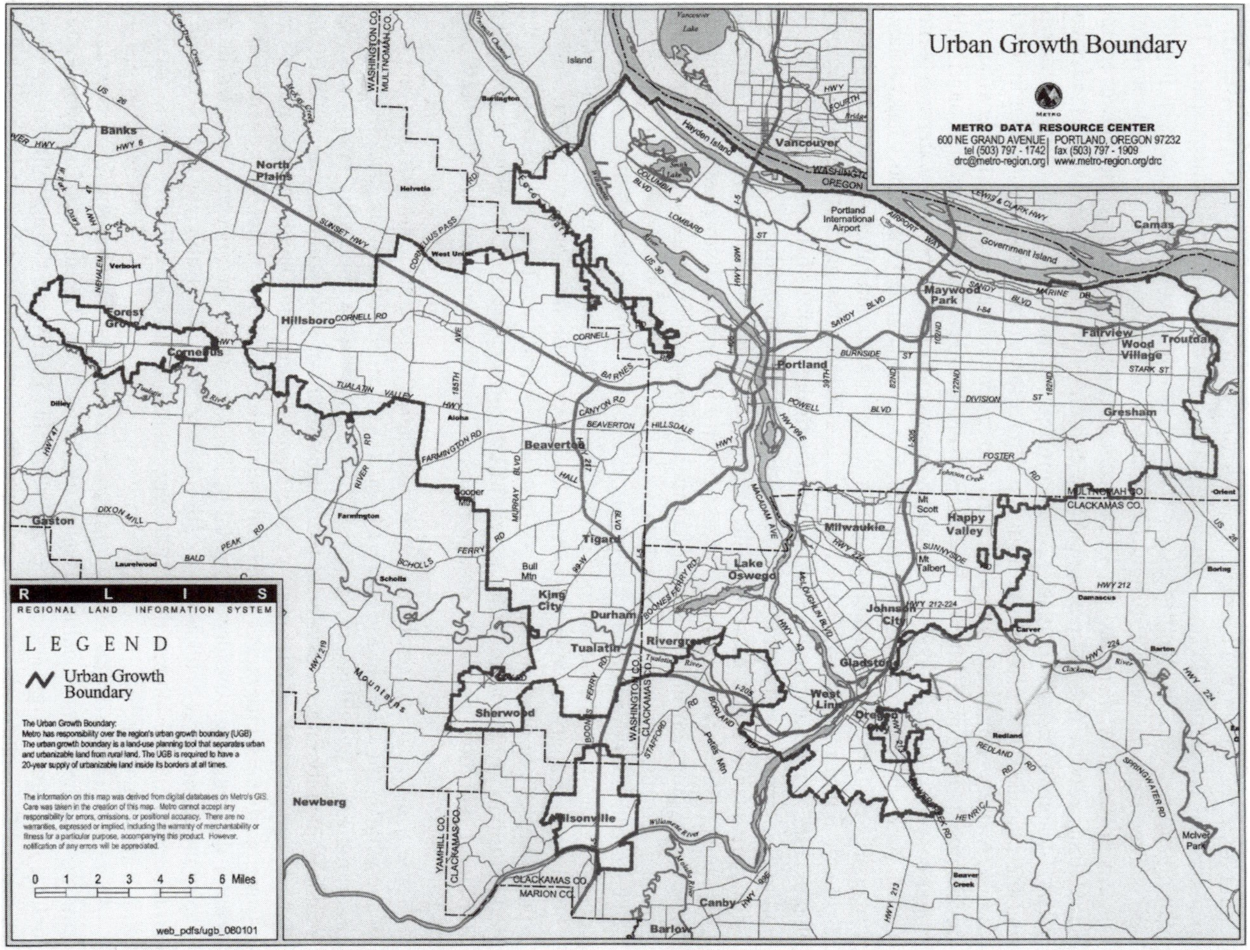

Urban Growth Boundary
METRO
METRO DATA RESOURCE CENTER
600 NE GRAND AVENUE | PORTLAND, OREGON 97232
tel (503) 797 - 1742 | fax (503) 797 - 1909
drc@metro-region.org | www.metro-region.org/drc
R L I S
REGIONAL LAND INFORMATION SYSTEM
LEGEND
Urban Growth Boundary
The Urban Growth Boundary:
Metro has responsibility over the region's urban growth boundary (UGB)
The urban growth boundary is a land-use planning tool that separates urban and urbanizable land from rural land. The UGB is required to have a 20-year supply of urbanizable land inside its borders at all times.
The information on this map was derived from digital databases on Metro's GIS. Care was taken in the creation of this map. Metro cannot accept any responsibility for errors, omissions, or positional accuracy. There are no warranties, expressed or implied, including the warranty of merchantability or fitness for a particular purpose, accompanying this product. However, notification of any errors will be appreciated.
0 1 2 3 4 5 6 Miles
web_pdfs/ugb_080101
Banks
North Plains
Hillsboro
Forest Grove
Cornelius
Beaverton
Tigard
King City
Durham
Tualatin
Sherwood
Wilsonville
Portland
Vancouver
Gresham
Fairview
Wood Village
Troutdale
Maywood Park
Milwaukie
Lake Oswego
Happy Valley
Johnson City
Gladstone
West Linn
Oregon City
Rivergrove
Canby
Barlow
Newberg
Gaston
Camas
Portland International Airport
WASHINGTON CO.
MULTNOMAH CO.
CLACKAMAS CO.
YAMHILL CO.
MARION CO.
WASHINGTON
OREGON

Figure 7.7 Portland's urban growth boundary provides certainty with respect to growth limits and ensures that economically prosperous farming remains close to the city.

that has followed urban development in city regions that do not have urban growth boundaries. But the most important impact of the UGB is that it maintains rural land values at substantially lower levels than urban land values, since development is not permitted outside it. Ethan Seltzer notes that in 2002, 86 per cent of Oregon's nursery products industry was in the metro area, and nursery products were the primary agricultural commodity in the state.[43] And prime farmland is a ten-minute drive to downtown Portland. The success of the UGB in controlling urban growth begs several connected questions. First, what is the role of natural landscapes *within* the urban growth boundary? Second, although the presence of the boundary promotes compact higher density development, does this also ensure sustainability when measured by high-quality urban design, liveability and sense of community?

Protecting natural landscapes within the boundary

A glance at Portland's urban growth boundary map (Plate 1 of colour plate section) tells us that it was drawn with no reference to natural systems or watersheds. In a simplistic sense, outside the boundary is where farming, forestry and rural low-density residential areas are protected. Inside the boundary is for urban growth. The land-use programme in the early 1980s, in fact, did not contemplate protection of natural features inside the boundary.[44] At the same time, all comprehensive plans, urban or rural, are required to address goals for the preservation of significant natural, historic and cultural features, although their protection was not a requirement. This resulted in many streams being buried, and others declared polluted by the Oregon Department of Environmental Quality.[45] Thus, a classic conflict existed within the urban boundary with respect to the protection of natural resources, particularly stream corridors and the banks of the Willamette River. On the one hand, their protection is imperative if biodiversity and the conservation of threatened species such as trout and salmon are to be maintained. On the other hand, it implies decreasing what planning has set aside as land that can be built on. To compound the 'problem' recent mapping of natural resources within the UGB showed a much higher level of bio-diversity than earlier mapping had revealed.[46]

Study of how to protect natural landscapes within the UGB began in 1994 with the Region 2040 growth management planning process. The vision for a regional growth strategy included the need to integrate urban, suburban and rural lands within a watershed context, to reduce downstream flooding, protect riparian corridors and wetlands and restore fisheries and wildlife habitat. Future development within urban reserves had to be sensitive to stormwater run-off, pollution and flooding of downstream communities. An integrated floodplain management strategy would recognize the multiple values of stream and river corridors as an interconnected biological system managed on an ecosystem basis. Metro, the Portland regional planning agency, declared over 16,000 acres off limits to development, an approach that subtracts sensitive lands from the regional plan before determining the region's capacity for development. To further ensure that natural landscapes are considered inside the UGB, Metro passed the 'Greenspaces Resolution' in 1996, a regional policy that encourages the provision of adequate parkland and fish and wildlife habitat protection and restoration, even if this necessitates expansion of the urban growth boundary. Other initiatives include a level of floodplain protection and protection of riparian and upland habitats.

The lack of protection for natural systems led to a grass-roots movement to press for a regional system of natural areas, trails and greenways for the Portland–Vancouver metropolitan region in late 1980s. The Audubon Society of Portland's urban conservation programme proposed the creation of a metropolitan wildlife refuge system. At the same time Metro initiated a regional park resources inventory, which led to the development of a new regional perspective for protecting and managing natural areas. This was incorporated into its Region 2040 growth management policies. It was argued that the Portland metropolitan area, since it was the agency responsible for regional growth, should assume the role of developing the strategy, and in 1992 the Metropolitan Green Spaces Master Plan was adopted. Its goals included:

- to create a cooperative regional system of natural areas, trails, wildlife greenways and people in the four-county metropolitan area;
- to protect and manage significant natural areas through partnerships with governments, non-profit organizations, land trusts and others;
- to preserve the diversity of plants and animals in the urban environment, using watersheds as the basis for ecological planning;
- to establish an interconnected system of trails, greenways and wildlife corridors.

By 2002, 40,000 hectares of parks and natural areas had been acquired.

New urbanism: giving form to design and liveability

The perceived formlessness, lack of identity, infrastructure and environmental costs of suburbia led to the promotion of an alternative to North American suburban form by the new urbanism movement. This replaces existing subdivision design with traditional forms of American city building. It borrows from local regional architectural forms and incorporates terraced housing in compact blocks, street grids and public spaces to give physical coherence to neighbourhoods. Mixed use and a range of housing choices replace conventional single-use zoning. Walkable neighourhoods and public transit replace going everywhere by car. The preservation of farmland, valleys and woods replace their obliteration under the bulldozer. Some of these principles are reflected in the Portland urban growth boundary. For instance, transportation planning in the Portland region has specific expectations for the role of walking and alternative means of transportation in specific areas of the city and in the region. Much attention has been given to public transit as a major structuring element of Portland, which has made it one of most walkable and appropriately scaled cities in North America. Transit usage is reported to have increased 143 per cent faster than the growth in population. It has also increased 31 per cent faster than growth in vehicle miles travelled since 1990.[47] There are also stated design expectations for main streets, town centres and regional centres, although 'the creation of clear and objective urban design standards is in its infancy'.[48] The new urbanism is also based on the premise that architectural design affects human behaviour and establishes a strong sense of place.[49] The urban design codes specified by new urbanism on street networks and their relationship to sidewalks and buildings, continuity of façade and public space are also a structural element for compact urban form. From a social and behavioural perspective, however, codes specifying the colour and design of siding, porches, fences, as the basis for building design, may be seen as highly prescriptive. There is, arguably, little difference between the new urbanism code from the standard municipal planning by-laws. The fundamental purpose of both is about control. The difference between what is structural from a planning or design perspective, such as engineering infrastructure or greenway systems, and what is design control over human behaviour, has considerable ramifications for the design of human communities.

There are several issues that bear on this situation. The first is about what shapes communities. From the new urbanism perspective, the design of the built environment influences behaviour, a view that has some validity. For instance, it has been found that if the distance between neighbours in a subdivision is small, visual contacts tend to be transformed into social ones.[50] Observation of how long-lived urban neighbourhoods have evolved and changed over many years to a high level of physical and social complexity have done so where a minimum of official controls have been applied. Toronto's Kensington community, for instance, began life rather sedately in the 1790s[51] and has evolved over 200 years into a colourful, vibrant neighbourhood and market, full of people and great architectural variety. House façades, porches, wall colours, the way front gardens are used, have all evolved in ways that meet the owner's individual needs and personality. Semi-detached houses may be neatly painted blue on one side and green on the other, signalling different owners. Additions to buildings on the market streets provide space for a shop with living quarters upstairs. 'Granny flat' additions have been built for a family relative over garages in the back lanes. In the tightly packed neighbourhoods of Kensington, gardens can be found on any structure with a flat roof. It has been called unplanned, chaotic and visually messy. It has also been called lively, socially and physically diverse, sensorily stimulating, safe, a place to find great characters, people who have lived there a long time, shops where one can buy almost anything grown anywhere, a place that has evolved largely on its own, within the physical and social infrastructure of the city but with a minimum of intervention from city by-laws. The transformation of Levittown in the USA over the last half century, from standard suburban housing to a physically and socially diverse community, is a testament to this phenomenon.[52] The by-laws and codes that are imposed on new housing developments tend to ensure conformity, inhibiting these social changes from evolving in the way they want. From a social perspective, the cultural diversity and richness of expression in the city ultimately comes about from the people who live in, and exert local control over, their own neighbourhoods within the overall structure of streets, open spaces and mixed use, and compact development. In terms of urban growth boundaries, the new urbanism brings with it the urbanity, spatial qualities, compact form, density and structuring elements necessary for urban life. The issue, however, is less about architecture than about the need for conditions that allow communities to develop on their own terms.

New urbanism principles recognize environmental criteria, such as protecting woodlands, river valleys, wetlands and other natural features, in its manifestos. While these are essential aspects of environmental design, there remains a basic need to recognize them on the ground. This involves an integrated approach to natural systems that links energy conservation, energy-efficient building, recycling of materials, water conservation, wind and solar energy production, natural ventilation and stormwater treatment, community gardens and the protection of local agriculture. While Village Homes in Davis, California reflects this model of sustainable design, its major limitation is its low density with 242 houses on twenty-eight hectares of land. Thus the challenge for smart growth lies in a holistic approach that incorporates the three interrelated facets of sustainability. First, flexible policies that permit communities to evolve in their own way. Second, an environmental approach to urban planning and design that seeks to contain sprawl, build compact higher density communities, protect local farmland and natural landscapes, reduce dependence on the automobile and restore air quality. Third, internalization of environmental thinking into all facets of urban growth.

Reflections on Portland's smart growth initiative

Portland's urban growth boundary is widely recognized as being successful in controlling urban sprawl. This is borne out in comments made by Congressman Earl Blumenauer, who represents the Portland area at a new urbanism conference:

(b)

Plate 7.2 (a) Kensington Market, Toronto. A traditional architectural and regional character on which new urbanism bases its new neighbourhood developments. Here, however, minimal controls and strong community spirit have given rise to numerous changes over its history in response to community needs. To some, it may appear disorganized and aesthetically messy, but it is a place that is full of life and vitality. (b) A new urbanism housing in Markham, Ontario. Strict urban design codes are integral to these developments. Both approaches to residential development are needed in the city, although how this place will evolve over time as a social community is unpredictable.

> you can see Sauvie's Island, prime farmland a ten minute drive to downtown Portland, flat and buildable. There is virtually the same amount of land in agriculture now as 25 years ago, which, but for our land use planning laws, would have all been lost. In Washington County next to Portland, despite the addition of 40,000 people between 1982 and 1992, annual farm income increased 57 percent. At the same time, neighbouring Clark County in Washington State, lost 6,000 acres and farm incomes rose only 2 percent per year.[53]

This key aspect of the urban growth boundary – protecting local agriculture – suggests that the *real* costs of food production and transportation – the ecological footprint – are greatly reduced (see Chapter 5). It also provides opportunities for establishing medium-density developments that are integrated into local small-scale market gardens, thus avoiding the classic incompatibility between suburban and farming values, where new developments edge out farming operations – the consequence of conflict between rural and urban cultures.

The concept of urban growth boundaries in North America as a strategy for controlling sprawl also provides the basis for protecting natural landscapes and biodiversity that can be integrated into development within urban boundaries. These, together with a compact urban form and densities that promote an efficient use of land and access to public transit, are all necessary steps towards a sustainable basis for living. They are also dependent on public involvement and support. How environmental policies are translated into reality will be the test of their success on the ground.

Nature as infrastructure

The town of Markham, Ontario

Thus far I have examined some of the issues of urban growth and how it can be contained or structured. The application of the greenbelt in Britain and the urban growth boundary in North America maintains countryside outside urbanized boundaries. Another way of understanding the issue of growth control has to do with the entire landscape: the river valley systems, forests, and wetlands and regional natural park systems and conservation areas, that must be understood within its bio-physical context. The concepts of linked park systems are planning strategies that focus first on these natural systems as a basis for organizing urban growth. The notion of nature as infrastructure is applicable, not only regionally, but also on a smaller urban scale. The following study, now incorporated into the official plan for a suburban township, examines landscape planning at the scale of a major municipality that has also recognized its regional and watershed contexts.[54]

The town of Markham is one of the fastest growing communities in the Greater Toronto region of southern Ontario. It is located about 8 kilometres from Lake Ontario and its northern boundary borders on the Oak Ridges Moraine. Its landscape is typical of the southern Ontario natural region – a flat till plain incised by forested river valleys. Three of these flow through the municipality, the most important being the ecologically significant Rouge valley system. The Rouge is one of the most significant remaining linked natural areas in the region. Much of it (4,250 hectares) was proposed, and later became, a park with the support of the Province of Ontario and the Federal Government.

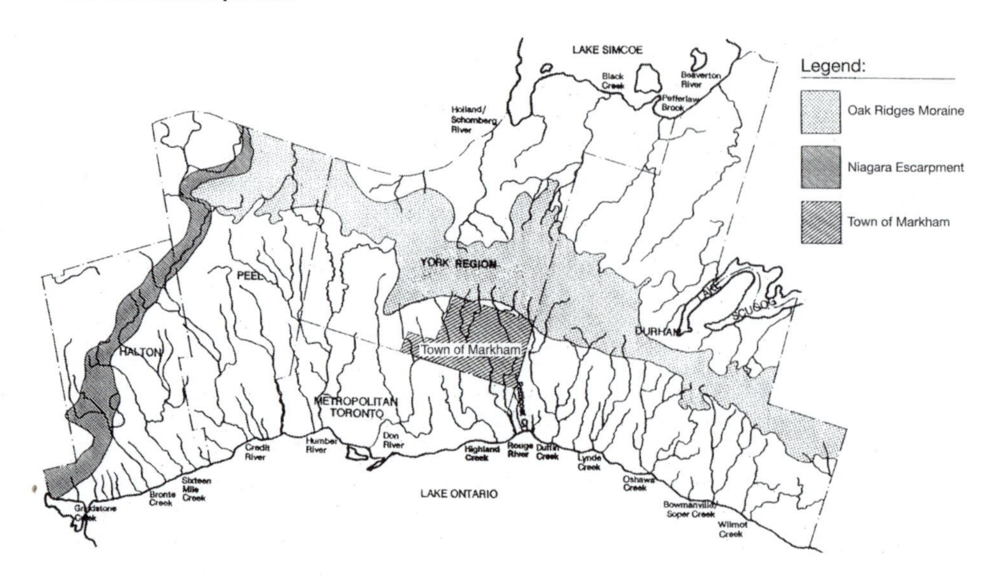

Figure 7.8 Location of the town of Markham within the greater Toronto bio-region. The town forms part of a remarkable regional system of rivers (see Colour Plate 2).

Source: Gore and Storrie Limited in association with Hough Stansbury Woodland Limited *et al.*

Markham's built-up area

The town's valley woodlands are all that remain in Markham after the tableland forests were cleared, first for agriculture and later for urban development. Today there are no more than a few scattered woodlots remaining outside the valleys. Overall, only 13.6 per cent of its land is naturally vegetated, most of which is anthropogenic (defined as naturally emerging vegetation on disturbed land, such as old field). On the tablelands, only 5 per cent remain in scattered woodlots. An accompanying low diversity of plants and animals is indicative of this fragmented and disturbed environment. The town had poor survey information on its remaining natural habitats, and incomplete planning controls to protect them, which meant that as development moved north, additional habitat losses would be almost inevitable.

The Natural Features study was initiated in 1993 as a response to a growing appreciation of environmental issues by the community at large, and a recognition that what remained in unimpaired valley lands had to be protected. Its overall purpose was to prepare a plan that would structure future growth in the town, develop a framework, or green infrastructure system, under the town's official plan to guide future development, and provide diverse and accessible recreational opportunities. Its objectives included preventing further losses of significant natural features, improving remaining ecological resources, and creating improved linkages between natural areas, creating large contiguous areas of natural communities, and increasing the total forested area in the town from 13.6 per cent to 21 per cent. Major natural nodes would be located to protect groundwater areas and as wildlife habitats. They would be a minimum of 100 hectares in size, spaced at 2.5 kilometres, together with interconnecting corridors 100 metres in width. The size of these areas was based on landscape ecology principles. A size of 100 hectares was selected for forested areas which will support viable populations of forest-dependent bird (interior) species and the potential to become self-sustaining ecosystems. A spacing of 2.5 kilometres apart is comparable to the spacing between nests of species with large home ranges, such as hawks, allowing each area to function as a discrete ecosystem without territorial disputes.

The plan for the Markham environment

The health and vitality of the natural systems depend primarily on dynamic interactions at several local levels: individual sites and habitats, sub-watersheds and watersheds, and the bio-region and the linkages between them. The plan also had to function within political and social spheres (see Plate 2, colour plate section). Since natural systems are interconnected, the policies, regulations and attitudes reflected in urban land uses had to be considered. And because ecosystems do not respect munipical boundaries, cooperation between local governments and agencies was necessary. Natural features also had to be accessible to the community and provide a platform for personal commitments to local places. The concept called for a framework that reconnects features *along* corridors; and connects valley corridors across tableland. Its intent was to protect and link isolated woodlots and marshes and establish new upland forest blocks, thus helping to restore the 95 per cent of forest cover lost over a 200-year period. The plan also proposed adopting new land-use and parks/recreation planning districts based on sub-watershed areas. This represented a basic shift to ecosystem planning within the municipal planning context. Strengthening potential habitat links between watersheds was also of considerable importance. The knee-jerk reaction to protecting valley systems in urban areas has traditionally accepted the notion that valley lands are for 'nature' and tablelands are for 'development'. From the perspective of biological diversity, however, cross-connections provide other travel route opportunities that are key factors in ecological networks. There was also the need to fully integrate the town's urban environment within its larger bio-region. This entailed substantial links southerly down all river valleys to Lake Ontario, northerly up the smaller streams to the Oak Ridges Moraine, and new east–west connections to adjacent watersheds.

The Natural Features study also identified a number of relatively undisturbed forested and wetland communities that contribute significantly to the bio-region by supplying cold water to cold water fish habitat, and conduits for migration and flows of water and nutrients. Protecting established natural communities in urbanizing regions was therefore of utmost importance in the plan. Several other factors were of significance to the strategy. First, there were areas of abandoned human activity where naturally succeeding habitats contribute to diversity and connections between undisturbed areas, and where active restoration could take place and fit into the patterns of habitat and linkage. Second, there were land uses that could be integrated into the planning framework as both natural linkages and multi-seasonal trails, serving a variety of recreational functions appropriate to each place. They included local parks, cemeteries, residential lands, such as backyards, golf-courses, agricultural and rural lands (see Chapter 3, pp. 123–4).

Implementation: official plan policies

The Natural Features study outlined ways in which the strategy could be implemented in the context of the town's official plan. It provided a strong set of policies for the environment that was consistent with the Ontario Planning Act, and how the official plan should be amended to protect natural features and establish a greenway system.

For instance, long-term goals for Markham included:

- to increase the extent and quality of natural areas;
- to protect and enhance natural features to achieve ecological diversity, recreational opportunities, healthy living environment, high-quality water and so on;

- to encourage development that respects the natural and cultural attributes of built up and rural areas, improvement of environmental quality;
- to incorporate the ecosystem approach into all facets of land-use planning.

Individual policies include:

- The greenway system. Includes valley lands, vegetation complexes, woodland parks and public open spaces, low-impact recreational trails. The system is intended to support natural processes, provide access to natural areas and link with the Rouge Valley Park, Oak Ridges Moraine and adjacent valleys in the region.
- Greenway corridors. Includes valley land corridors, golf-courses, parks, cemeteries.
- Greenway nodes. Includes large contiguous areas which, through restoration, will add significantly to Markham's natural diversity.
- Activity linkages. Serving to link built-up areas with nodes and greenway corridors.
- Watersheds and sub-watersheds. Involves recognition of the sub-watershed as the primary planning unit, protection of recharge and discharge areas, restoring water quality to support fish populations.
- Stewardship. Development of a comprehensive landowner stewardship programme including public education and involvement in decision-making.
- Buffers. Development on lands adjacent to the greenway system should be subject to appropriate controls that include metre buffers to protect its features and corridors These should be under public control and consistent with the Toronto and Region Conservation Authority definitions.

Developments since 1993

After the Natural Features study was approved by the Town Council, an amendment was introduced in 1997 to incorporate its proposed policies into the official plan, to protect the town's valley lands, corridors and buffers, for its urbanized lands and the lands not built on to the north.[55] The amendment provided 'a plan for the Environment and a framework for achieving ecologically linked, healthy and diverse natural features' (see Plate 7.3 and Plate 3 of colour plate section). Its goals were to prevent further losses of significant natural features, address the improvement of remaining natural features, create further forests on tableland and to create a sustainable linked open space system within the town with opportunities for connections with regional natural systems. The amendment's objectives addressed the need:

- to encourage awareness among residents, employees and landowners of their responsibilities for maintaining and enhancing all natural landscapes;
- to require development to acknowledge and respect these landscapes;
- to integrate the town's greenway system with broader inter-regional systems that perform a similar role;
- to retain existing wetlands, establish lowland vegetation and retain existing cold water fisheries.

Some of the amendment's requirements include:

- Sub-watershed studies should be undertaken in planning for new development.
- Town support of linked greenways to provide continuous trails linking the town's greenway system with the Rouge Park, Oak Ridges Moraine and the Don River valley.

Plate 7.3 Markham Centre, a new urbanism housing development, has responded to the Markham plan respecting Rouge valley edges and sensitive natural features (see colour Plate 3).

- Assistance in the public acquisition of tableland natural features.
- Consideration by the town for the retention of these natural features in private ownership.
- Permitting golf-courses in some protected areas if it can be demonstrated that their development will preserve and protect these lands, protect streams from pollution, and provide public paths needed to achieve greenway linkages.
- Design stormwater management systems to ensure quantity and quality control and to provide a net benefit to the environment.
- Establish reciprocal arrangement to accommodate stormwater flow across municipal boundaries and develop a master drainage agreement.
- Develop a linked trail throughout the town to connect natural features of local and regional significance.
- Encourage developers and builders to provide buyers with educational materials on the Open Space Master Plan to promote understanding of the importance of linked open space systems and the need to share responsibility for environmental management.

By the turn of the century, all not built on lands had been developed, an area of 2,400 hectares. A new compact town, Markham Centre, was proposed on lands in the built-up area – all developments following the ecological principles and requirements of the Natural Features study. And the Provincial Government had created the Rouge Park, the largest nature reserve in Southern Ontario. The long-term intent was to secure over 4,450 hectares of land in the Rouge River watershed, extending from Lake Ontario to the Oak Ridges Moraine.

Some concluding observations

The development of ecological criteria, enshrined in an official plan, for protecting natural corridors prior to development, represented a new standard for the town, made possible by a strong community and political will that has continued since the 1990s. This case study also raises a number of issues that have significance for sustainability in the regional landscape. First, the valley systems should be conceived of as 'green infrastructure', the overall structuring element that should influence the form and character of development, which suggests that built form and landscape are interdependent. The mandate for the Markham study, however, saw future development and the landscape as separate issues. Protecting valley landscapes involves strategies that integrate sustainable building design with natural processes.

Second, sustainable building design (green architecture) involves designs that protect and support natural systems, manage rainwater from hard surfaces at source, by infiltration, storage and green roofs. It is about appropriate energy technologies that do not contribute to the production of greenhouse gases.

Third, the relationship between communities and the natural valley landscape is about environmental learning, an idea that was incorporated, to some degree, into the official plan. It involves an awareness, on the part of local communities, of how to protect natural landscapes, and what plant species should, and should not, be incorporated into gardens associated with natural valley landscapes. It is about controlling pets from invading protected areas and the species that live in them, and about stewardship of the landscape. In effect these are about stewardship.

Fourth, green infrastructure is about community identity. The valley systems, tableland forest habitats and greenway linkages through the region form natural structuring elements. They determine where development may be located and provide the foundation for investing in local and regional places that arises when there are definable boundaries, varied landscapes, and access to parks and natural areas.

Regional planning and brownfields

Urban renewal in Germany's industrial Ruhr

The growing trend in cities to rebuild on 'brownfield' sites and avoid the 'greenfields' outside built-up areas, previously discussed, carries with it the task of revealing the 'special' in 'ordinary' places. It lies in recognizing the interrelationships between the physical and biological processes, the human histories that have made these places what they are, the human processes and social and economic needs that can ensure their future health. The following discussion of Germany's Ruhr Valley industrial region is about the renewal of brownfields, the significance of their industrial history, and the restoration of nature. These visions have been understood and acted on in ways that combine the will, political commitment, and the imagination to create quite remarkable places. The first implementation phase of the Emscher Project has been realized and can now be experienced.

Industrial renewal: combining ecology, heritage and economy

Strategies for the economic renewal of old industrial areas are the subject of intense discussion in Europe, the UK, North America and Japan. In Germany, the need for major restructuring has become most obvious in the Ruhr Valley where once prosperous industries based on coal and steel have become obsolete. They have left behind a ravaged landscape, large tracts of derelict land, massive subsidence from coal-mining, channellized rivers that carry both wastewater and rainwater, high unemployment and a depressed economy. In 1988 a decision was made to initiate economic and environmental renewal of the Emscher region within the Ruhr, an area of approximately 320 square kilometres with a population of 2 million. An 'International Building Exhibition' (IBA) (an implementation instrument with a long tradition in German architectural circles) was set up to initiate this process. The IBA saw its task as a coordinating agency rather than a planning authority, setting out development directions and targets, providing know-how, bringing potential partners together. It restricted itself to making interventions and alterations in selected places in the hope that this process would provide the seed developments in other locations.[56]

On a regional scale, the major projects include the ecological rebuilding of the Emscher River watershed, naturalizing channellized streams and establishing biodiversity where this can be realized. Six new decentralized sewage treatment plants were constructed throughout the watershed, as well as 320 kilometres of underground sewers, stormwater infiltration, and the separation of rainwater from wastewater. The programme also included residential and industrial developments, new social, cultural and sports activities, and the protection and creative reuse of a remarkable industrial heritage that has become a centre for theatre, art, recreation, tourism, and a focus for cultural institutions.[57]

Begun in 1989 and completed ten years later in 1999, the IBA brought together specialists and politicians to act as catalysts for change, to promote high-quality city planning, architecture and the arts, and provided a forum in which local authorities, private companies and citizens could develop projects. Many of these were financed jointly by the state, the cities and private enterprise, while others, such as the Emscher parks, were funded exclusively through public funds. Financial investment for the Emscher Park project over a ten-year period was some DM 2 billion, coming in equal measure from public and private sources.

Creating regional linkages

Between 1989 and 1999, the Emscher Park IBA played a crucial part in upgrading the region. It was established to provide the main unifying theme, the central core, of a new 'green' infrastructure for the region based on the principle that ecological upgrading is intended as the basis of fresh economic impetus.[58] The boom in coal and steel that took place early in the twentieth century subjected the region to a rapid and chaotic period of building in response to the need to house the workers who were pouring into the region. In response to this rapid industrialization, regional planning was introduced in 1920. Its purpose was to improve quality of life in the towns, improve traffic conditions, provide recreation areas and restore a badly degraded landscape. The Ruhr Coal Area Settlement Association, founded to begin this process, had legal powers to acquire and protect open space from development and in 1966 issued the first modern area development plan.[59] In 1976, the Association relinquished its planning powers to regional planning councils but retained its legal duty to acquire and conserve open areas.[60] The Emscher Park IBA continued this work, establishing the greenway system, as well as restoring vacant industrial land, preserving and finding new uses for the monuments to an industrial era.

The park, more than 320 square kilometres in size, connects seventeen towns and abandoned ironworks and coal-mines from Duisburg in the west to Bergkamen in the east. Its purpose is to achieve lasting improvements in the living and working environment

Figure 7.9 The Emscher Regional Park, a long-range strategy for the industrial region that will ultimately link seventeen towns and industrial focal points together to create an interconnected green infrastructure system. It covers an area of 320 square kilometres. Note the increase in size.

Source: IBA Emscher Park.

of more than two million inhabitants, by connecting isolated open spaces, restoring the landscape, and upgrading the ecological and aesthetic quality of the countryside. Its long-term aim is to create a connected area of high environmental value through the middle of the connurbation, accessible by public transport. Its main source of funding is the 'Ecology Programme in the Emscher-Lippe area' created specially for this purpose by the State of North-Rhine Westphalia. Politically, all these cities form a federation along the main artery of the Emscher River, the park acting as a spatial and conceptual entity unifying all IBA projects that encompasses housing developments, working and recreational areas. The preservation of the large industrial complexes was seen as an essential part of the overall strategy for several reasons. They played a role in local memory and have been put to a variety of new and highly creative uses. They also establish identity, a key function for articulating the terrain and assisting orientation, by creating a regular rhythm in different configurations, at an enormous scale, throughout a large region.

Regeneration

The greening, mostly through natural regeneration, of 10,000 hectares of industrial wasteland over twenty years, is a powerful expression of hope for the future. It is also a symbol of the passing of a twentieth-century era which built the great monuments to industrial production that will not be built again. The strategy in creating the Emscher Park is long term, 'not something to be achieved in a legislated period but a task for a generation'.[61] One might also say that this is 'a process begun but with no finishing post' (see Plate 7.4 and Plate 4 of colour plate section). The places completed in the

Plate 7.4 The Stone Dump, an enormous sculptural symbol of the industrial region. Situated on a created mound rising 52 metres above its surrounding landscape, it was constructed from blocks of scrap concrete from demolished sites in the Ruhr area and is itself 20 metres high.

Artist: Herman Prigann.

green network are growing together piece by piece – a regional cycleway, spectacular and creative reuse of the Thyssen steel works, the 'Stone Dump' atop a man-made hill of colliery spoil overlooking and visually linking the region to local places. These represent the new image of the Ruhr District and its mix of industrial artefact, culture and reclaimed countryside. The Emscher Park is a new landscape, its topography reshaped and transformed by industrial activity and now recreated by a combination of natural processes and cultural expressions of landscape design, architecture and the arts.

New life emerging

How does one reconstruct a previously ravaged landscape thousands of hectares in area? The answer, as the Germans discovered, was to leave it to nature. Since the cost of creating landscapes the conventional way would be prohibitive, doing nothing provided the 'possibility of creating a new type of open space' on a vast scale.[62] Economy of means is the appropriate principle for leaving the ground plain of derelict industrial land to reforest as it may – what was called the Leftover Land Project. The massive pioneer forests of birch, willow and poplar emerging in open land and within the foundations, steel piping, rail lines and support structures of these monuments to heavy industry leave one with vivid impressions (see Plate 7.5 and Plate 5 of colour plate section). There is an utter contrast and yet harmony between industrial building and emerging woodland. There is a sense of mystery between the decay of man-made works and biological renewal; of ancient Hindu temple ruins emerging from the jungle; of renewed life taking over from an industrial past; of animal, bird and insect life re-establishing itself in a naturalized habitat reinvented by human activity. This landscape is sufficiently large and interconnected, which is indeed the intent, to provide viable wildlife habitat and travel

Plate 7.5 The Emscher forests. At ground level, the visual complexity of industrial engineering and emerging vegetation is clearly apparent (see colour Plate 5).

corridors for a variety of common species as well as for people. But it is essentially an inhabited forest that will require a level of long-term management to maintain the emerging vegetation in relation to human activities. The Emscher Park is a regenerating landscape, its topography reshaped and transformed by mining and industrial activitity, and now recreated by a combination of natural process and cultural expression.

Industrial heritage: integrating preservation with the arts and economic renewal

The decommissioned ironworks coking plants associated with different towns provide orientation points for the region and are celebrated, for the dramatic scale of their industrial engineering, for their significance as monuments to the past, and for economic and social renewal. But they are seen as a great deal more: as venues for celebrating the arts, architecture, and landscape design and planning – celebrating ecological and social renewal, since technology, business, and service industries are now the economic basis for the future of the region.[63] The coal and steel industrial complexes have acted, and continue to act, as the seed for attracting other arts groups, musicians, associated businesses and tourists. For instance, a piano company decided in 2002 to set up its business in the Zollverein coke plant,[64] and other businesses associated with the arts will follow.

There are many other aspects of this project in urban renewal that are of considerable interest. Colliery tips and heights of land have become the focus for sculpture, celebrating panoramic views of the larger landscape, as regional orientation markers, and recalling, in their materials and form, the era of heavy industry and the region's future. Theatres have been built into factory buildings and outdoor courtyards for summer outdoor performance. Music is performed against a background of old industrial equipment and pipe systems. A museum has been built in an adapted gasometer. Night has become an art-form with evening cultural activities and factories dramatically lit. Approach roads to the Vollverein coke plant are street lit in different colours for orientation. In the ironworks at Meiderich the gasometer that once supplied energy to the industry has been filled with water and transformed into a deep-water scuba diving pool complete with a reef made of industrial objects. An industrial pond at ground level has been transformed into a wetland. Adaptations of industrial structures accommodate new uses but retain their essential industrial character and imposing scale (see Plates 6 and 7 of colour plate section). Whether they are theatre, music, dance, restaurant design or meeting place, the visitor is always intensely aware of the place.

While the political, economic, planning instruments and conditions that have established this project are unique to Germany, the basis on which it has been founded provides unique opportunities for the future of declining industrial areas. The geographic dimensions of this planning project are what impresses those from abroad. It is a special quality lacking in most North American revitalized brownfields. It has operated beyond boundaries and across different planning domains. The essential idea involves the notion that new economic impetus is based on ecological health.

What are the future directions? With the completion of the ten-year IBA, it became evident that funding by the region should continue until 2006 to maintain existing projects and build others. The seventeen town councils agreed that the Emscher Park should continue to be run as a state agency with a steering committee of local councils representing the Ruhr region. Planning would also determine what to invest in after 2006, and what should be regional and local responsibilities. The Ruhr remains a region in decline, with 6,000 hectares of brownfield sites.[65] Yet at the same time, it is the potential for positive change that has been created which is likely to continue to attract business, jobs and arts organizations to these centres of ecological, economic and artistic renewal. The fundamental green infrastructure of the Park is established at a regional scale. How development will emerge and grow over time is for the future to tell.

From islands to networks

Cities and protected areas in the macro-region

So far we have been examining the urban region, regeneration, containing growth and sustainable communities, and how the landscape can become an organizing framework for urban growth. We need now to expand the regional theme of this chapter to include cities and their relationship to wild areas of high ecological integrity in relation to their natural regions. In earlier editions of this book, I stressed that the perceptual separation between city, nature and countryside has long existed. Within urban regions today, there are indications that this exclusion has begun to break down. We are recognizing that nature and urbanism are beginning to merge. But when we turn to national parks and protected areas beyond the cities it becomes clear that the disconnections still exist. This is partly because these special places are mostly far from the cities. Getting there is limited to the yearly holiday or special event, reinforcing the perception that the parks are remote, untouched, pristine wilderness, protected for all time. But this is an image not supported by reality. Most are threatened by unsustainable numbers of visitors, water pollution from inefficient, worn-out infrastructure, tourism developments complete with green lawns, horticultural gardens and golf-courses. All these activities are urban in origin and are issues faced by the cities. Yet the value of these protected areas lies

Plate 7.6 The early National Parks in North America were established to protect dramatic mountain scenery, not ecosystems.

in their very ecological and social purpose, to protect unimpaired functioning natural systems, for what they can teach us about the very basis of life on earth. So meaningful links and outreach between cities and protected areas, no matter how far distant or close these may be, are crucial at every level of human experience. It is in this light that the remaining pages of this book will explore these relationships.

Habitat fragmentation and urban growth are among the most crucial issues facing both cities and protected areas. The early national parks in North America were established in the late nineteenth century at a time when little was known about ecological networks, ecological integrity, wildlife movement or natural corridor functions. Their boundaries were, in fact, originally drawn to protect scenery, not ecosystems or wilderness. It is not surprising, therefore, that the mountain parks in North America were established for their scenic beauty that continue to be highly prized and a major draw for international visitors. As Ruskin proclaimed over a hundred years ago, 'Mountains are the beginning and end of all natural scenery'.[66] After Banff was established in 1885, most people assumed that protected areas were safe for all time from the advancing tide of human development. The significance of this perception, and how wrong it has been shown to be today, was illustrated in a 1987 survey of major mammal populations throughout the western North American parks. It revealed how dramatically park boundaries are failing to protect the diversity of species originally found in the regions in which the parks were established.[67] Parks have become isolated islands in surrounding regions of conflicting land uses that include forestry, mining, urbanization and agricultural development across Canada and the USA.

In the late twentieth century, ecosystem-based management and the maintenance of ecological integrity brought with them a new way of thinking about managing protected areas. It was necessary to understand them in a regional context that emphasized the need to fully integrate protected areas into 'regional and local land-use planning and into all government land allocation processes'.[68] This would include national parks, wildlife sanctuaries, coastal areas, heritage river systems and other significant protected places in the larger regional landscape. In response to the overwhelming evidence that isolated protected areas cannot conserve species diversity by drawing lines on a map, environmental and non-governmental organizations today play a key role in initiating new citizen-led approaches to conservation, that focus on systems of protected areas, corridors and ecological linkages.

Protected area strategies

The year 2001 marked the international year of the volunteer that included many organizations initiating important initiatives. In Canada the national Endangered Spaces Campaign led by the World Wildlife Fund was a ten-year national strategy to complete a representative protected area network of ecologically significant sites. For instance, the Muskoka Heritage Foundation, a grass-roots community organization dedicated to protecting the natural and cultural values of the Muskoka region, established the world's first Dark Sky reserve. It was dedicated to protecting areas of wild land from light pollution, where pristine and unobstructed night skies are visible for star gazing and astronomy, and where the experience of night ecology and wildlife in remote areas may be had. In Britain, Plantlife, an organization dedicated to saving wild plants throughout Britain and Europe and a driving force behind the global strategy for plant conservation, reported that representatives from thirty-eight European countries would, after their 2001 conference, demonstrate a cohesive approach to plant conservation across the European continent. The idea of approaching plant conservation regionally, through the European Plant Conservation Strategy, then became a vital part of the global strategy.[69]

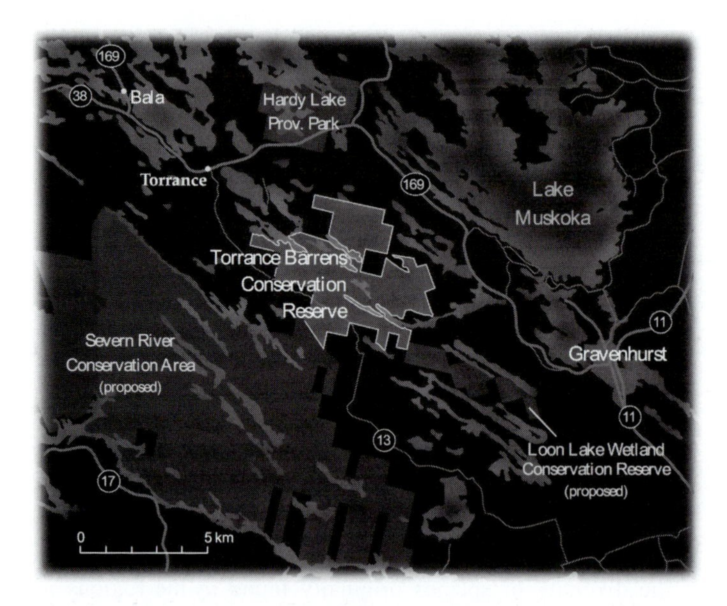

Figure 7.10 The Torrance Barrens, Dark Sky Reserve, in the District of Muskoka, Ontario, is protected for interpreting the ecology of the night and as a place to view the night stars in the absence of light pollution. The reserve was designated in 1997 and is the first of its kind in North America.

Source: Steven Fick, *Canadian Geographic.*

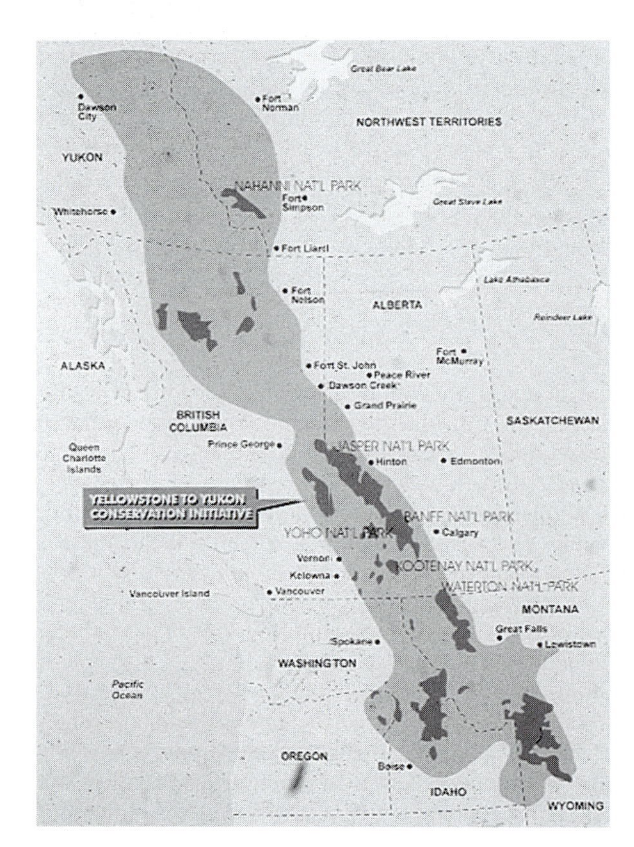

Figure 7.11 The Yellowstone to Yukon Conservation Initiative is a major community effort to establish a continuous corridor north–south across the Rocky Mountains that encompasses all the mountain parks. Non-profit conservation organizations such as this are making major contributions to creating networks of parks and protected areas.

Source: Rob Buffler, *Y2Y Initiative brochure.*

Networks of protected areas should be connected to each other through measures such as linking corridors and 'stepping stones' between core areas. The connectivity approach has emerged in the concept of the Pan-European Ecological Network (PEEN). In 2002 ministers from fifty-four countries in the UN–ECE region endorsed a resolve to establish the network by 2005.[70]

There are massive efforts by community groups in North America and Europe to protect natural areas, with the national parks acting as focal points of ecological integrity, many of which have set their sights on establishing connections on a very large scale. There is a recognition that an understanding of the range of different key indicator mammals, such as the wolf and grizzly bear, is the only way of protecting these species, using biology rather than politics to establish boundaries. Perhaps the largest and most complex effort is the Yellowstone to Yukon Conservation Initiative that belongs to a new global family of far-sighted, broad-based bio-diversity strategies that have arisen in response to the lessons of conservation biology. The Yellowstone to Yukon initiative is 'a vision for the future of the wild heart of North America, the vision of a bright green thread, uncut by political boundaries, stitching together 1800 contiguous miles of the Rocky, Columbia and Mackenzie Mountains', all the way from Yellowstone Park in the USA to the Yukon in Canada.[71] The 'Y2Y' initiative encompasses all the mountain parks within one corridor, forming one very large contiguous system. To protect bio-diversity it is necessary to protect much larger areas of habitat, to begin to think and act on a scale larger than ever before in the history of the North American conservation movement. The mission is to build and maintain a life-sustaining system of core protected areas and connecting wildlife movement corridors, both of which will be further insulated from the impacts of industrial development by transition zones. Existing national, state and provincial parks and wildlife areas will anchor the system, which the creation of protected areas and the conservation and restoration of critical segments of ecosystems will provide the cores, corridors and transition zones needed to complete it (see Fig. 7.11).[72]

Collaborative planning: linking urban and protected areas

These macro-regional networks of protected areas have great relevance to the urban regions. The cities themselves are increasingly developing networks of significant natural features and corridors, to protect wildlife, create linear recreation routes and structure urban growth. Their other function, given the context and opportunity, is to establish connections between these networks of protected areas and the urban regions via their river systems, major natural features and park systems, that act as a conduit for species migration and trail systems in and out of urban areas. Toronto's waterfront on Lake Ontario has species of birds not normally seen in cities, such as barred owl and Peregrine falcon, that travel through the river systems of the region.[73]

Regional landscape planning to establish ecological networks that can 'link urban areas' to protected areas is already being applied. For instance, Georgian Bay Islands National Park in Ontario is one of the smallest of Canada's parks, consisting of fifty-nine islands with an area of 25 square kilometres of spectacular scenery (see Fig. 7.12). It is also the closest national park to the Toronto centred region, the largest human population in the country. Due to the park's small size, maintaining ecological integrity can be achieved only by a collaborative approach of provincial agencies, community-based nature conservancies, cottage associations, and the District of Muskoka, all contributing to the bio-regional conservation network. Collaborative input into the municipality's land-use planning brought together existing, designated and proposed protected areas and corridors, all of which were incorporated into the district's official plan (see Figs 7.12 and. 7.13).[74]

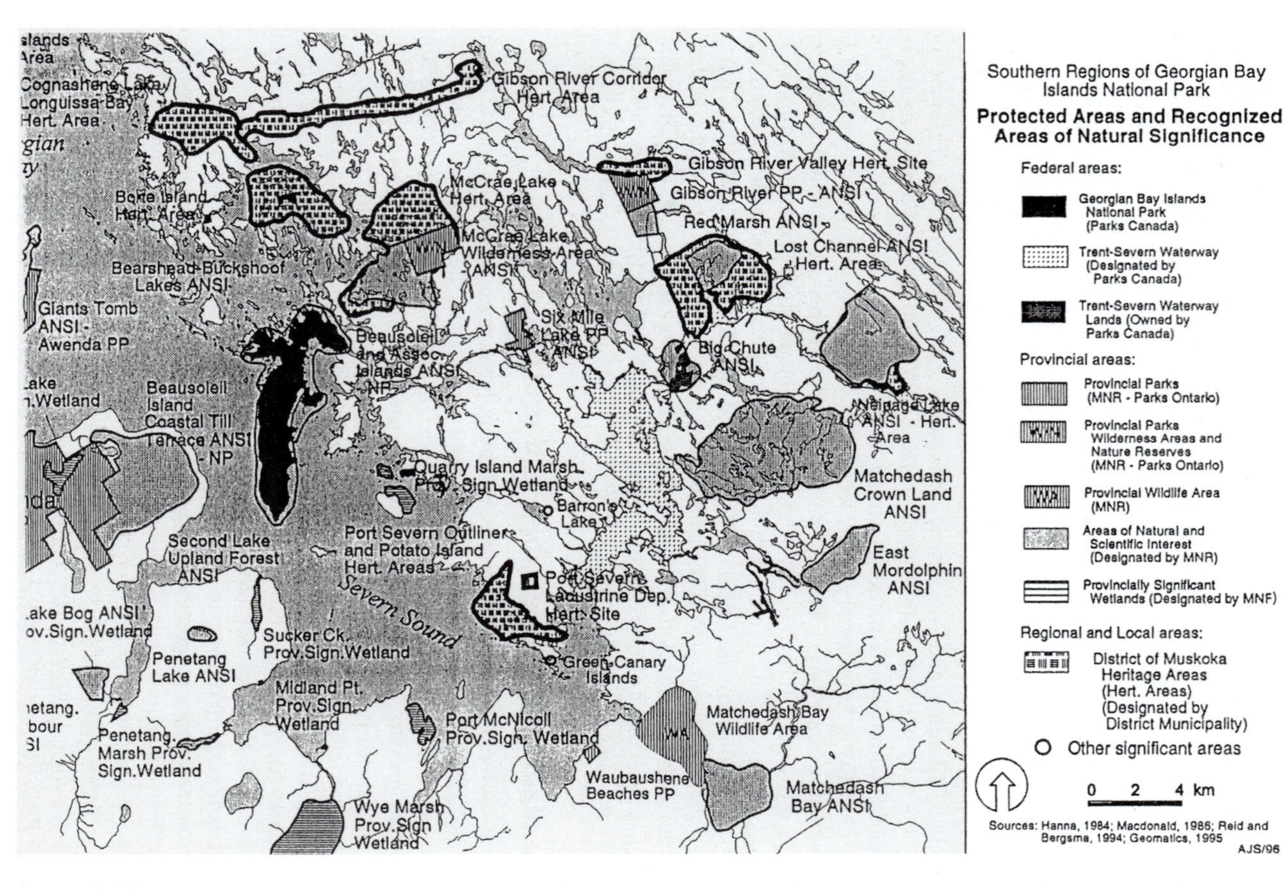

Figure 7.12 Individual protected areas and recognized areas of natural significance in the District of Muskoka. An example of where a rural district adjacent to a major urban region and the National Parks Agency have collaborated to develop a plan that links the small national parks and significant natural features in the region (see Fig. 7.13).

Source: Parks Canada Agency, District of Muskoka.

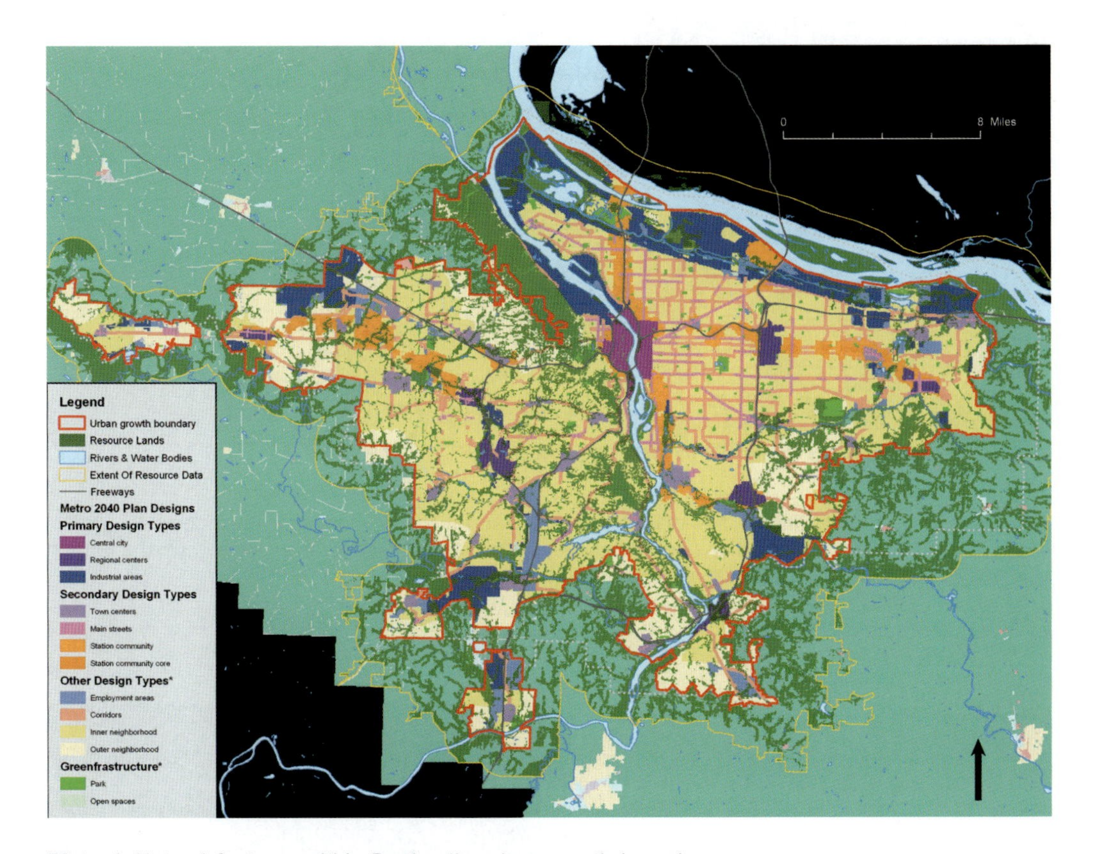

Plate 1 Natural features within Portland's urban growth boundary. The map shows the main built-up areas of Portland and the complex natural areas within (and beyond) the UGB to the west, south and east. When natural areas are incorporated into overall planning, the amount of land available for building is considerably less than the growth boundary lines drawn on the map in Figure 7.7. For example, within the 92,000 hectares overall area of the UGB, the natural areas have been calculated to be approximately 22,000 hectares. When the existing urbanized areas have been accounted for, the total buildable area of land in the UGB is about 12,000 hectares. Thus, when streams and valleys, forests and wetlands become part of the planning process that determines where development can be located and where it cannot, the amount of land appropriate for development shrinks considerably. It forms the basis for understanding the natural landscape as the infrastructure that shapes the urban environment, and a basis for quality of life and sustainability. This has been recognized by Portland's Metropolitan Region planner and was formally integrated into regional policy by adoption of the Metropolitan Green Spaces Revolution in 1996, although the regional government has yet to implement policies at the local level to ensure these areas are actually protected or restored.

Source: Justin C. Houk. Associate Regional Planner and GIS Analyst. Metropolitan Portland Planning Department. Mike Houck, Urban Naturalist, Audubon Society of Portland and Chair, Natural Resources Group.

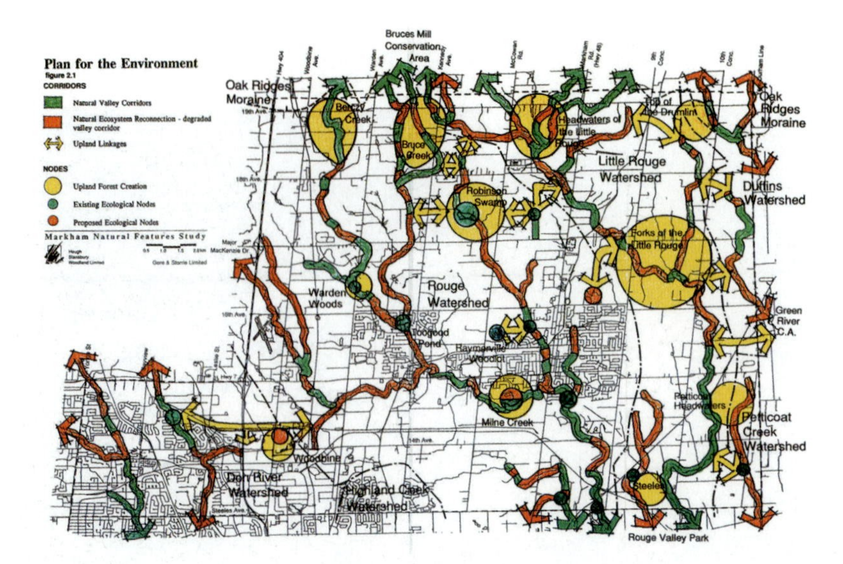

Plate 2 Markham's strategy plan for the environment. River corridors, cross-connections between valleys, large forest regeneration areas form the natural infrastructure for future urban growth.

Source: Gore and Storrie Limited in association with Hough Stansbury Woodland Limited *et al.*

Plate 3 A tributary of the Rouge River with the new development in the background, illustrating the effectiveness of the environmental strategy for the town. Avoiding the river forms only one component sustainable development, however.

Plate 4 Understanding the sense of the region from the local place. The steel mills and coal-mines of the region have created a totally different landscape from what was there before. Rather than attempting the impossible task of returning the landscape to its original 'natural state', it has celebrated the new industrial landscape for what it is, a symbol of regional renewal. The steel mills and coal-mines have been recognized as significant monuments to an industrial past. As local places they stand above the towns and regenerating forests, powerfully reinforcing the understanding of the region as a whole. Monuments such as this have also been created as sculptural elements using industrial materials.

Plate 5
The emerging forests penetrate industrial complexes and give a strong sense of renewed nature, of hope for the future, while we are reminded of the past. Pragmatically, this approach of allowing the forests to succeed naturally has greatly reduced the prohibitive costs of planting them.

Plate 6
The ‘Piazza Metallica’, one of the squares in the Meiderich Smelting Works, a place for festivities, theatre and meeting. No changes were made to the industrial surroundings. The floor was made with forty-nine industrial steel plates, each weighing 7 tons, that had once lined several foundry pits.

Source: Peter Leidtke.

Plate 7 A concert in the Meiderich power plant. Everywhere, different activities take place in industrial settings with no attempt to modify them beyond the necessities of the new use.

Source: Peter Leidtke.

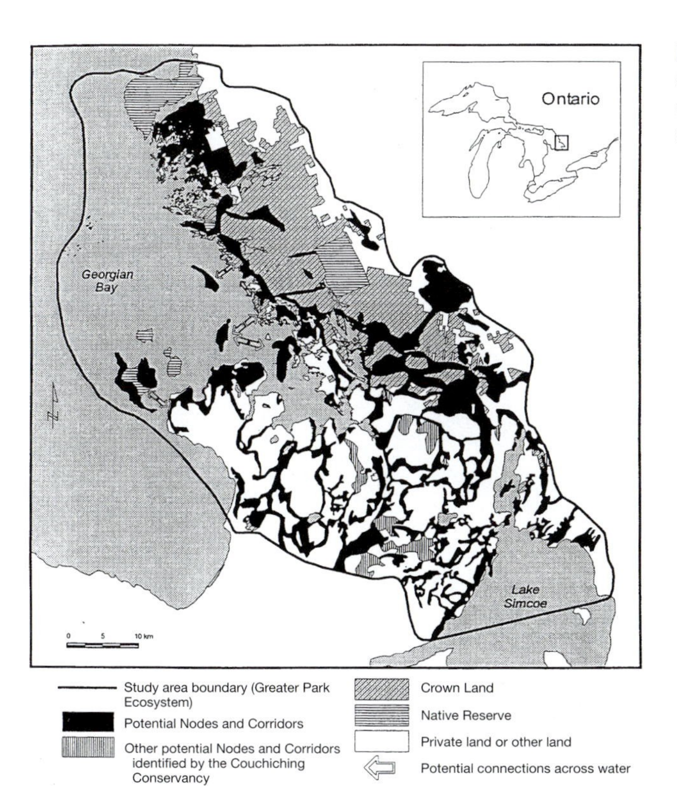

Figure 7.13 The potential connected system, greatly enhancing the ecological integrity of individual areas.

Environmental learning: linking cities and protected areas

In Chapter 1 I discussed the principle that environmental education begins at home. Learning about natural systems in the places where people live involves a continuous process of experiencing and exposure to one's local surroundings from which environmental understanding can emerge. Yet awareness of ecological integrity for visitors begins in the parks far from home. The once-yearly, or less frequent trip, to these special places is likely to leave a memory of an experience but little else. The memories of great mountains reflected in blue lakes, sharp, clear air and untouched forests are aesthetic impressions, which may be masking serious environmental problems. The town of Banff in the Canadian Rocky Mountains illustrates this issue. Banff was historically included within Banff National Park boundaries, and has subsequently grown into a major tourist town that receives five million visitors a year.[75] What is invisible to the average tourist enjoying the picturesque mountain scenery is the fact that the town sits astride a montane that provided crucial overwintering habitat and a movement corridor for many large mammals. By cutting off movement for predator and prey, the town's location resulted in habitat fragmentation and severe ecological impact. Infrastructure, such as sewage treatment systems, may, unknown to the visitor, be polluting lakes or streams. Wild animals get used to hand-outs from visitors on park highways, a dangerous habit that inevitably results in the animal becoming a hazard to visitors who may be unaware of the harm in feeding them.

The conventions of aesthetic values frequently become anachronistic when one starts thinking ecologically. While no one would argue against the values of an appropriate aesthetic, the conventions of formal design, as we saw in Chapter 1, have a way of

Plate 7.7 The five million visitors to the town of Banff are generally unaware of the ecological impacts on the Park when major migratory routes are blocked and impede wildlife movement. Following a shrinking of the town's boundaries wildlife passage began to be restored.

subverting genuine insights into what lies beneath the surface of the visible landscape. What makes one place beautiful and worthy of protected area status and another not, some landscapes special and others commonplace? Such value-laden questions suggest an inherent dilemma between those places we learn to admire and those we learn to ignore. The parallels between scenic urban landscapes and those of wild nature are evident. The conflict that arises between use and protection in the national parks and protected areas of North America is a constant threat to their purpose and goals – to protect ecological integrity for future generations. The management approaches being taken to protect the wild places of the continent are both relevant to the cities and crucial to natural science education. Aesthetic motivations, therefore, can become a powerful tool in the pursuit of environmental literacy. They provide a starting point to understanding how the scenic mountain landscape evolved over aeons of time and the biological species that live there, how natural forest fires are essential to the continuity of fire-dependent forest communities; exorcising Smoky the Bear's long-standing mythical message that fire is bad. The need to limit human activities during the spring migration and nesting season is part of a process of learning and public involvement in the maintenance of ecological health in special places.

How and in what context environmental knowledge is delivered is crucial to ecological understanding, and to patterns of behaviour that can be positively influenced by that knowledge. Conceptually, the way the natural sciences are taught externalizes our understanding of 'nature'. It implies that we are simply observers of natural processes but not a part of it. It is here that the overall theme of sustainability provides the foundation for environmental literacy. The infrastructure that keeps us alive and healthy, such as the need for clean water, wastewater systems, energy, recycling and composting, provides a foundation for linking people to these processes. From here it is a short step to understanding that the same processes also sustain national parks and wild places. Their varying climates, geology and geomorphology, hydrological processes, soils, plants and animals, unique natural and cultural settings – in effect, the forces that have created compelling and beautiful scenery, are the same as those that sustain life in the cities.

As we have seen in previous chapters, urban parks, indigenous or regenerating, face a wide range of conservation issues that provide important opportunities for interpreting natural processes. There are several approaches to making links between national parks and other types of protected areas with the cities. They include: first, establishing ecological networks and scientific knowledge through the internet; second, learning with school communities through involvement in the arts; and third, establishing direct links with urban parks systems.

The internet

As a communication medium the internet can provide instantaneous and interactive ecological information and dialogue to urban audiences locally, regionally and internationally. The website of Wapusk National Park on the west coast of Hudson Bay, for instance, contains information on the realities of the park's environment, on such topics as the rigorous climate and the danger that polar bears pose for park visitors. It relates these realities to 'appropriate visitor use'. This involves the necessity for developing codes of behaviour that respect the park's environment, its wildlife and visitor safety, and therefore its ecological integrity. An interactive internet programme for schools was initiated by the coastal British Columbia field unit of Parks Canada, and developed over a number of years as an urban outreach strategy. The intent was to reach urban audiences in Vancouver and Victoria with crucial messages about the parks, covering such information as their floral and faunal communities, their significance in their natural regions, management problems and issues. A pilot project to reach youth in the Victoria school system linked Parks Canada's website with a 'KidsKare' website that contains interactive components to educate students on the national parks, national historic sites and marine conservation areas, initially in British Columbia and eventually across the country. After some years, however, this initiative was abandoned due to lack of funding.

Learning with school communities and the arts

Parks Canada subsequently supported another, quite different, environmental initiative, a 'musician in residence' programme in elementary schools that began in 2000 in British Columbia. It focuses on the Salish Sea Strait off the Pacific Ocean that is divided between the USA and Canada. Working with a local environmentalist folk-singer and songwriter, marine conservation messages were presented to schools with the full participation of the children. After the first year, the programme was extended to other school districts on Vancouver Island and then to schools in the lower mainland. The students learn a number of songs with environmental messages, a CD is cut, and at the end of the year a schools concert is held where all the children get their chance on stage with parents and relatives as the audience. As the Parks Canada representative noted: 'after 30 years of working in the [environmental] field, it blows me away to see and hear a whole school singing about saving the environment . . . over the years we have connected with literally thousands of families.'[76] This led to the production of a teacher's environmental handbook called *Salish Sea: A Handbook For Educators* which is used to reach a larger audience of teachers and students. The handbook includes notes and stories on the environment by the author, songs about the sea and its inhabitants, advice on how to be prepared for a field trip, what to look for and how to behave. An example from the 'Salish Sea field trip to the Douglas Fir forest' taken from the handbook illustrates the general theme:

> The coastal Douglas Fir forest is the main type of forest growing around the Salish Sea. In Canada, it is a unique and endangered ecosystem and only exists around the Salish Sea. It is referred to as a rain-shadow forest because it lies in the shadow of the mountains and doesn't receive much rain. As clouds blow off the open ocean they hit the Olympic Mountains and Vancouver Island mountains first, dropping most of their precipitation on the western side. . . . Many areas around the Salish Sea experience 70 centimetres of rain or less each year.
>
> There is a saying that fish grow on forests and forests grow on fish. This saying refers to the important links between the sea and the forest. The forests of the Salish Sea are 'nature's umbrella' over the land. They provide shade along the shoreline and stabilise the banks of rivers and shorelines. Native shrubs like Salal, Oregon-grape and Oceanspray stop runoff and filter out pollution and sediments that can hurt intertidal life. . . . The Douglas fir forests once grew prolifically but since the mid 1800s logging and the spread of towns has threatened them.[77]

The programme received corporate sponsorship and funding from federal government partners that included the Departments of Fisheries and Oceans, Environment Canada, British Columbia Parks and Communications Canada, led by Parks Canada. As another way of learning environmentally as a community, this experience is both novel and involving.[78] Of particular significance is the clear focus on students being prepared and knowledgeable about this place *before* they visit, and the linkages that are made between fish and forest, and between the impacts of logging – and urbanization – an approach that is valid for all types of visitors to natural areas.

Links with urban parks

While communicating via the internet has great potential for today's technologically sophisticated schoolchildren, there are other ways of linking distant protected areas to the city, and that is through direct experience. For lasting educational value a number of things are needed. First, environmental messages need to be close to home where awareness of natural processes can be a daily event. Second, local urban parks that are ecologically diverse as well as heavily used, such as regenerating landscapes, provide the best opportunity for environmental messages to be heard and understood as much through direct experience as through interpretation. Third, similarities between the habitat characteristics of the local park and those of the natural regions represented in the national parks can make comparisons clearly relevant. Fourth, a national parks presence in the urban parks provides the basis for meaningful learning on site and connections between the two. Community support also lies in strong citizen organizations with sound knowledge of environmental issues and commitments to the ideals of protected areas. Such conditions may often be absent in rural communities. For example:

- A joint interpretive programme for the Rouge Park within the Toronto region's protected valley and ravine system was initiated by the Federal Government at the beginning of the second millennium. Because of its relatively untouched wild areas, the park retains a high level of bio-diversity with plants and animal species not usually seen in urban areas. It is operated by the Toronto and Region Conservation

Authority and provides great potential for studying the implications of human impacts on a natural system that still retains a high level of ecological integrity as an urban park.

- The city of Calgary in Alberta is located in the grasslands region of Canada. The Nose Hill City Park is a major open space in the middle of the city that has been dedicated to celebrating the short grass prairie ecosystem in this natural region. It has great value for interpreting how such a plant community, now rare in North America, can be protected and managed. There is also the potential for learning how prescribed burns can maintain the bio-diversity of the plant community in the city park, a management strategy being used in the Federal Grasslands National Park. The relatively small size of both the local and the national park also has important lessons for connecting habitat fragmentation with the reduction of native species and the increase of alien species.
- Restoration of a black oak savannah community with prescribed burn management and removal of exotic and invasive vegetation, caused by lack natural fires, has been underway since 1993 in High Park, Toronto's largest and most used woodland park. A major burn was carried out in the year 2000. The following spring native lupins appeared again for the first time in may years of traditional mowing. The surrounding community accepted this process of vegetation management within the urban area which provides excellent comparisons with similar plant associations in the mid-west of the continent. The links between national parks fire regeneration management and the urban park make for compelling lessons in environmental education.

(a)

(b)

Plate 7.8 (a) Helping urban residents understand the need for prescribed burns in urban parks – in this case restoring a black oak savannah habitat in Toronto's High Park – can increase awareness and acceptance for similar active management in national parks and protected areas far from the cities. Other issues such as measures to protect natural areas from human impacts are best learned in urban parks. (b) The result of fire management for this kind of habitat – the emergence of the wild lupins, native to this natural community, the following year. Environmental education, in its regional context, is best begun at home.

Source: (a) Karen Yukich and (b) Gera Dillon, both with City of Toronto Parks Department.

- The Outer Harbour Headland (see Chapter 4, pp. 139–42) on the Toronto waterfront is one of the most ecologically diverse naturally regenerated urban habitats in Canada. Its qualities for interpreting Great Lakes coastal plant communities and for natural and biophysical regenerative processes in relation to intensive human use are rich in potential. There is the pressing issue of human impact on the environment that affects both city places and the national parks. The Spit may be seen as a learning experiment on how human impacts on plants and wildlife can be minimized, through physical management of pedestrian circulation, controls on access in sensitive areas and through environmental education and awareness of personal behaviour in relation to natural habitats. To some extent this is already in progress; for instance, during the spring nesting season people are asked not to visit these areas. The lessons learned in the city's natural areas have direct relevance to the national parks and protected areas, at the scale of the Great Lakes region and beyond.

Some final comments

Connecting urban areas to the countryside, at a local scale and the natural regions, is about interdependence and a necessary facet of sustainability. Connecting cities in a seamless network of protected places, maintaining bio-diversity and acting as an organizing framework for growth, is part of the work that must be done to contribute to a sustainable future in the twenty-first century. In 1992, however, the World Conservation Union (IUCN) made the point that:

> despite the growing recognition of the importance of national parks and other categories of protected areas worldwide, less than five percent of the world's surface is afforded protection under IUCN categories. The distribution of these areas is not biogeographically balanced, with some key ecosystems – such as tropical dry forests, fresh waters, temperate rainforests, temperate grasslands, Mediterranean-climate areas, and oceanic islands – being under represented.

Correcting this problem will require: developing an internationally recognized set of guidelines for evaluation of the present coverage of protected areas; identifying major gaps in this coverage; and setting targets to fill these gaps (such as establishing at least one viable representative site in each ecological unit). Priorities for expansion of the network of terrestrial and coastal marine protected areas worldwide should be based on the following criteria:

- inclusion in protected areas of all biological species, ecosystems, communities and habitats, including varieties and genotypes of economic value;
- ability to provide sustainability, including essential processes (such as migration);
- variety of geomorphological and geological formations and historically significant cultural landscapes;
- degrees of endemism, irreplaceability, natural rarity and the presence of threatened species, habitats or formations;
- visibility in relation to local social and economic factors as well as benefits provided to people;
- site selection so as to achieve maximum possible sustainability coverage of biological and geomorphological diversity.[79]

In 2003, the IUCN, announcing the World Parks Congress, had this to say:

> Protected areas have long been a corner stone of global conservation efforts. They represent some of humankind's earliest efforts to conserve the natural resources on which our survival and well-being depend. The century we have just departed has been a period of unprecedented change, escalating development and destructive impact on natural resources. But it has also seen an impressive growth in the number of protected areas, from a mere handful in 1900, to over 44,000 covering more than 10 percent of the earth's land surface at the end of the past century. Though the problems we have experienced in establishing and managing protected areas are considerable and challenging, these areas have never been needed more. Whether it be as reservoirs of biological diversity, sources of clean air and water, buffers from storms, sinks for carbon or places to escape and reconnect with nature, protected areas are vitally important to our individual and collective futures.[80]

In this and previous chapters I have tried to show how, by bringing urban and natural processes together at a local level, a new integrated design language emerges that has significance for the evolving form of the city. The forces that have shaped North America's urban regions have been governed by unlimited energy resources and by attitudes that have paid little heed to the ecological footprint of human activities, the necessity for a sustainable future or the relevance of nature to cities. We begin to see evidence of change, however, with the adoption in some cities of controls to growth and principles of sustainability. Such initiatives are making sense from environmental, economic and quality-of-life points of view. Similarly, working with natural processes as the basis for shaping urban environments has always made sense intuitively, but we are now beginning to prove it economically – a necessary component of a sustainability perspective. Emerging environmental values in the new century are contributing to notions of renewal, citizen empowerment and action. It is also evident that change, and to a great extent unpredictability, are inherent to cities as they are to all life processes. So, as in the community vision to restore the Don River, conceptual and practical ways of achieving sustainability in urban life have no finishing post. Borrowing from the natural sciences, this involves adaptive management, a process of learning by doing.

Notes

Introduction

1 Boardman, Philip. *The Worlds of Patrick Geddes*. London: Routledge & Kegan Paul, 1978.

1 Urban ecology: a basis for shaping cities

1 World Business Council on Sustainable Development. http://www.wbcsd.ch/

2 Wackernagel, Mathis and Rees, William. *Our Ecological Footprint*. Gabriola Island, BC: New Society Publishers, 1996.

3 I have borrowed the word 'pedigreed' from Bernard Rudofsky's *Architecture Without Architects*. New York, NY: Museum of Modern Art, 1964. Rudofsky uses it to describe the formal architecture of cities that express power and wealth: in this case the formal designed urban landscape.

4 Hough, Michael. *Out of Place: Restoring Identity to the Regional Landscape*. New Haven, CN: Yale University Press, 1990.

5 Carver, Norman. *Italian Hill Towns*. Kalamazoo, MI: Documen Press, 1979.

6 Mumford, Lewis. *The City in History*. New York, NY: Harcourt, Bruce & World, 1961.

7 Ibid.

8 Rudofsky. *Architecture Without Architects*.

9 Greenberg, Ken. Personal communication.

10 Giedion, Sigfried. *Space, Time and Architecture*. Oxford: Oxford University Press, 1952.

11 Laurie, Michael. 'Nature and City Planning in the Nineteenth Century'. In Ian C. Laurie (ed.) *Nature in Cities*. New York, NY: John Wiley, 1979.

12 McHarg, Ian L. *Design with Nature*. Garden City, NY: Natural History Press, 1969.

13 Lang, Reg and Armour, Audrey. *Environmental Planning Resource Book*. Montreal, PQ: Lands Directorate, Environment Canada, 1980.

14 Smith, Peter F. *Architecture in a Climate of Change*. Oxford: Architectural Press, 2001.

15 Ibid.

16 Steadman, Philip. *Energy, Environment and Building*. Cambridge: Cambridge University Press, 1977.

17 Nudds, Thomas D. 'Adaptive Management and the Conservation of Biodiversity'. In *Unimpaired for Future Generations? Panel on the Ecological Integrity of Canada's National Parks*. Ottawa, ON: Parks Canada Agency, 2000.

18 Lowenthal, David. 'Daniel Boone is Dead'. *Natural History*. American Geographical Society, August–September 1968.

19 Ibid.

20 Hoskins, W. G. *English Landscapes: How to Read the Man-made Scenery of England*. London: BBC, 1979.

21 *Jakarta Post*, 12 May 1992.

22 Odum, Eugene P. 'The Strategy of Ecosystem Development'. *Science*, vol. 164, April 1969.

23 Commoner, Barry. *The Closing Circle*. New York, NY: Knopf, 1971.

24 Royal Commission on the Future of the Toronto Waterfront. *Watershed*. Report, 1990.
25 Fowles, John. 'Seeing Nature Whole'. *Harpers*, vol. 259, no. 1554, November 1979.
26 Ibid.
27 Deelstra, Tjeerd. 'Enforcing Environmental Urban Management – New Strategies and Approaches'. Conference material for UN Conference on Environment and Development (UNCED), Berlin, 6 February 1992.
28 Ibid.

2 Water

1 Bellamy, David. *Bellamy's Europe*. London: BBC, 1976.
2 Environment Canada. *Water – Nature's Most Versatile Substance*. Ottawa, ON: Inland Waterways Directorate, 1976.
3 Ibid.
4 Petawawa Forest Experiment Station. *Water Trail*. Chalk River, ON: Ontario Public Awareness Program, Canadian Forest Service, 1978.
5 Golding, D. L. *Forests and Water*. Fact Sheet. Ottawa, ON: Department of Fisheries and Environment, n.d.
6 Ibid.
7 Hough Stansbury and Associates. 'Water Quality and Recreational Use in Inland Lakes'. Prepared for the Ontario Ministry of the Environment, SE Region, May 1977.
8 Vallentyne, J. R. *The Algal Bowl*. Miscellaneous Special Publication 22, Ottawa, ON: Department of the Environment, Fisheries and Marine Service, 1974.
9 Royal Commission on the Future of the Toronto Waterfront. *Watershed*. Interim Report. Toronto, ON, August 1990.
10 Lull, Howard W. and Sopper, William E. *Hydrological Effects from Urbanization of Forested Watersheds in the NE*. USDA Forest Service Research Paper NE 146, Washington, DC: US Department of Agriculture, 1969.
11 Ontario Ministry of the Environment. 'Evaluation of the Magnitude and Significance of Pollution Loadings from Urban Stormwater Run-off in Ontario'. Research Report no. 81.
12 Leopold, Luna B. *Hydrology for Urban Land Planning. A Guide Book on the Hydrologic Effects of Urban Land Use*. Geological survey circular 554. Washington, DC: US Department of the Interior, 1968.
13 Ferguson, Bruce K. *Introduction to Stormwater*. New York, NY: John Wiley, 1998.
14 Ontario Ministry of the Environment. *Modern Concepts of Urban Drainage*. Conference Proceedings no. 5, A Canada, Ontario Agreement on Great Lakes Water Quality, Toronto, ON, 1977.
15 Ibid.
16 Lull and Sopper. *Hydrological Effects*.
17 Forbes, R. J. 'Mesopotamian and Egyptian Technology'. In Melvin Kranzberg and Carroll W. Pursell Jr (eds). *Technology in Western Civilization*, vol. 1. Oxford: Oxford Unversity Press, 1967.
18 Drachmann, A. G. 'The Classical Civilization'. In Kranzberg and Pursell (eds) *Technology in Western Civilization*, vol. 1.
19 Mumford, Lewis. *The City in History*. New York, NY: Harcourt, Brace & World, 1961.
20 Wright, Lawrence. *Clean and Decent*. London: Routledge & Kegan Paul, 1960.
21 Mathews, Leslie S. *The Antiques of Perfume*. London: G. Bell, 1973.
22 Wright. *Clean and Decent*.
23 Mumford. *The City in History*.
24 Goldstein, Jerome. *Sensible Sludge*. Emmaus, PA: Rodale Press, 1977.
25 Ibid.
26 Wood, L. B. *The Rehabilitation of the Tidal River Thames*. Unpublished paper, London: Thames Water, n.d.
27 Ibid.
28 Royal Commission on the Future of the Toronto Waterfront. *Watershed*.

29 Ibid.
30 Deelstra, Tjeerd. 'Enforcing Environmental Urban Management – New Strategies and Approaches'. Conference material for UN Conference of Environment and Development (UNCED), Berlin, 6 February 1992.
31 Wood. *The Rehabilitation of the Tidal River Thames.*
32 Harrison, Jeffery and Grant, Peter. *The Thames Transformed.* London: Andre Deutsch, 1976.
33 Vallentyne. *The Algal Bowl.*
34 Task Force to Bring Back the Don. *Bringing Back the Don.* Toronto, ON: Hough Stansbury Woodland, Prime Consultants, in association with Gore and Storrie Ltd, Dr Robert Newbury, The Kirkland Partnership, 1991.
35 Ibid.
36 Newbury, Robert. Personal communication.
37 Task Force to Bring Back the Don. *Bringing Back the Don.*
38 Ibid.
39 Don Watershed Regeneration Council. *A Time for Bold Steps: The Don Watershed Report Card 2000.* Report February 2000, Downsview, ON: Toronto and Region Conservation Authority.
40 Asaduzzaman, Md. Abdur Rob. 'Urbanization of Dhaka City'. *Dhaka: the mappa*, November 1997.
41 Hai, Arifa. 'Designing a City of Water: A Sustainable Future for the Urban Landscape of Dhaka'. Major Paper, spring 2002, Toronto, ON: Faculty of Environmental Studies, York University.
42 Salar Khan, M. *et al.* (eds) *Wetlands of Bangladesh.* Dhaka: Bangladesh Centre for Advanced Studies, May 1994.
43 Asaduzzaman. 'Urbanization of Dhaka City'.
44 Hai. 'Designing a City of Water'.
45 Asaduzzman. 'Urbanization of Dhaka City'.
46 Ibid.
47 Hai. 'Designing a City of Water'.
48 Salar Khan. *Wetlands of Bangladesh.*
49 Asaduzzaman. 'Urbanization of Dhaka City'.
50 Hofer, Thomas. *Floods in Bangladesh: A Highland–Lowland Interaction.* Ph.D. thesis, Bern: Geographisches Institut, 1997.
51 Ibid.
52 Lerner, Steve and Poole, William. *The Economic Benefits of Parks and Open Space.* San Francisco, CA: The Trust for Public Land, 1999 (www.tpl.org).
53 Furedy, C. *Wastes and the Urban Environment: Perspectives on People, Animals, and their Wastes.* Calcutta, Institution of Public Health Engineers, with WEDC, Loughborough University of Technology and British Deputy High Commissioner, 1985.
54 Baylin, Frank. 'Solar Sewage Treatment'. *Popular Science*, May 1979.
55 Burke, William K. 'From Sewer to Swamp'. *E Magazine*, July–August. 1991, pp. 39–42 and 67.
56 Caroll, Debra. 'Greenhouses that Grow Clean Water'. *Sunworld*, vol. 14, no. 3, 1990, pp. 71–4.
57 Ibid.
58 Burke. 'From Sewer to Swamp'.
59 Cited in Burke. 'From Sewer to Swamp'.
60 Deelstra, Tjeerd. 'Ecological Approaches to Wastewater Management in Urban Regions in the Netherlands'. In *Ecological Engineering for Wastewater Treatment. Proceedings of the International Conference.* Trosa: Stensund Folk College, 1991.
61 Ibid.
62 Ferguson, Bruce K. *Introduction to Stormwater.* New York, NY: John Wiley & Sons, 1998.
63 Ibid.
64 Ibid.
65 Hough Stansbury Woodland Limited. *Naturalization/Restoration of Parks and Open Spaces.* Report for the Parks and Recreation Department, City of Kitchener, December 1990.

66 William Sleeth. Kitchener Parks and Recreation Department. Personal Communication.
67 Ontario Ministry of the Environment. *Stormwater Quality Best Management Practices*. Toronto, ON: Queen's Printer for Ontario, 2002.
68 Mark Taylor and Associates. *Constructed Wetlands for Stormwater Management; A Review*. Report Prepared for Water Resources Branch, Ontario: Ministry of the Environment and Metropolitan Toronto and Region Conservation Authority, April 1992.
69 Ibid.
70 Bunyard, Peter. 'Sewage Treatment in a Swedish Sculpture Garden'. *Ecologist*, no. 1, January–February 1978.
71 Hess, Alan. 'Technology Exposed'. *Landscape Architecture*, vol. 82, no. 5, May 1992.
72 Hough Stansbury and Associates. *LeBreton Flats Landscape Development*. Unpublished report, Ottawa, ON: LeBreton Flats Office, Central Mortgage and Housing Corporation, January 1979.

3 Plants and plant communities

1 Odum, Eugene P. *Ecology*. New York, NY: Holt, Rinehart and Winston, 1963.
2 Odum, Eugene P. 'The Strategy of Ecosystem Development'. *Science*, vol. 164, April 1969.
3 Farb, Peter. *The Forest*. New York, NY: Time-Life Books, 1963.
4 Elias, Thomas S. and Irwin, Howard S. ' Urban Trees'. *Scientific American*, November 1976.
5 Jorgensen, Erik, L. *Forestry: Some Problems and Proposals*. Toronto, ON: Faculty of Forestry, University of Toronto, September 1967.
6 Laurie, Michael. 'Nature and City Planning in the Nineteenth Century'. In Ian C. Laurie (ed.) *Nature in Cities*. New York, NY: John Wiley, 1979.
7 Fairbrother, Nan. *New Lives, New Landscapes*. London: Architectural Press, 1970.
8 Sukopp, Herbert, Blume, Hans-Peter and Kumick, Wolfram. L. 'The Soil, Flora and Vegetation of Berlin's Wastelands'. In Laurie. *Nature in Cities*.
9 Livingston, John A. *Canada*. Toronto, ON: Jack McLelland, 1970.
10 Sukopp, Blume and Kunick. 'Soil, Flora and Vegetation'. In Laurie. *Nature in Cities*.
11 Bellamy, David. *Bellamy's Europe*. London: BBC, 1976.
12 Mollison, Bill. *Permaculture (1 and 2)*. Stanley, Tasmania: Tagari Publishers, 1979.
13 Hammond, Herb. *Seeing the Forest Among the Trees*. Vancouver, BC: Polestar, 1991.
14 Teagle, W. G. *The Endless Village*. Shrewsbury: Nature Conservancy Council, 1978.
15 Sukopp, Blume and Kunick. 'Soil, Flora and Vegetation.' In Laurie. *Nature in Cities*.
16 Dorney, R. F. and Eagles, Paul E. J. *et al.* 'Ecosystem Planning, Analysis and Design in Ontario as Applied to Environmentally Sensitive Areas'. Paper presented to American Association for the Advancement of Science Meeting. Toronto, ON, January 1981.
17 Hough, M. and Barrett, Suzanne. *People and City Landscapes*. Toronto, ON: Conservation Council of Ontario, 1987.
18 National Urban Forestry Unit. http://www.nufu.org.uk/urban.htm
19 Hough Stansbury Woodland Ltd. *Naturalization Project*. Unpublished report. Ottawa, ON: National Capital Commission, 1982; and *Naturalization Project: Five Year Test Plot Evaluation*. Ottawa, ON: National Capital Commission, 1989.
20 Hough Stansbury Woodland Ltd. *Naturalization Project*.
21 Hough Stansbury Woodland Ltd. *Naturalization Project. Five Year Test Plot Evaluation*.
22 Ibid.
23 Ibid.
24 Bos, H. J. and Mol, J. L. 'The Dutch Example: Native Planting in Holland'. In Laurie. *Nature in Cities*.
25 Ibid.
26 Jacobs, Ton. 'The Gilles Estate in Delft'. In Tjeerd Deelstra (ed.) *Shaping Nature in Cities*.
27 Ibid.
28 Tjallingii, Sybrand. Personal communication.
29 Ruff, Allan R. *Holland and the Ecological Landscapes 1973–1987*. Delft: Delft University Press, 1987.

30 Jacobs. 'The Gilles Estate in Delft'. In Deelstra. *Shaping Nature in Cities.*
31 McPherson, E., Gregory, David J. and Rowntree, R. A. (eds) *Chicago's Urban Forest Ecosystem: Results of the Chicago Urban Forest Climate Project*. General Technical Report NE-186 Northeastern Forest Experimental Station: Forest Service, US Department of Agriculture, 1994.
32 American Forests. *Urban Ecosystem Analysis for the Houston Gulf Coast Region*. Washington, DC: December 2000 (wwwamericanforests.org).
33 American Forests. *Regional Ecosystem Analysis for the Willamette/Lower Columbia Region of Northwestern Oregon and Southwestern Washington State*. Washington, DC: October 2001 (www.americanforests.org).
34 Ibid.
35 Simpson, James R. 'Urban Forest Impacts on Regional Heating Energy Use: Sacramento County Case Study'. *Journal of Arboriculture*, vol. 24, no. 4, July 1998.
36 American Forests. *Regional Ecosystem Analysis Chesapeake Bay Region and the Baltimore–Washington Corridor*. Washington, DC: 1999 (www.americanforests.org).
37 Ibid.
38 Ibid.
39 City of Toronto Planning and Development Department. *A Working Guide for Planning and Designing Safer Urban Environments*. Toronto, ON: Safe Cities Committee, City Hall, Toronto, 1992. Quoted from *Community Safety in Nottingham City Centre*. Report of the Steering Group. Nottingham Safer Cities Project, October 1990.
40 Ibid.
41 City of Toronto Planning and Development Department. *A Working Guide*.
42 Frolich, Emil. Forest Engineer, Stadtforstamt, Zurich. Personal communication.
43 *New York Times*, 5 May 1991.
44 Pollan, Michael. 'Why Mow? The Case Against Lawns'. *New York Times Magazine*, 28 May 1989.
45 Ibid.
46 Ibid.
47 Ibid.
48 Morrison, W. O. *et al. Avi Fauna Survey of Vacant Grasslands*. Ottawa, ON: National Capital Commission, 1981.
49 Smith, R. A. H. and Bradshaw, A. D. 'Use of Tolerant Plant Populations for the Reclamation of Metalliferous Wastes'. *Applied Ecology*, no. 16, 1979.
50 Cole, Lyndis. *Conservancy in Urban Areas*. Report for the Nature Conservancy Council. Land Use Consultants. May 1978.
51 Lowday, J. E. and Wells, T. C. E. *The Management of Grasslands and Heathlands in Country Parks*. Report to the Countryside Commission by the Institute of Terrestrial Ecology, Shrewsbury: Countryside Commission, West Midlands Region, 1977.
52 Laurie, Ian C. 'Urban Commons'. In Laurie. *Nature in Cities*.
53 Weaver Liquifuels. Downsview, ON. Personal communication.
54 van Leeuwen, Willem. Rotterdam Parks Department. Personal communication, 2002.
55 Hough Stansbury and Associates. *LeBreton Flats Landscape Development*. Unpublished report for Central Mortgage and Housing Corporation, Ottawa, ON: January 1979.
56 Jacobs, Jane. *The Death and Life of Great American Cities*. New York, NY: Random House, 1961.
57 Brower, Sidney. 'Street Front and Sidewalk'. *Landscape Architecture*, July 1973.
58 Francis, Mark. 'The Making of Democratic Streets'. In Anne Verner Moudon (ed.) *Public Streets for Public Use*. New York, NY: Van Nostrand Reinhold, 1987.
59 Ibid.
60 Goulty, George. 'Landscape Electric'. *Landscape Design*, August 1986.
61 International Brownfields Exchange. *The Nature of Possibility; Experiences in Risk-based Decision Making*. Toronto, ON: Waterfront Regeneration Trust, 2000/02.
62 Ibid.
63 Jones, Brad. 'Comparisons of the Washington State and Ontario Regulatory Frameworks'. Paper in Proceedings of the Lower Don Lands: Site Remediation Workshop, Toronto, ON: May 1993.

4 Wildlife

1 Livingston, John A. *Rogue Primate*. Toronto, ON: Key Porter Books, 1994.
2 Farb, Peter. *The Forest*. New York, NY: Time-Life Books, 1963.
3 Ibid.
4 Tofts, Richard and Clements, David. 'Hedgerows as Wildlife Habitat'. *Landscape Design Extra*, January 1992.
5 Baines, Chris. 'The Really Wild Show Comes to Town'. *The Times*, 17 March 2001.
6 Geis, Aelred K. 'Effects of Urbanization and Types of Urban Development on Bird Populations'. In *Wildlife in an Urbanizing Environment Symposium*. Co-operative Extension Service, Amherst, MA: University of Massachusetts, June 1974.
7 Wilcove, D. S. 'Forest Fragmentation as a Wildlife Management Issue in the Eastern United States.' Quoted in Hough Stansbury Woodland/Gore and Storrie Ltd. *Ecological Analysis of the Greenbelt*. Ottawa, ON: National Capital Commission, 1991.
8 Livingston, John A. *Canada*. Toronto, ON: Natural Science of Canada, 1970.
9 Ibid.
10 Bertrand, Nick (ed.) *Deptford Creek: Life on the Edge*. London: Creekside Environmental Project, 2002.
11 Sukopp, Herbert. 'Urban Ecosystems'. *Journal of Natural History Museum Institute*, Chiba, vol. 2, no. 1, March 1992, pp. 53–62.
12 Karstad, Lars. 'Disease Problems of Urban Wildlife'. In David Euler *et al.* (eds) *Wildlife in Urban Canada Symposium*. Guelph, ON: University of Guelph, 1975.
13 Ibid.
14 More, Thomas A. 'An Analysis of Wildlife in Children's Stories'. In *Children, Nature and the Urban Environment*. Symposium proceedings. USDA Forest Service, General Technical Report, NE 30, Washington, DC: US Department of Agriculture, 1977.
15 Worster, Donald. *Nature's Economy*. Garden City, NY: Anchor Press/Doubleday, 1979.
16 Livingston. *Rogue Primate*.
17 Eisenberg, Evan. *The Ecology of Eden*. Toronto, ON: Random House of Canada, 1998.
18 Ibid.
19 Forman, Richard, T. T. and Gordon, Michel. *Landscape Ecology*. New York, NY: John Wiley, 1986.
20 Livingston. *Canada*.
21 Hamel, Peter J. 'Wastewater Treatment and Shoreline Ecology in the Regional Municipality of Ottawa-Carlton'. Unpublished student paper, Toronto, ON: Faculty of Environmental Studies, York University, 1976.
22 Carling, Paul M., McIntosh, Karen L. and McKay, Sheila M. 'The Vascular Plants of the Leslie Street Headland'. *Ontario Field Biology*, vol. 31, no. 1, 1977 (Toronto and Region Conservation Authority).
23 Ibid.
24 Metropolitan Toronto and Region Conservation Authority. *Tommy Thompson Park. Master Plan and Environmental Assessment*. Addendum report, December 1992.
25 Ibid.
26 Metropolitan Toronto and Region Conservation Authority. *Tommy Thompson Park Newsletter*, n.d.
27 Blokpoel, H. and Haymes, G. T. 'How the Birds Took Over the Leslie Street Spit' *Canadian Geographic*, April–May 1978.
28 The Friends of the Spit. *The Eastern Headland and Aquatic Park*. Unpublished report, Toronto, ON: The Friends of the Spit, n.d.
29 Johnson, Bob. *Familiar Amphibians and Reptiles of Ontario*. Toronto, ON: Natural Heritage/Natural History Inc, 1989.
30 Ibid.
31 Dougan, J. 'The Fate of ESAs in Urban Environments: Two Case Histories in Peel and Halton'. *Plant Press*, vol. 2, nos. 7–9, 1984. Quoted in Hough Stansbury Woodland/Gore and Storrie Ltd. *Lake Ontario Greenway Strategy: Restoration Ecology*. Unpublished report, Toronto, ON: Waterfront Regeneration Trust, May 1994.

32 Robinson, Steve. 'Environmental Implications of Golf Courses'. *Landscape Design Extra*, April 1992.
33 Tiner, Tim. 'Green Space or Green Waste?' *Seasons*, Federation of Ontario Naturalists, summer 1991.
34 Etchells, Jon. 'Golf Answers Back'. *Landscape Design*, no. 210, May 1992.
35 Baines, Chris. *How to Make a Wildlife Garden*. London: Elm Tree Books, 1985.
36 Ibid.
37 O'Connor, Glenn A. *Establishing and Maintaining Low Maintenance Landscapes for Southern Ontario*. Unpublished research report, prepared for the Faculty of Architecture and Landscape Architecture, University of Toronto, 1984.
38 Laut, Jamie. Environmental Officer, Lakeshore Refinery, Mississauga. Personal communication.
39 *The Black Redstarts of Deptford Creek*. Pamphlet, London: Creekside Education Trust (ecs.lewisham.gov.uk./cet).
40 *CET Creekside Environmental Project*. London 2002.
41 Creekside Education Trust. *Deptford: 'Life on the Edge'*. London, 2002.
42 Bertrand (ed.). *Deptford Creek: Life on the Edge*.
43 Hough, Michael. 'The Port Lands: The Significance of the Ordinary'. In Betty I. Roots (Editor in Chief) *Special Places: The Changing Ecosystems of the Toronto Region*. Vancouver, BC: UBC Press, 1999.
44 Steele, Jess (ed.). *Deptford Creek: Surviving Regeneration*. Deptford Forum Publishing, 1999.
45 Ibid.
46 Bertrand. *Deptford Creek: 'Life on the Edge'*.
47 Wheater, C. Philip. 'Urban Habitats'. Quoting C. E. J. Kennedy and T. R. E. Southwood. 'The Number of Species of Insects Associated with British Trees: A Re-analysis'. *Journal of Animal Ecology*, 1999, no. 531, pp. 455–78.
48 Vanstone, Ellen. 'Racoons. The Bandit in the Attic'. *Toronto Globe and Mail*, 20 February 1993.
49 Ibid.
50 Barrett, Suzanne and Kidd, Joanna. *Pathways: Towards an Ecosystem Approach*. Royal Commission on the Future of the Toronto Waterfront, Toronto, ON: Minister of Supply and Services Canada, 1991.
51 Duffey, Eric. 'Effects of Industrial Air Pollution on Wildlife'. *Biological Conservation*, vol. 15, 1979, pp. 181–90.
52 Ibid.

5 City farming

1 Heady, Earl O. 'The Agriculture of the U.S.'. *Scientific American*, (Issue on Food and Agriculture), September 1976.
2 Commoner, Barry. *The Closing Circle*. New York, NY: Knopf, 1971.
3 Geno, Barbara J. and Geno, Larry M. 'Food Production in the Canadian Environment'. *Perceptions 3*. Ottawa, ON: Science Council of Canada, 1976.
4 The New Alchemist Institute. 'Modern Agriculture: A Wasteland Technology'. *Journal of the New Alchemists*, 1874.
5 Sinclair, Geoffrey. 'Upland Landscape Study'. Unpublished address to the Employment in the Countryide Symposium, Devon: New Mills Study Centre, May 1980.
6 Wackernagel, Mathis and Rees, William. *Our Ecological Footprint*. Gabriola Island, BC: New Society Publishers, 1996.
7 Ibid.
8 Murray, Alex. Personal communication.
9 Murray, Alex L. and Kraus, Eric. 'The Ecological Footprint of Food Transportation'. P. 84. Proceedings from 'Moving the Economy'. An International Conference on Economic Opportunities in Sustainable Transportation. Toronto, ON: 1998.

10 Green, John. *Memorandum on Domestic Livestock Keeping in Urban Areas*. Unpublished report, 1943.
11 Animal Control Sub Committee. *Proposals for Animal Control*. Toronto, ON: City of Toronto, 1981.
12 Mumford, Lewis. *The City in History*. New York, NY: Harcourt, Brace & World, 1961.
13 Ibid.
14 Michell, W. R. 'The Cow Keepers'. *The Dalesman*, vol. 43, no. 11, 1980.
15 Urban Agriculture Magazine Editorial. *The Economics of Urban Agriculture*, no. 7, August 2002 (www.ruaf.org).
16 Ibid.
17 Darmajanti, Irwina. 'Integrating Informal City Farming Practices into Green Space Management'. Major paper, Faculty of Environmental Studies, York University 1994.
18 Personal communication with the author.
19 *Urban Agriculture Magazine*, no. 7, August 2002.
20 Ibid.
21 Green. *Memorandum on Domestic Livestock Keeping*.
22 Ibid.
23 Green, John. Personal communication.
24 *USA Today, International Edition*. 14 August 1993.
25 Ibid.
26 Ibid.
27 Wade, Isabel. 'Urban Self-Reliance in the Third World: Developing Strategies for Food and Fuel Productivity'. World Futures Conference, Toronto, ON, 1980.
28 Ibid.
29 Hassan, Mehmood U. I. 'Maximising Private and Social Gains of Wastewater. Agriculture in Haroonabad'. *Urban Agriculture Magazine*, no. 7, August 2002.
30 Ibid.
31 *Toronto Star*, 12 February 1994.
32 Novo, Mario Gonzalez. 'A Real Effort in the City of Havana'. *Urban Agriculture Magazine*, no. 6, April 2002.
33 *Jakarta Post*, 12 May 1992.
34 Ibid.
35 Poerbo, Hasan. 'Urban Solid Waste Management in Bandung: Towards an Integrated Recovery System'. *Environment and Urbanization: Rethinking Local Government – Views From the Third World*, vol. 3, no. 1, April 1991, pp. 60–9.
36 Ibid.
37 Brown, Lester R. *et al*. *State of the World 1992*. New York, NY: Worldwatch Institute, 1992.
38 Thompson, J. William. 'San Francisco's Gardens of Diversity'. *Landscape Architecture*, vol. 83, no. 1, January 1993.
39 Thompson, J. William. 'Reconsidering South Central'. *Landscape Architecture*, vol. 83, no. 1, January 1993.
40 Cited in Thompson. 'San Francisco's Gardens of Diversity'.
41 Cited in ibid.
42 City of Toronto Compost Recycling Program. http://www.city.toronto.ono.ca/composting
43 The Farallones Institute. *The Integrated Urban House. Self Reliant Living in the City*. San Francisco, CA: Sierra Club Books, 1979.
44 Hough, Michael. 'Bottom Line Horticulture'. *Harrowsmith*, no. 56, August 1984.
45 Unpublished calculations by the author.
46 Riley, Peter. *'Economic Growth': The Allotments Campaign Guide*. London: Friends of the Earth, 1979.
47 Hayes, Denis. 'Energy. The Case for Conservation'. *Worldwatch Paper*, 4, Washington, DC: Worldwatch Institute, January 1976.
48 BBC Television. *Tomorrow's World*, 21 February 1980.
49 Mumford. *The City in History*.
50 Lumley-Smith, Ruth. 'The Road to Utopia'. *New Ecologist*, no. 1, January–February 1978.
51 *Toronto Star*, 20 November 1979.

52 Creasy, Rosalind. *The Complete Book of Edible Landscaping*. San Francisco, CA: Sierra Club Books, 1982.
53 Laurie, Michael. 'Nature and City Planning in the Nineteenth Century'. In Ian C. Laurie (ed.) *Nature in Cities*. New York, NY: John Wiley, 1979.
54 Cited in Fein, Albert. *Landscape into Cityscape*. Ithaca, NY: Cornell University Press, 1968.
55 Iles, Jeremy. Director, Federation of City Farms and Community Gardens. Personal communication.
56 'Promoting Community Gardens and City Farms'. *Briefing Sheet*, November 2001 (www.farmgarden.org.uk).
57 Ibid.
58 Federation of City Farms and Community Gardens. *Future Directions for City Farms in London*. Study funded by Mayor of London, October 2002.
59 'Promoting Community Gardens and City Farms'. Briefing Sheet.
60 Magennis, Mick. Education Worker. Personal communication.
61 Mayor of London. *Connecting with London's Nature*. The Mayor's Bio-diversity Strategy, London: Greater London Authority, 2002.
62 Ibid.

6 Climate

1 Meiss, Michael. 'The Climate of Cities'. In Ian C. Laurie (ed.) *Nature in Cities*. New York, NY: John Wiley, 1979.
2 Chandler, T. J. *The Climate of London*. London: Hutchinson, 1975.
3 Meiss. 'The Climate of Cities'. In Laurie. *Nature in Cities*.
4 Rocky Mountain Institute. 'Climate: Greenhouse Gases and Where They Came From'. *Spring Newsletter*, 2000.
5 Ibid.
6 Ford Foundation. *Exploring Energy Choices: A Preliminary Report of the Ford Foundation's Energy Policy Project*. Washington, DC: The Ford Foundation, 1974.
7 Ibid.
8 McPherson, E. Gregory, *et al. Chicago's Evolving Urban Forest: Initial Report of the Chicago Urban Forest Climate Project*. General Technical Report NE-169, Northeastern Forest Experiment Station, Washington, DC: Forest Service, U.S. Department of Agriculture, 1993.
9 Landsburg, H. E. 'The Climate of Towns'. In William L. Thomas Jr (ed.) *Man's Role in Changing the Face of the Earth*, vol. 2. Chicago, IL: University of Chicago Press, 1956.
10 Illinois Environmental Protection Agency. *Illinois 1990 Annual Air Quality Report*. IEPA/APC/91–92, Springfield, IL, 1991. Reported in McPherson, *et al. Chicago's Evolving Urban Forest*.
11 *Guardian*, 26 March 1993.
12 Oke, T. R. *Boundary Layer Climates*. New York, NY: Methuen, 1987. Quoted in McPherson, *et al. Chicago's Evolving Urban Forest*.
13 Lowry, William P. 'The Climate of Cities'. *Scientific American*, August 1967.
14 Akbari, H. *et al.* 'The Urban Heat Island: Causes and Impacts'. In *Cooling Our Communities: A Guidebook on Tree Planting and Light-Coloured Surfacing*. Washington, DC: U.S. Environmental Protection Agency. Quoted in McPherson, *et al. Chicago's Evolving Urban Forest*.
15 McPherson, *et al. Chicago's Evolving Urban Forest*.
16 Burke, James. *Connections*. London: Macmillan, 1978.
17 Rotsch, Melvin M. 'The Home Environment'. In Melvin Kranzberg and Carroll W. Pursell Jr (eds) *Technology in Western Civilization*, vol. 2. Oxford: Oxford University Press, 1967.
18 Carter, George F. *Man and the Land, A Cultural Geography*. New York, NY: Holt, Reinhart & Winston, 1966.
19 Cain, Allan, *et al.* 'Traditional Cooling Systems in the Third World'. *Ecologist*, vol. 6, no. 2, 1976.

20 Ibid.
21 Ibid.
22 Rudofsky, Bernard. *Architecture Without Architects*. New York, NY: The Museum of Modern Art, 1964.
23 Bahadori, Medhi N. 'Passive Cooling Systems in Iranian Architecture'. *Scientific American*, February 1978.
24 Turner, T. H. D. 'The Design of Open Space'. In Timothy Cochrane and Jane Brown (eds) *Landscape Design for the Middle East*. London: RIBA Publications, 1978.
25 Robinette, Gary O. *Landscape Planning for Energy Conservation*. Reston, VA: Environmental Design Press for the American Society of Landscape Architects Foundation, 1977.
26 Ibid.
27 Meiss. 'The Climate of Cities'. In Laurie. *Nature in Cities*.
28 Doernach, Rudolf. 'Uber den Nutzen von biotektonischen Grunsystemen'. *Garten + Landschaft*, no. 6, 1979.
29 Ibid.
30 Falk, Trillitzsch. 'Anregungen zum Thema Dachgarten'. *Garten + Landschaft*, no. 6, 1979.
31 http://www.roofmeadow.com or, http:///www.roofscapes.com
32 Ibid.
33 Federer, C. A. 'Effects of Trees in Modifying Urban Microclimate'. In *Trees and Forests in an Urbanizing Environment Symposium*. Amherst: Co-operative Extension Service, University of Massachusetts, 1970.
34 Ibid.
35 Ibid.
36 Schmid, James A. *Urban Vegetation*. Research Paper no. 161, Department of Geography, University of Chicago, 1975.
37 Philleo, Jerryne. *The Davis Energy Handbook*. Davis, CA: City of Davis, 1981.
38 Ibid.
39 Novak, D. J. 'Urban Forest Structure and the Functions of Hydrocarbon Emissions and Carbon Storage'. In McPherson, *et al. Chicago's Evolving Urban Forest*.
40 Akbari, H. 'The Impact of Summer Heat Islands on Cooling Energy Consumption and CO2 Emissions'. In McPherson, *et al. Chicago's Evolving Urban Forest*.
41 Tucker, John. 'Trees for People'. *Landscape Design*, no. 215, November 1992, pp. 41–3.
42 Meiss. 'The Climate of Cities'. In Laurie *Nature in Cities*.
43 Oke, T. R. 'The Significance of the Atmosphere in Planning Human Settlements'. In E. B. Wiken and G. Ironside (eds) *Ecological (Biophysical) Land Classification in Urban Areas*. Ecological Land Classification Series, no. 3. Ottawa, ON: Environment Canada. 1977.
44 Robinette. *Landscape Planning*.
45 Olgay, Victor G. *Design with Climate*. Princeton: Princeton University Press, 1963.
46 Robinette. *Landscape Planning for Energy Conservation*.
47 McPherson, *et al. Chicago's Evolving Urban Forest*.
48 Ontario Ministry of Housing. *Residential Site Design and Energy Conservation, Part 1. 'General Report'*. Toronto, ON: Government of Ontario, 1980.
49 Salvatore, Fidenzio. *The Potential Role of Vegetation in Improving the Urban Air Quality, A Study of Preventative Medicine*. Willowdale, ON: Mork-Toronto Lung Association, 1982.
50 McPherson, *et al. Chicago's Evolving Urban Forest*.
51 Schmid. *Urban Vegetation*.
52 Wainwright, C.W.K. and Wilson, M.J.G. 'Atmospheric Pollution in a London Park'. *International Journal of Air and Water Pollution*, vol. 6, 1962.
53 Salvatore. *The Potential Role of Vegetation*.
54 Schmid. *Urban Vegetation*.
55 Ibid.
56 Energy Probe. *Facts on Energy Conservation*. Toronto, ON: Energy Probe, n.d.
57 Brown, Lester and Jacobson, Jodi L. *The Future of Urbanization: Facing the Ecological and Economic Constraints*. Washington, DC: Worldwatch Institute 1987.

58 Paehlke, Robert. 'The Environmental Effects of Intensification'. Prepared for Municipal Planning Policy Branch, Ministry of Municipal Affairs, 1991. Quoted from Canadian Urban Institute. *Housing Intensification, Policies, Constraints and Challenges*. Background Paper. Toronto, ON: Canadian Urban Institute, 1990.
59 Ibid.
60 Jacobs, Jane. *The Death and Life of Great American Cities*. New York, NY: Random House, 1961.
61 Meiss. 'The Climate of Cities'. In Laurie *Nature in Cities*.
62 Ibid.
63 Schmid. *Urban Vegetation*.
64 Bernatzky, Aloys. 'The Effects of Trees on the Climate of Towns'. In Shirley E. Wright *et al.* (eds) *Tree Growth in the Landscape*. London: Department of Horticulture, Wye College, University of London, 1974.
65 Salvatore. *The Potential Role of Vegetation*.
66 Diamond, A. J. and Myres, Barton. *University of Alberta, Long Range Development Plan*. Consultant Report, June 1969.
67 Hough Stansbury and Associates. *University of Alberta, Long Range Landscape Development Plan*. Consultant report, June 1971.
68 Baumuller, Juergen. Director of Urban Climatology, Office for Environmental Protection. Personal communication, 2002.
69 Baumuller. Personal communication.
70 Department of Parks and Cemeteries with Departments of Environonmental Protection, Urban Planning and Press and Information. *Stuttgart, A Livable Community*. Stuttgart, 2001.
71 Ibid.
72 Baumuller. Personal communication.

7 The regional landscape: a framework for shaping urban form

1 Boardman, Philip. *The Worlds of Patrick Geddes*. London: Routledge & Kegan Paul, 1978.
2 Ibid.
3 Hall, Peter. *Cities of Tomorrow*. Oxford: Blackwell Publishing, 2002.
4 Lewis, Philip H. Jr. *Tomorrow by Design*. New York, NY: John Wiley, 1996.
5 Hall. *Cities of Tomorrow*.
6 Ibid.
7 Office of the Deputy Prime Minister. 'Planning Policy Guidance', Note 2 (http:///www.planning.odpm.gov.uk//ppg//ppg2/01.htm).
8 Office of the Deputy Prime Minister. 'Planning Policy Guidance', Note 2.
9 Department of the Environment, Transport and the Regions. *The Countryside – Environmental Quality and Economic and Social Development* (PPG7) (http://www.planning.dtir.gov.uk/ppg/ppg701.htm).
10 County of Dorset Structure Plan to 2011. Examination in Public, October to November 1996. Report to the Panel.
11 www.tcpa.org.uk.
12 *Royal Town Planning Institute Journal*. 'The Future of the Green Belt, What Are Available Options?' 7 March 2003.
13 The Countryside Commission. *Linking Town and Countryside*. Pamphlet, Cheltenham, 1999.
14 Caffyn, Alison and Dahlstrom, Margarita with Revell, Kerry. *Connecting Town and Country*. London: Countryside Agency, September 2001.
15 'Fringe Benefits: Realizing the Potential of the Countryside Closest to Towns and Cities' Paper, Countryside Agency, April 2002.
16 Department of the Environment, Transport and the Regions. *The Countryside*.
17 The Countryside Agency. *A Strategy for Sustainable Management in England*. London, June 2001.
18 Ibid.
19 Ibid.

20 The Royal Society for the Protection of Birds. *Futurescapes: Large-scale Habitat Restoration for Wildlife and People*. Sandy, 2001.
21 Cornwell, Sue. Countryside Agency. Personal communication, 1999.
22 *The Times*, 17 March 2001.
23 *The Old Bradfieldian*, spring 2002.
24 Nick Tremlett. Quoted in *The Old Bradfieldian*, spring 2002 (obsociety@Bradfield college.org.uk).
25 Hough Stansbury Woodland and Gore and Storrie. *Ecological Analysis of the Greenbelt*. Unpublished report prepared for the National Capital Commission, October 1991.
26 Forman, Richard T. T. and Godron, Michel. *Landscape Ecology*. New York, NY: John Wiley, 1986.
27 Hough Stansbury Woodland and Gore and Storrie. *Ecological Analysis of the Greenbelt*.
28 Ibid.
29 The National Capital Commission. *Greenbelt Master Plan*. Report, 1996.
30 Ibid.
31 Ibid.
32 Hough, Michael. 'Changing Roles of Urban Parks – an Environmental View'. *Environments*, vol. 17, no. 2, 1985.
33 Ahern, Jack. 'Greenways as a Planning Strategy'. In *Landscape and Urban Planning*, vol. 33, 1995, pp. 131–55.
34 *Globe and Mail*, 16 October 1991.
35 Grove, Noel. 'Greenways: Paths to the Future'. *National Geographic*, June 1990.
36 Waterfront Regeneration Trust. *Regeneration*. Toronto, ON: Minister of Supply and Services, 1992.
37 Waterfront Regeneration Trust. *Lake Ontario Greenway Strategy*. Toronto, ON: Ministry of Supply and Services, 1995.
38 William Fulon *et al. Who Sprawls Most? How Growth Patterns Differ Across the US*. Survey Series, Washington, DC: Brookings Institution, July 2001.
39 Houck, Mike. 'The Humane Metropolis; People and Nature in the 21st Century'. Address to New York University, 6–7 June 2002.
40 Sprawl Atlanta. Environmental Justice Resource Centre, Atlanta, GA (http://wwwl.ejrc. cau.edu).
41 Nelson, Arthur C. *et al. The Link Between Growth Management and Housing Affordability: The Academic Evidence*. Washington, DC: Brookings Institution Center on Urban and Metropolitan Policy, February 2002.
42 http://www.uoregon.edu/-pppm/llanduse/UGB.html.
43 Seltzer, Ethan. Director, Institute of Portland Metropolitan Studies, Portland State University. Personal communication, 2002.
44 Houck. 'The Humane Metropolis'.
45 Houck, Mike. Personal communication.
46 Ibid.
47 Blumenauer, Congressman. Address 'Portland Ground Zero in the Liveable Communities Debate'. Congress for New Urbanism, June 2000.
48 Seltzer, Ethan. Personal communication.
49 Drawn from the Smart Growth Network ICMA (http://smartgrowth.org).
50 Mayo, James M. 'Effects of Street Forms on Suburban Neighbourhood Behaviour'. *Environment and Behaviour*, vol. 11, no. 3, September 1979.
51 DeLaurentilis, Joanne. *Kensington Roots*. Toronto, ON: St Stephen's Community House, 1980.
52 Kelly, Barbara M. *Expanding the American Dream: Building and Refurbishing Levittown*. New York, NY: University of New York Press, 1993.
53 Blumenauer, Congressman. Address 'Portland Ground Zero in the Liveable Communities Debate'.
54 Gore & Storrie Ltd in association with Hough Stansbury Woodland Ltd *et al. Town of Markham Natural Features Study, Phase 2*. Report, March 1993.

55 Town of Markham. Official Plan Amendment No. 52. *Policies to Protect, Enhance and Restore Natural Features* (Urban Development Area), 1997.
56 *Topos*. 26 March 1999.
57 The Emscher Park International Building Exhibition. *An Institution of the State of North-Rhine Westphalia*. Information Brochure, September 1991.
58 *Topos*. 26 March 1999.
59 The Ruhrgebiet. *Facts and Figures*. IBA, n.d.
60 Ibid.
61 *Topos*. 26 March 1999.
62 Ibid.
63 The Ruhrgebeit. *Facts and Figures*.
64 Rodrian, M. Schwarze. Director of the State Agency for the project. Personal communication.
65 Ibid.
66 Ruskin, John. *Modern Painters*, vol. 4, pt 5, Orpington: George Allen, 1888.
67 McNamee, Kevin A. 'Fighting for the Wild in the Wilderness'. In Monte Hummel (gen. ed.) *Endangered Spaces*. Toronto, ON: Key Porter Books, 1989.
68 Panel on the Ecological Integrity of Canada's National Parks. 'Unimpaired for Future Generations?'. Ottawa, ON: Parks Canada Agency, 2000, p. 9.2.
69 *Plantlife Annual Review 2001–2002*.
70 Planta Europa. Planta Europa Secretariat, c/o Plantlife – The Wild Plant Conservation Charity, June 2002.
71 Adapted from the *Y2Y* brochure.
72 Ibid.
73 Barrett, Suzanne and Kidd, Joanna. *Pathways: Towards an Ecosystem Approach*. Prepared for the Royal Commission on the Future of the Toronto Waterfront. Ministry of Supply and Services, Canada, 1991.
74 Panel on the Ecological Integrity of Canada's National Parks. 'Unimpaired for Future Generations?'
75 Banff-Bow Valley Study. *At the Crossroads*. Report, Toronto, ON: Ministry of Supply and Services, Canada, 1996.
76 Barlow, Jim. Parks Canada Superintendent. Personal communication.
77 Arntzen, Holley. *Salish Sea. A Handbook for Educators*. Victoria, BC: Parks Canada, 2001.
78 Barlow, Supt. Jim. Personal communication.
79 *Recommendation 16, Expanding the Global Network of Protected Areas*. IUCN report of the IVth World Congress on National Parks and Protected Areas, February 1992.
80 IUCN World Parks Congress.

Select bibliography

This bibliography is a small selection of references that has been compiled with two purposes in mind: (1) to provide further readings on the various subjects the book covers; and (2) to provide up-to-date reference material for those subjects where the literature has expanded significantly since the first edition was published.

Foundations of environmental thought

Evernden, Neil. *The Natural Alien*. Toronto, ON: University of Toronto Press, 1985. The author contends that the environmental movement's ability to achieve its goals is in doubt, despite the publicity given environmental issues today. He reviews the inherent assumptions of Western industrial societies; assumptions that work in opposition to environmental causes. The existence of differing approaches to human/environmental relations is, however, evidence of a basic characteristic of humanity – that of flexibility, and it is this that provides the basis for a new understanding of how as a species we relate to the non-human world. Only through this understanding, the author argues, can the impulse to environmental advocacy and crisis be understood and addressed.

Livingston, John A. *Rogue Primate: An Exploration of Human Domestication*. Toronto, ON: Key Porter Books, 1994. This book challenges most conventional ideas about the relationship between people and the natural world. The author contends that the first domesticated animal was neither dog nor goat, but human. Humans cut themselves adrift from the real world by becoming entirely dependent on ideas, and technical ideas have provided our species with the power to manipulate nature. The natural world itself has consequently been drawn into the service of humans and their belief systems. Our understanding of nature is informed by an ideological insistence that domination is somehow 'natural', a social theory that lay behind Charles Darwin's explanation of evolution as the 'struggle for existence'. This book is necessary reading for anyone wishing to understand the impact of people on nature.

Seddon, George. *Landprints: Reflections on Place and Landscape*. Cambridge: Cambridge University Press, 1997. This book is a series of essays about landscapes. It covers a wide range of ideas, from a discussion of the nature of Nature, the evolution of perceptual attitudes, the *genius loci* of the Australian landscape, to the garden as Paradise. The book is important reading, for its insights and acute observations about the way we use and abuse land and the moral and environmental aspects of history.

General reading

Boardman, Philip. *The Worlds of Patrick Geddes*. London: Routledge & Kegan Paul, 1978. A biography of Geddes' life, ideas and achievements as a biologist, town planner, sociologist, educator. The book is an excellent introduction to this great pioneer of twentieth-century thought, and a precursor to Geddes' *Cities in Evolution* (1915) republished by Rutgers University Press (1972).

Hoskins, W. G. *English Landscapes*. London: BBC, 1979. This book is about the English landscape; how people have made their mark on the land over thousands of years. As the author says, 'it has been written upon over and over again in a kind of code'. In chapters on settlement, colonization, boundaries, churches, transport and industry the author shows us how to read these immensely varied landscapes and understand how they came about.

Hough, Michael, Benson, Beth and Evenson, Jeff. *Greening the Toronto Portlands*. Waterfront Regeneration Trust, October 1997. This publication examines a green infrastructure system as a framework for future development on the Portlands area of the Toronto waterfront. It proposes key principles that consider regional as well as local issues, precedents from economic as well as environmental points of view, functions and performance criteria, and implementation as this affects public policy, current initiatives, ownership and public involvement.

Jacobs, Jane. *The Death and Life of Great American Cities*. New York, NY: Random House, 1961 (latest edn). Jane Jacobs asks: What makes cities work? Why are some neighbourhoods full of things to do and see and why are others dull? Over thirty years after its first publication, the observations of the city she made in the early 1960s are as relevant in the 1990s. She taught people to see places as they really are.

Johnson, Bart R. and Hill, Kristina. *Ecology and Design: Frameworks for Learning*. Washington, DC: Island Press, 2002. This book is about teaching ecology to designers. It includes contributions by a number of authors. It is concerned with the need to integrate ecological principles into environmental design and planning education and outlines priorities and approaches for incorporating ecological principles in the teaching of landscape design and planning. The book represents an important signpost and source of ideas for faculty, students and professionals in the science and design professions.

Laurie, Ian C. (ed.). *Nature in Cities*. New York, NY: John Wiley, 1979. A series of informative chapters on urban climate, plants and urban ecosystems, nature and city planning in the nineteenth century, and alternative 'naturalized' approaches to parks and open spaces in modern cities. Remains useful and relevant today.

McHarg, Ian L. *Design with Nature*. New York, NY: Natural History Press, 1969 (2nd edn 1992). A seminal book that has had a major influence on modern environmental planning. McHarg's central theme is that of ecological determinism. The biophysical processes that shape the physical landscape are deterministic; they respond to natural laws and are form-giving to human adaptations. Nature is a fundamental force that will determine the morphology of modern cities and human endeavours.

Miller, Helen. *Patrick Geddes: Social Evolutionist and City Planner*. London: Routledge, 1990. Another excellent biography of Geddes with full bibliographic lists.

Mumford, Lewis. *The City in History*. New York, NY: Harcourt, Brace & World, 1961. This classic exploration of the city – its origins, transformations and prospects – remains essential reading for understanding cities from all points of view.

National Greening Australia, Conference Proceedings. *A Vision for a Greener City: The Role of Vegetation in Urban Environments*. Freemantle, Western Australia: Greening Australia Ltd, October 1994. (Available from Greening Australia Limited, GPO Box 9868, Canberra, ACT 2601.) This publication examines problems and issues of Australia's greening programme in relation to cities. It includes papers on the environment of urban areas, planning for conservation and development, management of urban environments with respect to vegetation, economic and social issues, and community involvement, and bio-diversity. Numerous case studies are included.

Papanek, Victor. *Design for the Real World*. New York, NY: Pantheon Books, Random House, 1971 (2nd rev. edn 1985). This book focuses on the integration of industrial design and natural processes. Its underlying message and themes, however, have direct application to environmental design and urban ecology. Useful reading.

Platt, Rutherford H., Rowntree, Rowan A. and Muick, Pamela C. (eds). *The Ecological City: Preserving and Restoring Urban Biodiversity*. Amherst, MA: The University of Massachusetts Press, 1994. A series of chapters on cities and natural environments. The book explores issues of geography, ecology, landscape architecture, urban forestry and environmental education, from broad overviews of problems to detailed case studies.

Royal Commission on the Future of the Toronto Waterfront. *Watershed.* Interim Report, August 1990. Toronto, ON. The first of a series of reports published by the Toronto Waterfront Commission. Outlines an integrative approach to planning based on the 'ecosystem approach' that links natural processes, people, human activities and the city. Suggests a series of principles for a green waterfront that flow from this approach and outlines prescriptions for regenerating the city.

Royal Commission on the Future of the Toronto Waterfront. *Lake Ontario Greenway Strategy.* Report, May 1995. This publication encompasses the lands and waters that show direct ecological, cultural or economic connections to the waterfront It extends across the entire Ontario north shore and south US side of the lake including the river systems that flow south from the Oak Ridges Moraine to the lake. It provides an analysis of the objectives and actions necessary to realize the waterfront of tomorrow and an overview of implementation mechanisms and roles of the trail. This report provides a regional view of the topic and is a valuable addition to the literature.

Thayer, Robert L. Jr. *Gray World, Green Heart: Technology, Nature, and the Sustainable Landscape.* New York, NY: John Wiley & Sons, 1994. This book identifies three conflicting forces in each of us that the author believes lie at the heart of the environmental crisis: love of land and nature; dependence on and affection for technology; and fear of technology's negative side effects. They give rise to a sense of guilt that is expressed in the way we shape our living places; for instance, air-conditioners hidden behind fences and bushes, or the use of consumptive technologies to build fake waterfalls. The author suggests an alternative theoretical framework upon which to build a future, one in which technologies support rather than dominate nature. Many concrete examples are given to illustrate the theory.

Wackernagel, Mathis and Rees, William. *Our Ecological Footprint.* Gabriola Island, BC: New Society Publishers, 1996.

Waterfront Regeneration Trust. *Regeneration: Toronto's Waterfront and the Sustainable City.* Final report, December 1991. The report summarizes the Commission's work over three years. Includes an analysis of the ecosystem approach; a suggested framework for ecosystem-based planning; and environmental imperatives that link water and the state of the Great Lakes, shore line issues, greenways and the winter waterfront, healing an urban watershed (the story of the Don River), places within the Toronto bio-region, regeneration and strategies for recovery.

On water

Stream hydrology/storm drainage/wastewater

Burke, William K. 'From Sewer to Swamp: the Green Answer to Dirty Water'. *E Magazine,* July/August 1991. An investigation into 'natural' sewage treatment through the use of aquatic plants and fauna in the USA. Discusses the need and applications of John Todd's approach, and the difficulties in gaining acceptance of such alternatives.

Ferguson, Bruce K. *Introduction to Stormwater.* New York, NY: John Wiley, 1998. An authorative introduction to site engineering, up-to-date approaches to urban stormwater management, including constructed wetlands, groundwater recharge, stream restoration. The book includes a wide range of methods for site drainage, control and enhancement of water quality and quantity.

Gover, Nancy. 'Greenhouse Technology Treats Wastewater Naturally'. *Small Flows,* vol. 7, no. 2, April 1993. Another useful introduction to John Todd's work in solar aquatics. Describes the process, gives several examples and contact for further information.

Hongkong Standard. 'The Politics of Rivers'. Special section on global affairs prepared by the *Hongkong Standard,* November 1994. Discusses world issues of rivers, the dependency on rivers for existence in the Middle East, how water is shared among nations, national security and related political problems. Interesting for its larger national focus in comparison to the watershed issues under discussion at more local levels in North America.

Hough Stansbury Woodland *et al. Bringing Back the Don.* Toronto, ON: Task Force to Bring Back the Don (City Hall, Toronto, Ontario), August 1991. The complete report on the restoration strategy for Toronto's Don River. Includes history, context and issues, potential strategies for restoration and a vision for the Lower Don.

National Committees of the Netherlands and Germany for the International Hydrological Program of UNESO. *Hydropolis Reader: the Role of Water in Urban Planning*. Proceedings of an international workshop on hydrology, held in Wageningen, the Netherlands, March 1993. A series of papers by workshop participants that focus on three themes: urban water systems, experiences and views of technicians and managers; water in urban development, experiences of planners and designers; the role of water in decision-making and planning procedures. Useful case studies from the USA and Canada, China, Australia, Ireland, Turkey and Europe. Contact: Section OCC (IAC), Box 88, 6700 AB Wageningen, the Netherlands.

Newbury, Robert W. and Gaboury, Marc N. *Stream Analysis and Fish Habitat Design*. Privately published by Newbury Hydraulics Ltd (newbury@cablean.net), Okenagen, British Columbia, 1993. A very useful and authoritative technical field manual and guide to stream habitat projects. Includes planning, field exploration, evaluation of stream behaviour, design and construction of stream habitat works.

Tamminga, Kenneth R. 'Land Application of Treated Municipal Effluent: Literature Review of Post Implementation Evaluation'. Department of Landscape Architecture, Pennsylvania State University. Unpublished paper, n.d. A selective survey and literature review of the overall environmental impacts and general efficiency of wastewater irrigation. Paper focuses on the infiltration capacity of soils and vegetation on sprayed, treated wastewater. It also addresses the larger framework of wastewater reuse for agricultural purposes. Includes the work of William Sopper (the Living Filter concept) and other field researchers in the USA and developing countries.

Wright, Lawrence. *Clean and Decent*. London: Routledge & Kegan Paul, 1960. A small but highly informative and timeless book on the social history of water throughout the ages – how it was used and disposed of, historical traditions in keeping clean, and the evolution of the bathroom.

On plants

General/naturalization/restoration

American Forests (sponsored by the USDA Forest Service and the Houston Green Coalition). A series of research reports that cover the economic costs and benefits of vegetation on the urbanizing environment. The organization's numerous study reports cover US states such as the Regional Ecosystem Analysis Canton-Akron Metropolitan Area, Ohio, 1999, and the Regional Ecosystem Analysis for Metropolitan Denver and Cities of the Northern Front Range, Colorado, 2001.

Bellamy, David. 'Doge City'. In *Bellamy's Europe*. London: BBC, 1976. This book was the product of a BBC television series and the chapter 'Doge City' is about the wide variety of plants one can find growing fortuitously in Venice. Very useful for a keen sense of observation and understanding of urban nature.

Bray, Paul M. 'The City as a Park'. *American Land Forum Magazine*, winter 1985. The idea of a park as an oasis amidst the dense bricks and mortar of the city is thoroughly ingrained in our thinking. This thoughtful article suggests that all parts of the city – its work spaces, living quarters, streets and heritage – have equal aesthetic and recreational potential, qualities that are not confined only to the park, but must be seen as part of the entire structure of the urban environment.

Buckley, G. P. (ed.). *Biological Habitat Reconstruction*. London: Belhaven Press, 1989. Excellent reference book on habitat reconstruction in England, addressing principles of and opportunities for habitat reconstruction, the creation of new habitats and enhancing existing habitats. Both the urban and rural contexts are explored.

Dramstad, Wenche E., Olson, James D. and Forman, Richard T. T. *Landscape Ecology Principles in Landscape Architecture and Land-Use Planning*. Washington, DC: Island Press, 1996. A non-technical book, it illustrates the practical principles that can be applied to landscape planning. Included are sections on foundations, principles for patches, edges, corridors and connectivity, mosaics, and practical applications.

Forman, R. T. T. and Godron, M. *Landscape Ecology*. New York, NY: John Wiley, 1986. This book on the study of the landscape is written from an ecologist's perspective, but is intended to be readily understandable to the non-scientist. It is based, first, on direct observation of the visual landscape (its patterns and characteristics); second, by a scientific exploration of what made that landscape the way it is (its structure); third, how it behaves (its function). The book explores the natural science, artistic, social and economic dimensions of the landscape, that include terrestrial and aquatic environments, populations, ecological communities, energy flows and matter. Very helpful for a basic understanding of how landscapes work.

Hammond, Herb. *Seeing the Forest among the Trees: The Case for Holistic Forest Use*. Vancouver, BC: Polestar Press, 1991. A valuable book for understanding forests as dynamic biological entities in relation to the practice and politics of forestry. The author discusses forest ecology and provides practical solutions to forestry problems in this context. Comprehensive and well illustrated, this, book is important to an understanding of urban forests.

Harker, D., Evans, S., Evans, M. and Harker, K. *Landscape Restoration Handbook*. Boca Raton, FL: Lewis Publishers, 1993. A good reference book on landscape restoration with chapters on naturalizing the managed landscape, planning approaches for naturalizing the landscape, summary of ecological principles relating to bio-diversity, conservation biology and ecosystem restoration, principles and practices of natural landscaping, bird and butterfly gardens, a detailed directory of natural regions in the USA and a listing of the plant communities with the primary woody and herbaceous species.

Hough Stansbury Woodland. *Naturalization Project*. Ottawa, ON: National Capital Commission, 1982. A review of the most economical techniques for restoring native woodland in urban areas. Alternative site preparation techniques, tree and shrub species and mulching methods tried on a number of test plots.

Hough Stansbury Woodland. *Naturalization Project: Five Year Test Plot Evaluation*. Ottawa, ON: National Capital Commission, 1989. A follow-up study to review the success and failure of the various techniques employed in the 1982 Naturalization Project. Evaluation of the test plots, the survival of different species and the success of mulching techniques.

Hough Stansbury Woodland Naylor Dance, with Gore and Storrie Ltd. *Ecological Restoration Opportunities for the Lake Ontario Greenway*. Toronto, ON: Report prepared for the Waterfront Regeneration Trust, June 1994. Survey of ecological restoration and implementation strategies in practice as of 1994. Includes chapters on site selection, assessment of ecosystem functions, implementation techniques, vegetation communities and the urban landscape. Extensive annotated bibliography.

Kirkwood, Niall (ed.). *Manufactured Sites*. London: Spon Press, 2001. This book is about the legacy of industrial activities and pollutants on the contemporary landscape and their influence on new scientific research, innovative site technologies and progressive site design. The twenty-first century will have to address the legacy of now past industrialization, and this area of reseach and redevelopment will be increasingly important as urban growth moves away from 'greenfields'and towards development on 'brownfield' sites. The book presents innovative environmental, engineering and design approaches that include phytoremediation, landfill capping and bioengineering that focus on multi-discipinary and collaborative work.

Kirt, R. B. 'Quantitative Trends in Progression Toward a Prairie State by Use of Seed Broadcast and Seedling Transplant Methods'. In *Abstracts of the 12th N.A. Prairie Conference, 1990* (see D. D. Smith and C. A. Jacobs (eds) 'Recapturing a Vanishing Heritage'. Proceedings of the twelfth North American Prairie Conference. University of Northern Iowa. Cedar Falls, IA, 1990). On the basis of a four-year study, results showed that prairie species could be transplanted equally well by broadcasting seeds or transplanting seedlings.

Lerner, Steve and Poole, William. *The Economic Benefits of Parks and Open Space*. The Trust For Public Land, 1999 (www.tpl.org). This is a case-book that presents data and examples of the economics of parks and open spaces and documents accumulating evidence of how their conservation can make economic sense. The book covers topics that include attracting investment, revitalizing cities, preventing flood damage and safeguarding the environment.

Luken, J. O. *Directing Ecological Succession*. New York, NY: Chapman and Hall, 1990. Book concerned with the consequences of deliberately disturbing vegetation: to achieve a desired goal through manipulating the plant community by disturbance. Deals with methods of managing succession including plant and plant part removal, managing succession by changing resource availability, changing propagule availability, and animals and succession.

Plantlife. The Wild-Plant Conservation Charity, London. An organization journal dedicated to protecting wild plants in Britain. The charity carries out practical conservation work, influences relevant policy and legislation and collaborates widely to promote the cause of wild plant conservation (www.plantlife.org.uk).

Plant Europa. European Plant Conservation Strategy. A developing network of organizations (government and non-government) working for plant conservation in Europe. It focuses on preservation of plants and their habitats (www.planteuropa.org).

Rappaport, Bret. 'Weed Laws: A Historical Review and Recommendations'. *Natural Areas Journal*, vol. 12, no. 4, 1992, pp. 216–17. Discussion of weed laws and ordinances that allow the establishment of natural areas on public and private lands while addressing the concerns of neighbours.

Ruff, Allan R. *Holland and the Ecological Landscapes 1973–1987*. Delft, the Netherlands: Delft University Press, 1987. Since his first book *Holland and the Ecological Landscape* which described the successional approach to planting developed by the Dutch, the author has revisited many of the Dutch housing projects he examined in 1979, and presents an analysis of how they have fared over some twenty years.

Tanacredi, J. J. 'Management Strategies for Increasing Habitat and Species Diversity in an Urban National Park'. In J. J. Berger (ed.) *Environmental Restoration: Science and Strategies for Restoring the Earth*. Proceedings of the Conference on Environmental Restoration, Berkeley, CA, January 1988. Washington, DC: Island Press, 1990. Since 1972, restoration programmes have been implemented in the 10,522 hectare Gateway National Recreation Area lands within and adjacent to New York City. Specific activities include active and passive revegetation of impacted sites, restoration of grassland habitat utilized by regionally rare grassland-dependent bird species, creation of freshwater habitat, and a transplant programme to restore populations of native amphibians and reptiles.

On wildlife

Baines, Chris. *The Wild Side of Town*. London: Elm Tree Books and BBC Publications, 1986. This is a book about urban wildlife and experiencing wild places, the places in which wildlife lives and how people can begin to value wildlife. The author asks, 'Children choose to play in wild, unofficial landscapes, and can't avoid contact with nature. Why do grown ups dismiss these landscapes simply as "untidy"?' Beautifully and simply written, there is much to learn from this publication from mapping neighbourhood habitat networks, to making habitats and nesting boxes and getting everyone involved.

Duffey, Eric. 'Effects of Industrial Air Pollution on Wildlife'. *Biology Conservation*, no. 15, 1979. This article deals with the little-studied effects of air pollutants on the worldwide decline of vertebrate wildlife, and the changes of distribution of certain wildlife species.

Forest Service USDA and Pinchot Institute of Environmental Forestry Research. *Children, Nature, and the Urban Environment. Symposium Proceedings*. USDA Forest Service, General Technical Report, NE-30, USDA, Washington DC, 1977. Papers on the natural environment and human development, research on urban children and the natural environment, and the community and institutional response to nature education.

McClanahan, T. R. and Wolfe, R. W. 'Accelerating Forest Succession in a Fragmented Landscape: The Role of Birds and Perches'. *Conservation Biology*, vol. 7, no. 2, June 1993, pp. 279–86. Reviews the effectiveness of bird perches as a tool to accelerate ecological succession. Study of a mined site in central Florida revealed that seed fall beneath perches had a higher diversity of seed genera and seed numbers (up to 150 times higher). While species composition included both early-successional and late-successional species, the harsh conditions of the site favoured the establishment of the early-successional species.

Noss, Reed E. 'Wildlife Corridors'. In Daniel S. Smith and Paul Cawood Hellmund (eds) *Ecology of Greenways, Design and Function of Linear Conservation Areas*. Minneapolis, MN: University of Minnesota Press, pp. 43–68, n.d. Discussion of wildlife corridors, their function, predation and design issues, including corridor quality and width. Provides guidelines for wildlife corridor design.

Royal Society for the Protection of Birds, Sandy, UK. *Futurescapes*. A publication about the future of wildlife in the English countryside – a blueprint for restoring wildlife habitat. It includes information on what has been lost, the vision for restoration, where to begin a statement of the benefits to Britain's countryside. The publication outlines the habitat types from salt-marshes to mountain and heath land, and includes an excellent map of their location across the country.

Wheater, Philip C. *Urban Habitats*. London: Routledge, 1999. This book is a practical guide to the wide range of urban habitats and the flora and fauna that live within them. It also examines the important conservation and management issues that are relevant to urban areas.

Cultural/behaviour issues

Francis, Mark. 'Control as a Dimension of Public-Space Quality'. In I. Altman and E. Zube (eds) *Public Spaces and Places*. New York, NY: Plenum, 1989. This paper examines the form and meaning of public space; how it can support public culture and outdoor life, how public spaces can be designed and managed to satisfy human needs and expectations, and enable people to control how they use them. Several aspects of user control are examined that include the evolving nature of American public life, and the nature of control as psychological construct and participation concept.

Marcus, Clare Cooper and Francis, Carolyn (eds). *People Places: Design Guidelines for Urban Open Spaces*. New York, NY: Van Nostrand Reinhold, 1998.

Wekerle, Gerda and Whitzman, Carolyn. *Safe Cities: Guidelines for Planning, Design and Management*. New York, NY: Van Nostrand Reinhold, 1994. This book addresses problems of violence and fear of crime in cities: where and why crime occurs; what makes some places unsafe and feared; and what politicians, planners, social agencies and communities can do to make them safer. The book stresses the relationship between the physical environment and social problems, and how solutions must be multi-faceted to be effective. Practical solutions and numerous examples of similar experiences in many cities in North America and Europe.

On urban farming/recycling

American Community Gardening Association. *National Community Gardening Survey* (preliminary report), 1990. This report contains preliminary results of the first year of a two-year survey by the American Community Gardening Association and its members' activities in cities and counties across America. (Information available from American Community Gardening Association (ACGA), 325 Walnut Street, Philadelphia, PA 19106.)

Chowdhury, Tasneem and Furedy, Christine. *Urban Sustainability in the Third World: A Review of the Literature*. Winnipeg, MB: The University of Winnipeg, Institute of Urban Studies, 1994. An examination of the literature on issues of poverty, and the relationship of sustainability to development in the context of Third World cities. It includes a conceptual analysis of 'sustainable development' that highlights the differences between developed and developing countries; studies that deal with problems of services such as water, sanitation and health and how their absence impacts on the urban poor; urban resources such as urban agriculture and waste management/recycling that might be effectively tapped; planning and policy that addresses the various mechanisms of urban management and the changes needed in international cooperation and assistance.

Creasy, Rosalind. *The Complete Book of Edible Landscaping*. San Francisco, CA: Sierra Club Books, 1982. A useful book that integrates productive plants with design. It deals with pesticide issues, costs, water and soil conservation, history landscape gardening, planning and design for small gardens. Well illustrated with detailed plant layouts and gardening techniques.

European Federation of City Farms. Guide to city farms in various countries in Europe. Published by the NFCF Brussels, Belgium, n.d. Contains history, current activities and addresses of the city farms in Europe, including those in the UK. For further information contact: National Federation of City Farms, Avon Environmental Centre, Junction Road, Brislington, Bristol BS4 3JP, UK.

Feder, Natasha. *Population and Global Sustainability*. Toronto, ON: Conservation Council of Ontario, May 1992. A summary report on global sustainability in relation to population. The earth's carrying capacity with respect to soil degradation, food production and distribution, forests, and wastes, is examined in addition to other issues such as acid rain, energy, family planning and policy. Useful reference guide to world issues.

Freeman, Don. *City of Farmers*. Montreal, PQ: McGill/Queens University Press, 1991. A study of urban farmers and farming in Nairobi, Kenya. It covers the context within which the study was made, the nature, background, distribution and practices of the farmers, and the the role and significance of urban agriculture in Nairobi at the levels of the families, the community and the nation.

Furedy, Christine. 'Wastes and the Urban Environment'. *WEDC 12th Conference: Water and Sanitation at Mid-decade*, Calcutta, 1986. This paper examines resource recovery and recycling in Asian societies. Even though recycling of urban wastes occurs extensively in Asian cities, it argues that a fresh concept of the city itself is required – its essential functioning as a natural system.

Mollison, Bill and Holmgren, David. *Permaculture*. Australia: Taragi Publications, vol. 1, 1978 (2nd edn 1982); vol. 2, 1979. Two well-known books that describe permaculture as a philosophy of working with, rather than against, nature; an agricultural system that combines landscape design with perennial plants and animals to make a safe and sustainable resource for town and country.

Sachs, Ignacy and Silk, Dana. *Food and Energy: Strategies for Sustainable Development*. Paris: United Nations University, 1989. Publication based on the activities of the Food–Energy Nexus Program (FEN) of the United Nations which took place between 1983 and 1988. Discusses the contribution urban agriculture can make using the survival skills of rural refugees to grow food, its contribution to the 'hungry season' between major harvests, types of crops and obstacles to growing food gardens.

Strut, Jac and Nasr, Joe. 'Urban Agriculture for Sustainable Cities: Using Wastes and Idle Land and Water Bodies as Resources'. *Environment and Urbanization*, vol. 4, no. 2, October 1992. The paper describes how cities can be transformed from being only consumers of food and agricultural products into important resource-conserving, health-improving, sustainable generators of these products.

Tinker, Irene (ed.). 'Urban Food Production – Neglected Resource for Food and Jobs'. *Hunger Notes*, vol. 18, no. 2, autumn 1992. A series of articles on food production in cities including Africa (Kampala) and the USA (San Francisco).

Urban Agriculture Magazine. An international journal that deals with a wide range of issues from nutrition and health to economics, linked to urban farming, with a focus on developing nations (www.ruaf.org).

Wackernagel, Mathis and Rees, William. *Our Ecological Footprint*. Gabriola Island, BC: New Society Publishers, 1996. This book outlines an approach to determining the human impact of growing food on the land. It presents tools for measuring and visualizing the resources required to sustain human activities and the limits of carrying capacity.

On climate and energy

Back, Boudewijn and Pressman, Norman. *Climate Sensitive Urban Space: Concepts and Tools for Humanizing Cities*. Delft: Colophon, 1992. A practical book on environmentally sensitive approaches to the design of urban spaces with climate in mind. Includes the social significance of public spaces, mobility and liveability, historic precedents, traffic calming, improvements for pedestrians and cyclists, and recommendations for wind and solar design in hot and cold climates.

McPherson, E. Gregory. 'Economic Modeling for Large-scale Tree Plantings'. *Proceedings of the ACEEE 1990 Summer Study on Energy Efficiency in Buildings*, vol. 4, American Council for an Energy-Efficient Economy, Washington DC. Large-scale urban tree planting is advocated to conserve energy and improve environmental quality, yet few data exist to evaluate its economic and ecologic implications. This paper describes an economic-ecologic model applied to the Trees for Tucson/Global reLeaf reforestation programme.

McPherson, E. Gregory, Simpson, James and Livingston, Margaret. 'Effects of Three Landscape Treatments on Residential Energy and Water use in Tucson, Arizona'. *Energy and Buildings*, no. 13, Elsevier Sequoia, 1989, pp. 127–38. This paper reports on research on the role of vegetation in reducing the cooling loads of buildings in hot, arid climates. Three similar model buildings were surrounded by different types of landscapes: turf, rock mulch with shrub foundation planting and rock mulch with no planting. Irrigation water use and electricity required to power air-conditioners and interior lights were measured for two week-long periods, and energy use measured. Preliminary findings suggest that localized effects of vegetation on building micro-climate may be more significant than boundary layer effects in hot, arid regions.

McPherson, E. Gregory, Novak, David J. and Rowntree, Rowan A. *Chicago's Evolving Urban Forest Ecosystem: Results of the Chicago Urban Forest Climate Project*. Northeastern Forest Experiment Station, General Technical Report NE-169, USDA, Forest Service, 1993. This highly informative publication includes the role of vegetation in the city, urban forest structure, the influence on wind and air temperatures, air pollution and atmospheric carbon reduction and the energy-saving potential of trees.

Manty, Jorma and Pressman, Norman (eds). *Cities Designed for Winter*. Helsinki: Building Books Ltd, 1988. Paints a broad picture of politics, planning and design issues for northern cities. The book highlights the essential concerns of winter cities and documents useful experience by international experts. Case studies from China, Finland, Norway, Canada and other countries with similar climates. An important document for local authorities, urban analysts, designers and town planners, and concerned citizens.

Oke, T. R. 'The Significance of the Atmosphere in Planning Human Settlements'. In E. B. Wiken and G. Ironside (eds) *Ecological (Bio-physical) Land Classification in Urban Areas*. Ecological Land Classification Series No. 3. Environment Canada, Ottawa, 1977. This paper outlines the importance of incorporating atmospheric considerations into the design of settlements. It stresses the interaction between the atmosphere and the built environment. The regional climate imposes constraints upon the nature of the settlement, and the nature of the settlement evokes a change in the local climate. Some of the fundamental aspects of these interactions are examined and how they may be incorporated into the design of urban areas.

Paehlke, Robert. *The Environmental Effects of Urban Intensification*. Report prepared for Municipal Planning Policy Branch, Ministry of Municipal Affairs, Government of Ontario, Canada, March 1991. This report deals with urban sprawl as a key environmental concern in Ontario that has resulted in loss of agricultural land, damage to habitat and biological diversity. The dominant role of the automobile in urbanization is associated with high energy uses, and air quality problems. The report examines the positive as well as potential negative impacts of intensification in ameliorating urban problems. Very useful information on energy use, air quality and environmental issues in relation to populations and urban form. Extensive annotated bibliography.

Philleo, Jerryne. *The Davis Energy Book*. Davis, CA: City of Davis, March 1981. This publication describes the development of an energy conservation strategy for the City of Davis, California, that included public involvement, climate research, energy conservation, building code and city conservation plans, transportation, trees and related energy policy issues. There have subsequently been numerous changes and improvements to these policies.

On regional issues

The Countryside Agency, London, England. *A Strategy for Sustainable Land Management in England*. Distributed by Countryside Agency Publications, PO Box 125, Wetherby, West Yorkshire LS23 7EP. A strategy for sustainable practices for England's farms.

Hall, Peter. *Cities of Tomorrow*. (3rd edn) Oxford: Blackwell Publishers, 2002. This is a critical essay of planning in theory and practice in the twentieth century, as well as of the social and economic problems and opportunities that gave rise to it. The book is both very readable and useful as an overview of this subject.

Panel on the Ecological Integrity of Canada's National Parks. *Unimpaired for Future Generations?*, vol. 2, Parks Canada Agency, 2000. This report is the result of eighteen months of travel across Canada and an examination of Parks Canada's approach to ecological integrity. Its recommendations for improvements were presented to the Minister of Canadian Heritage in November 1988.

Smart Growth Network International City/County Management Association (ICMA) (http://smartgrowth.org). Smart Growth involves a network of organizations that is concerned with alternative approaches to urban sprawl (e.g. new urbanism, urban growth boundaries) and has been prevalent in North America for over fifty years. This document identifies 100 policies for implementation to that end.

Topos. European Landscape Magazine, 26 March 1999. Issue on the Emscher Park (in English). The transformation of industrial steelworks and coal-mines within the Emscher region of the Ruhr Valley. The magazine includes discussion of its primary environmental, economic and social goals in its planning and implementation of parts of the project after 1999.

Urban Growth Boundaries. Fact sheet prepared the Centre of Excellence for Sustainable Development and the Green Alliance, 1998. Information on UGBs includes the cities where they have been adopted, what they are, how they work and where to find further information (http://user.gru.net/domz/ugb.htm).

Index

Page references for figures, tables and plates are in *italics*

www.study-guides.com
www.study-guides.com
www.study-guides.com
www.study-guides.com